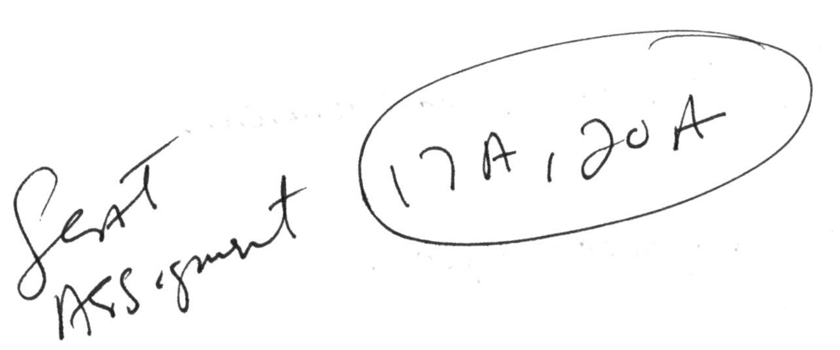

THE BODY TEMPLE

Next Week - Liver / Gall Bladder

Following wk - Digestive System
2 wks

Following 2 wks - Endocrine System

After that 2 wks - Nervous System

After that - Immune

THE BODY TEMPLE

Written and edited by

R. Duane McEndree, N.D., PhD.
and
Nancy L. McEndree, N.D., PhD.

TEACH SERVICES, INC.
New York

PRINTED IN
THE UNITED STATES OF AMERICA

2010 11 12 13 14 15 · 5 4 3 2 1

ISBN-13: 978-1-57258-589-8
ISBN-10: 1-57258-589-7
Library of Congress Control Number: 2009938537

Published by

TEACH Services, Inc.
www.TEACHServices.com

Acknowledgements and Heartfelt Appreciation

The authors of The Body Temple wish to acknowledge the following individuals for their contributions, prayers & many hours spent in compiling *The Body Temple.*

In Memoriam: **Bob & Gwen McEndree**—Duane's Mother and Father who helped us see God's handiwork in nature.

With Love to: **Don & Esther Minett**—Nancy's "special" Mom & Dad. Thank you for your encouragement and being there for us in hard challenging times.

Acknowledgements:

Dr. Bill Palmer—Our Colleague and contributor for many of the ideas and concepts found in this book.

Fran Fisher—For her hard work and contributions.

Ron & Linda Christianson—Thank you for investing in God's work.

Dr. Carol Donovan –Thank you for being a pioneer & teacher of naturopathy.

Dr. Jim & Elisa Sharps—Thank you for carrying on with IIOM, the School we founded.

In Appreciation to:

Our children: Cheryl, Christopher, Charles, Terri, Caryl, William, Lori, Sean, Amy, Lara, and Brian. A very special thanks to Charles & Terri for their support, encouragement and investment. Also a special thanks to Lara Green Roering *(Lu Lu)* who was given to us by God to help us grow in Christian character.

Our "adoptive children" and colleagues: Peter & Christine Rose; Tim & Brenda Doyle; Marie Fish; Lia Fu Wang, Danny Campbell; Doris Hynek.

Special Thanks:

A special thanks to Christopher McEndree for his time and design for our book cover.

Preface

True Education

From age to age the search for the "Tree of Knowledge" and the pursuit for riches and superiority has led men and women to seek higher education. Human ambition seeks for knowledge, degrees and diplomas that will bring to themselves glory, self-exaltation, and supremacy. Self-exaltation always brings unhappiness and poor relationships with others.

Satan has artfully woven his dogmas, his false religion, his false theories, and social mores into the instruction given in worldly universities and colleges. From the podiums of these "trees of knowledge," satanic agencies, worldly manuals, and New Age course work speak the most seductive flattery and bogus science. Thousands are misled and are partaking of the fruits of these trees, but it only leads to further degradation. After graduating from these "Halls of Hades," the student is awarded a Babylonian degree that the State declares equips them to perpetuate the education of their progenitors with education that the teacher of lies began centuries ago. God's Holy Word teaches that men "*spend money for that which is not bread.*" (Isaiah 55:2) These students use their heaven-entrusted talents and finances to secure an education that God pronounces as foolishness.

The eating of the "Tree of Knowledge of Good and Evil" caused the ruination of our first parents, and the amalgamation of mingling good and evil is ruining men and women today. Modern education that concentrates the minds of the young upon hedonistic amenities such as competitive sports, scholastic mind games, honor rolls and sororities and fraternities leads the student to depart further and further from "The True Source of Wisdom" and closer and closer to the concepts of perdition. When the word and principles of GOD are laid aside for worldly books, games, honors, and sports—confusion of understanding pertaining to the Kingdom of Heaven will become evident. Those who seek this education that the world esteems so highly will become *"educated worldlings."* They have chosen to accept what the world calls knowledge in the place of what God has revealed to us through His Word and His Son.

All true education comes from a *personal* knowledge and *diligent* study of the Scriptures. This education will **renew** and even **expand** the mind; it will TRANSFORM the character and will RESTORE the image of God of every child of the Almighty. God's Holy Word will enable the learner to understand the Voice of God. The diligent student of this course of study will then become a co-worker with the Savior. He will become a vessel used by "The Potter" to dispel moral darkness about him and bring light and *true knowledge* to men. No other education is equal to Gospel education which imparts to the student the character of God.

As you, Dear Reader, study "The Body Temple," by R. Duane McEndree, N.D., PhD. and Nancy L. McEndree, N.D., PhD, please acknowledge and understand that the human body was fashioned by God to replicate His perfect body. He created man's body to be the dwelling place of HIS HOLY SPIRIT. In order to understand the science of anatomy and physiology we need to understand the difference between the true and false science and the difference between true and false education. Sometimes our ideas of education take too narrow a sphere. True education means much more than "parroting" back a certain course of study. It means more than a preparation for this life on Earth. True education prepares the student for a humble life of service to mankind in this short sphere of time and the HIGHER realm of joy and service for the world to come.

All TRUE EDUCATION has its source in the knowledge, love, and service to God and mankind. The mind of man is brought into communion with the mind of God, the finite with the Infinite. As you, Dear Student study to reach God's ideal you will earn an education that is high as Heaven and as broad as the Universe. This education cannot be completed on Earth. This science will be continued in the world to come. This true education, ordained by God, will secure the successful student his passing diploma from the preparatory school on Earth to the Celestial School of higher learning above. AMEN!

frontal (x-axis)

In this coronal view from Grey's Anatomy, 1901 Edition, page 670 of the third and fourth ventricles of the human brain plainly reveals the angel wings over "the ark of the covenant" and the dwelling place of our Heavenly Creator.

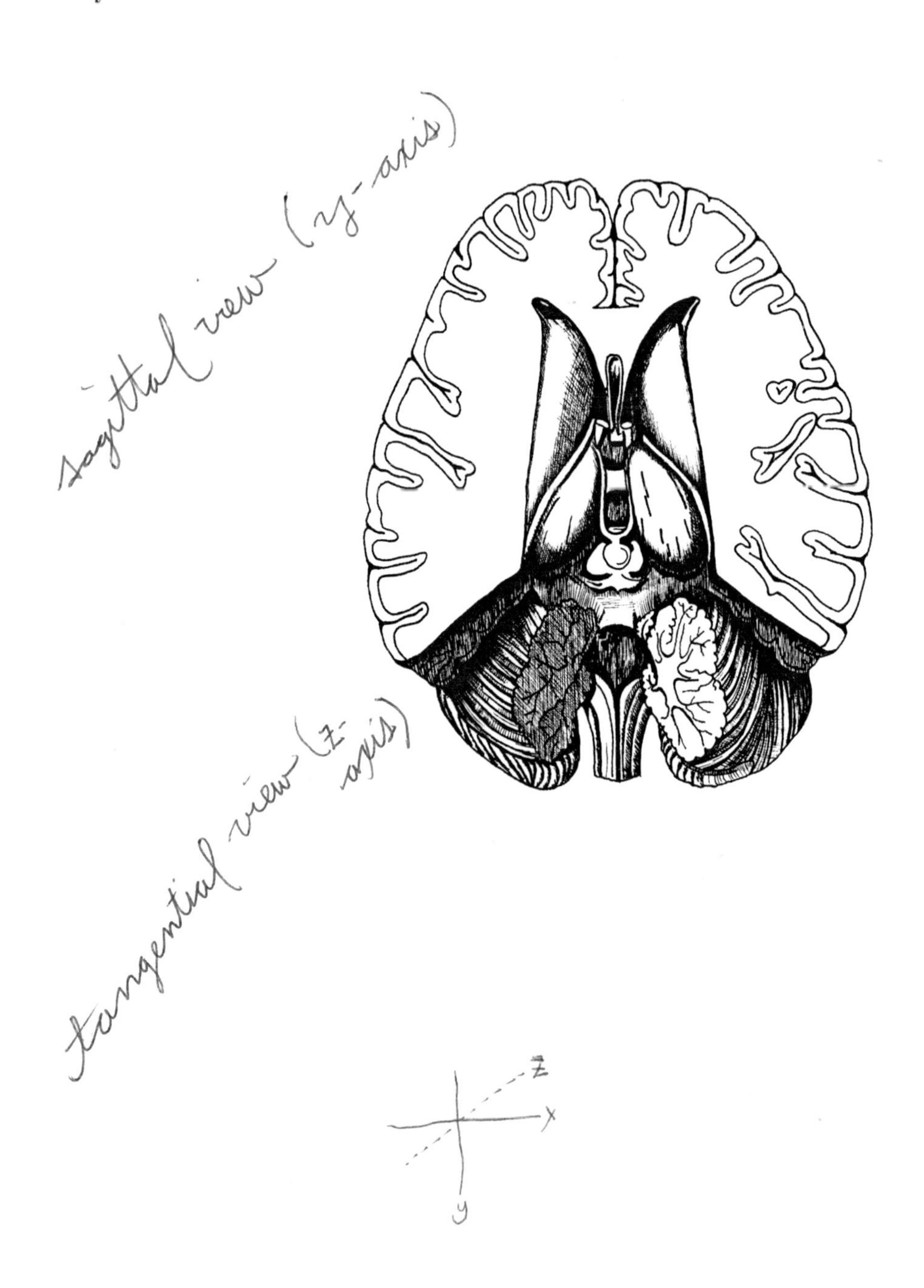

The Body Temple

By: Nancy L. McEndree - Feb. 13, 2007

This temple of my body and soul
Is to be kept pure and whole
Given to me by God to use
To guard and keep and never abuse.

How do I care for my body temple?
In the Book of Genesis is the perfect example
God has a plan for health and soul
Obedience to His word; make that your goal.

I stand in awe before your throne
This aging body can only moan
You alone, Lord, can make old new
When we co-operate with you

Lord God, I ask you to consecrate
This body to you I dedicate
Accept this flesh of human hands
Eternal One, your Temple stands!

"Know ye not that ye are the temple of God, and [that] the Spirit of God dwelleth in you?" 1 Corinthians 3:16

You have a choice—either God or Satan will dwell in your mind. God wants you to choose Him. In the coronal view of the brain from Grey's Anatomy, the form of angel's wings are apparent. This is God's throne room!

Table of Contents

Chapter One

"A BODY THOU HAST PREPARED ME"

Genesis 1:26, 27

"And God said, Let us make man in our image, after our likeness: and let them have dominion over the fish of the sea, and over the fowl of the air, and over the cattle, and over all the earth, and over every creeping thing that creepeth upon the earth. So God created man in His [own] image, in the image of God created He him; male and female created He them."

God not only created mankind but He gave us an "owner's manual," THE HOLY BIBLE. He said, *"Trust in the Lord with all thine heart; and lean not unto thine own understanding. In all thy ways acknowledge Him and He shall direct thy paths."*—Proverbs 3:5,6. God knows how the body functions, He is the CREATOR. When we live according to "God's Plan," our bodies function quite nicely. In the first three chapters of the Bible one will find the ten health principles or laws. These ten principles are:

1. Godly Trust
2. Open Air
3. Daily Exercise
4. Sunshine
5. Proper Rest
6. Lots of Water
7. Always Temperate
8. Nutrition
9. Attitude of Gratitude
10. Benevolence.

When sickness comes, the very best place we can turn is to the GREAT PHYSICIAN and the Manual that He provided. We are told in God's Manual that OBEDIENCE to these health laws brings health: *"And said, If thou wilt diligently hearken to the voice of the LORD thy God, and wilt do that which is right in his sight, and wilt give ear to his commandments, and keep all his statutes, I will put none of these diseases upon thee, which I have brought upon the Egyptians: for I [am] the LORD that healeth thee."*—Exodus 15:26. DISOBEDIENCE results in sickness and sin: *"But it shall come to pass, if thou wilt not hearken unto the voice of the LORD thy God, to observe to do all his commandments and his statutes which I command thee this day; that all these curses shall come upon thee, and overtake thee."*—Deuteronomy 28:15. *"The LORD shall make the pestilence cleave unto thee, until He has consumed thee from off the land, whither thou goest to possess it."*—Deuteronomy 28:21. *"The LORD shall smite thee with a consumption, and with a fever, and with an inflammation, and with an extreme burning, and with the sword, and with blasting, and with mildew; and they shall pursue thee until thou perish."*—Deuteronomy 28:22.

He also gave us the lesser light in E. G. White's *Counsels on Health,* p. 38: "*So closely is health related to our happiness that we cannot have the latter without the former. A practical knowledge of the science of human life is necessary in order to glorify God in our bodies. It is therefore of the highest importance, that among the studies selected for childhood, physiology should occupy the first place. How few know anything about the structure and functions of their own bodies, and of nature's laws! Many are drifting about without knowledge, like a ship at sea without compass or anchor; and what is more, they are not interested to learn how to keep their bodies in a healthy condition and prevent disease."*

Thousands of books have been written on the subject of human anatomy and physiology. The great majority of these books have attempted to break down the human body into a long list of integral body parts. These body parts have been given names. Many of these names are nearly impossible to pronounce and certainly impossible to remember. In **THE BODY TEMPLE** we shall approach the study of anatomy and physiology from an entirely different perspective than perhaps you have ever contemplated.

What is the study of anatomy?

Anatomy is the branch of science that deals with the ***structure*** of the body or organism. The study of human anatomy is of course, absolutely fundamental to an understanding of the human organism. We can better understand *who* we are when we understand *what* we are.

What is the study of physiology?

The study of physiology is the science of the ***functions*** of the living organism and its components and the chemical and physical processes involved. In other words physiology is the study of the mental, physical and spiritual bodily functions.

Chapter Two

MATTER, ELEMENTS, AND ATOMS

II Peter 3:10-13

"But the day of the Lord will come as a thief in the night; in the which the heavens shall pass away with a great noise, and the elements shall melt with fervent heat, the earth also and the works that are therein shall be burned up. [Seeing] then [that] all these things shall be dissolved, what manner [of persons] ought ye to be in [all] holy conversation and godliness, looking for and hasting unto the coming of the day of God, wherein the heavens being on fire shall be dissolved, and the elements shall melt with fervent heat? Nevertheless we, according to his promise, look for new heavens and a new earth, wherein dwelleth righteousness."

Anything that occupies space and has mass is called matter. Everything in the universe is composed of matter. Mass refers to the amount of matter in an object. Matter is composed of elements which are chemical substances that cannot be broken down into simpler substances by ordinary chemical means. Elements are substances in which all the atoms are similar and on the same frequency. There are about 106 elements listed on the periodic table of elements. The human body is made of 84 of these elements. *"And are built upon the foundation of the apostles and prophets, Jesus Christ Himself being the chief corner stone: In whom all the building fitly framed together groweth unto an holy temple in the LORD: In whom ye also are builded together for an habitation of God through the Spirit."*—Ephesians 2:20-22. Four of these elements account for 96% of the body's weight: Oxygen—65%; Carbon—18%; Hydrogen—10%; and Nitrogen—3%. The rest, such as calcium, potassium, sulfur, sodium, magnesium, are all around 1% or less.

Each element is composed of its own chemically distinct kind of atom. An atom is one of the smallest units of an element that retains the chemical characteristics of that element. Atoms are so small, that if you filled a child's balloon with hydrogen it would contain about 100 million, billion, billion hydrogen atoms. A drop of water contains more than 100 billion, billion hydrogen and oxygen atoms. If an atom were as large as the head of a pin, all the atoms in one grain of sand would fill a cube measuring one mile high.

Atoms are not solid bits of matter. In fact even atoms of the densest substances consist mostly of space. Everything in the universe has basically the same patterns. As planets in our galaxy orbit around the sun, likewise the nucleus of an atom like the planets has small particles orbiting around it. Think of the nucleus (the middle of the atom) as the sun, and the small particles (electrons) circling around the nucleus as the planets. Relative to the difference in size, electrons are as far away from the nucleus as the planets are from the sun.

Each electron travels in its own circle called a shell (7 of these shells are known to exist). Each shell can hold only a certain number of electrons. There are never more than 2 electrons in the shell that is nearest the nucleus.

Every human being has basically the same structure, but this structure is arranged differently. Some people are short, some tall, some stocky, some slender, some dark, some light, etc. We identify each other by how we look or sound. We identify atoms by number and weight. The number of identification is the number of protons in the nucleus. The weight is determined by how many neutrons are in the nucleus. The hydrogen atom is the only one without any neutrons in the nucleus.

When two or more of these elements are joined together they form a compound. Some examples of these are: Potassium Chloride; Ammonium Chloride, water, etc. For instance water is called H2O because two hydrogen atoms have joined with one oxygen atom to make a water molecule. Matter is constantly changing and by so doing emits small amounts of energy, that are measured by positive (+) or negative (-) emissions.

The word "KINDS" is mentioned in the first chapter of Genesis. Each kind, plant or animal, has a unique frequency. Foods also have frequency. We will learn later how a knowledge of these individual frequencies will enable you to know the proper diet necessary to bring plant or animals back into perfect health. Frequency has to do with how the various elements are stacked together in any substance. From the frequency we obtain the energy needed to live. *"As the living Father hath sent me, and I live by the Father: so he that eateth Me, even he shall live by Me."*—John 6:57.

In the economy of the human body, there are two forms of energy. One is called <u>reserve energy</u> and the other is called <u>usable energy</u>. Reserve energy is compared to a <u>savings account.</u> It tells you how much accumulative life force there actually is to use. The useable energy is compared to a <u>checking account</u>; it is the energy that is used to fuel the body throughout the day.

WE DO NOT LIVE OFF THE FOOD WE EAT, BUT INSTEAD OFF THE <u>ENERGY</u> FROM THE FOOD WE EAT.

Illness is defined as a ***<u>loss of reserve energy</u>***.

ANY DAY THAT ANYONE DOES NOT TAKE <u>IN</u> MORE ENERGY THAN THEY USE <u>UP</u>—THAT IS THE FIRST DAY OF ILLNESS.

Chapter Three

PROTONS, ELECTRONS, ANIONS AND CATIONS

Ephesians 2:20-22

"And are built upon the foundation of the apostles and prophets, Jesus Christ Himself being the chief corner stone: In whom all the building fitly framed together groweth unto an holy temple in the LORD: In whom ye also are builded together for an habitation of God through the Spirit."

The ten Biblical principles of **fresh air, rest, sunshine, exercise, pure water, temperance, proper diet, trust in God, benevolence and gratitude** are given to guard against an excessive loss of energy and to increase our gain of energy.

In high school and college chemistry we were taught that an ***atom*** has equal number of protons and electrons. We were taught that "opposites attract," therefore it was understandable why there had to be equal charges in the basic atom for it to be stable. However, the student needs to understand *why* this was thought to be true. The atom is an ***electromagnetic structure.*** Yet only the *electrical* relations were being addressed, while the *magnetic* aspect was ignored. Anywhere there are electrical relations there are magnetic relations also. And in reality, what has been considered as the attraction of opposite charges in electricity, is actually the alignment of current flow causing ***magnetism.*** The rule of magnetism is defined as currents flowing in the same direction pulling together and currents flowing in opposite directions pushing apart.

Nature has a law which says "electricity will always follow the line of least resistance." This electrical flow potential, or resistance level, can be measured for each element in the body. This is defined by deductive reasoning or deducing that something that is true of all instances must be true of individual instances.

There is a principle of the structure of the atom that traditional basic chemistry does not address. These concepts relate to the proton and electron ratio within the atom. If an atom has less negatively charged electrons than positively charged protons it is called a **cation** (CAT-e-on). If it has fewer protons than electrons it is negative and is called an **anion** (AN-e-on). (*Ana* means up. *Cata* means down.) We know that the line of least resistance (a measure of an amount of resistance) is different for an atom having a 1:1 ratio of anions and cations than it would be for an atom having a 1:40 ratio of anions and cations. This being true, there is then a difference in the ease with which electrical current is able to flow in an atom of 1:1 ratio than there is for one of a different ratio. As the *ratio* varies the *resistance* will vary. As the *resistance* varies, the *electrical flow* will vary. Then as the *electrical flow* varies the *magnetism* will vary. And finally as the chain relationship is completed, it brings about the ***electromagnetic frequency.***

Frequency is the ratio of anions to cations. The atomic weight of an element is the ***frequency.*** Therefore, the atomic weight is a number that gives us the ratio of anions and cations. The ratio affects the electrical flow. The importance of this will be understood as we continue in our study.

When 2 atoms of different elements join they make **compounds.** This is called chemically bonding. This happens in 3 different ways: **ionic** bonding, **covalent** bonding, and **hydrogen** bonding.

The combining of two or more atoms, or the breaking of the bonds, causes a chemical change called a *chemical reaction*. This produces electricity within the body. Chemical reactions are part of the overall process of **metabolism** which includes all the chemical activities that go on in the body. Metabolism builds up substances (**synthesis**) or breaks down substances (**decomposition**). The synthesis process is called *anabolism* and the decomposition process is called *catabolism*. (*Ana* means up, and *Cata* means down.) All chemical reactions either release or absorb energy. When chemical bonds are being *broken*, they release energy. When chemical bonds are being *formed*, they require energy.

Organic compounds always contain *carbon* and *hydrogen*. They, along with water, are the main materials of the body. The **organic chemicals** of the body are the essential body nutrients: carbohydrates, lipids (fats), proteins, and nucleic acids (the hereditary material).

Homeostasis (ho-mee-oh-STAY-siss; Gr. Homois, *same* + stasis, *standing still*) is a state of inner balance and stability in the body. This is a process that entails thousands of chemical reactions going on at the same time usually on a sub-conscious level. We are not aware of all this activity unless something goes wrong. *"And if any man think that he knoweth any thing, he knoweth nothing yet as he ought to know."*—I Corinthians 8:2.

When **hydrolysis** (Gr. - *water loosening*) takes place, a molecule of water breaks up the bonds in a compound into smaller and different molecules. This is an important process in the body because large molecules of proteins, nucleic acids, and fats are ***broken down*** into simpler, smaller, and more usable molecules. When the opposite reactions occur, small molecules are ***united*** into larger molecules, and one or more molecules of water are eliminated. It is called **dehydration synthesis**.

Chapter Four

ENERGY

I Chronicles 29:11

"Thine, O LORD, [is] the greatness, and the power, and the glory, and the victory, and the majesty: for all [that is] in the heaven and in the earth [is thine]; thine [is] the kingdom, O LORD, and thou art exalted as head above all."

God is the source of all our energy (power). Without this energy we would cease to exist. Physical chemistry must therefore be concerned with both matter and energy.

What is energy? Unlike matter it does not have properties that we can easily sense like color or odor, but it can be defined in terms of what its presence allows to be done. **Energy** is possessed by a body if it has the ability to change the position or arrangement of a sample of matter; that is, if it has the ability to do work. When work is done, energy is transferred from the object doing the work to the object being moved whether this reference is on the very small scale of molecules or the larger scale of our Universe. *"By the word of the LORD were the heavens made; and all the host of them by the breath of his mouth. He gathereth the waters of the sea together as an heap: he layeth up the depth in storehouses. Let all the earth fear the LORD: let all the inhabitants of the world stand in awe of him."*—Psalms 33:6-8.

An example of how energy works is as follows: Suppose you allow an ice cube to melt in your hand. As the ice begins to melt, your hand becomes cold. Both effects are due to a single occurrence. This effect is caused by the flow of energy in the form of heat out of your hand and into the ice cube. The water molecules are moved from the fixed positions in the ice crystals and now become free to move about in the liquid. Energy is absorbed and molecules are moved. The removal of heat energy from your hand which you sense as a cooling sensation causes changes in the arrangement of the molecules in the tissue. If the cooling were allowed to continue, damage not unlike that from frost bite would occur—all because energy is being taken away.

The Law of Conservation of Energy states that: Although energy can be transferred from one place to another, and can change form, it can be neither created nor destroyed in the process. In any process, then, from the melting of ice to a nuclear explosion, it is possible to account for all the energy involved in it. Consider both the law of conservation of energy and the law of conservation of mass. It is apparent that with ordinary chemical and physical changes neither mass nor energy is created or destroyed even though either may change form. It is a fundamental rule of nature that we cannot use heat energy with 100% efficiency. Only a fraction of the energy consumed ever goes into doing work. Our best heat engines operate at about a 40% efficiency. This means that 60% of the energy is transferred to the environment while only 40% is used to do work.

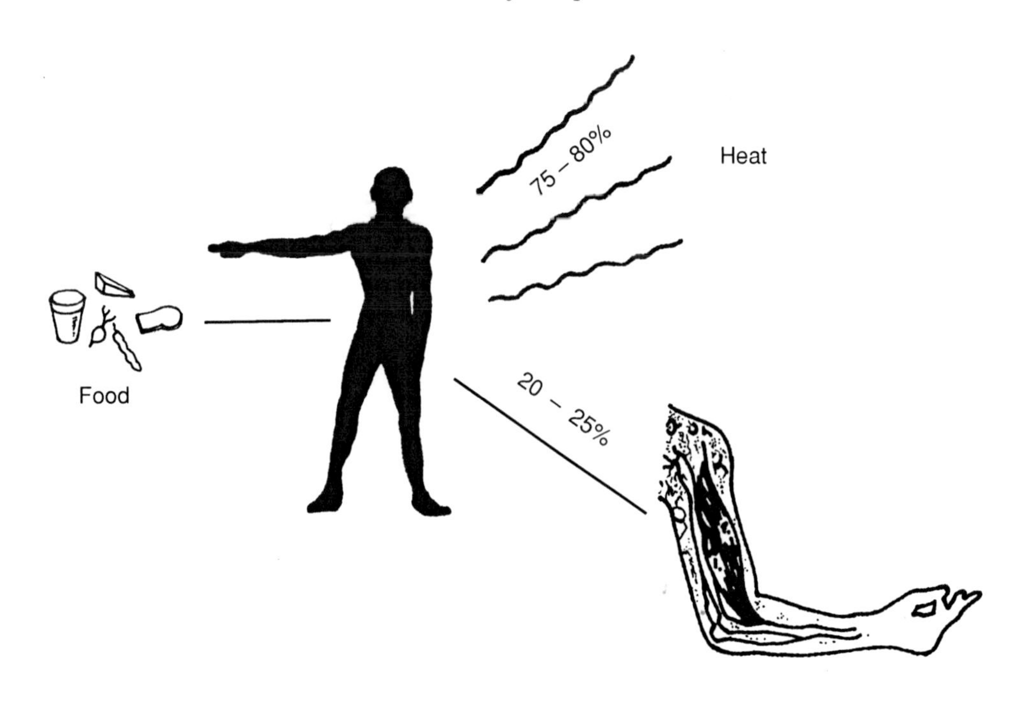

When we speak of energy we also need to speak of work since as when energy is used, the two become linked together. We can recognize that work is being done if we see something move, like the rustling leaves of a tree, a fast moving train, or a soaring baseball. Work is also done by our bodies molecularly. Biological work consumes energy just as any other kind of work does. The energy for this biological work is supplied by the foods we consume. As food is digested it brakes down into smaller molecules. This food fuel is transported across the intestinal membrane and eventually sent to other parts of the body. Often the movement of digested material to cells and tissues must be done against a resistance to the flow. This requires a use of energy since *transport work* must be done to move against the resistance. This movement is known as active transport. Large molecules known as proteins are used for cellular growth. Cells are continuously being synthesized from smaller molecules. This *biosynthesis work* requires energy. When a muscle contracts energy is being used as the muscle tissues draw together. As you move your arms, legs, or eyes, *mechanical work* is being done, which also requires energy.

FORMS OF ENERGY

Though energy cannot be created or destroyed, it can change form. As the light bulb burns, electrical energy is converted into heat energy and light energy. Since we use energy in all its forms, it would be useful to briefly describe some of them.

1. ***Heat energy*** is obtained from the burning of fuels and also from the sun. Most energy is ultimately lost to the environment as heat. Heat is used in the motion of atoms and molecules in matter.
2. ***Mechanical energy*** is that of machinery in motion. A moving arm, leg, or car possesses mechanical energy.
3. ***Electrical energy*** is generated by friction. This energy is used by the motion of electrical conducting materials in a magnetic field and in certain chemical reactions.
4. ***Chemical energy*** is used by a compound when certain kinds and arrangement of atoms are found in its molecules. Upon reaction some of this energy can be released if the products of the reaction possess lesser amounts of energy.
5. ***Light energy or electromagnetic energy*** can be thought of as energy traveling as waves through space at a speed of 186,000 miles per second. We commonly call this the "speed of light."

All matter is but the unique combination of heat and electricity. To put it another way, we might say if all the electricity were suddenly removed from the human body it would turn to dust. *"In the sweat of thy face shalt thou eat bread, till thou return unto the ground; for out of it wast thou taken: for dust thou [art], and unto dust shalt thou return."*—Genesis 3:19.

Electricity is the part of energy that produces magnetic forces. When a current flows in a wire there is an accompanying magnetic field around the wire and also at perpendicular or at right angles to it. Magnetism does not work separately from electricity, but it is an integral part of it. Of course the body cannot be as simplified as an electric wire. However the changing electric fields that are a result of the pulsations within the basic anion and cation relations do have the effect of producing magnetic fields. When there is no electrical current flow there is no magnetism.

Each one of the atoms in our bodies is an electrical generator. They are also magnetic generators. As they come in contact with each other, the building and bonding process becomes one of electro-magnetic plating. This action is similar to the process of silver or gold plating. In order to increase our state of health or wellness we need to increase our Reserve Mineral Energy (vital force).

Each individual born into this world is given a limited supply of vital force at birth. This supply, like the energy supply in your new battery, has to last a lifetime. The faster you use up your energy supply the shorter your life. A great deal of our energy is wasted haphazardly throughout life. The habit of cooking our food and eating it processed with chemicals and the use of alcohol, drugs, and junk food all draw out tremendous quantities of energy from our limited supply.

"Intemperance in eating and drinking, intemperance in labor, intemperance in almost everything, exists on every hand. Those who make great exertions to accomplish just so much work in a given time, and continue to labor when their judgment tells them they should rest, are never gainers. They are living on borrowed capital. They are expending the vital force which they will need at a future time. And when the energy they have so recklessly used is demanded, they fail for want of it. The physical strength is gone, the mental powers fail. They realize that they have met with a loss, but do not know what it is. Their time of need has come, but their physical resources are exhausted. Every one who violates the laws of health must some time be a sufferer to a greater or less degree. God has provided us with constitutional force, which will be needed at different periods of our lives. If we recklessly exhaust this force by continual overtaxation, we shall sometime be losers. Our usefulness will be lessened, if not our life itself destroyed."
—E. G. White, Counsels on Education, p. 166.

When mineral energy is lost out of our basic molecular structure of the human body the line of resistance (vital force) will be altered at that point. This in turn alters the electrical flow. When the electric flow is altered, the magnetism will also be altered. High level wellness is reduced and the person will be subject to a greater and greater number of disease symptoms.

We cannot receive energy from dead food. This is why God gave us the diet that would be the most beneficial to us in Genesis 1:29: *"And God said, Behold, I have given you every herb bearing seed, which [is] upon the face of all the earth, and every tree, in the which [is] the fruit of a tree yielding seed; to you it shall be for meat."* Food, prepared in the most simple manner and preserving its natural properties as much as possible, and avoiding flesh meats, grease, and all spices, is the best food to promote wellness.

In the digestion of live foods, anions and cations work and rub against each other causing resistance. When this happens energy is given off as a result of this resistance. The energy given off from the digestion is what the human lives on. He does not live off of the food he eats, but rather he lives off the energy given off as the cationic food is resisted by the anionic digestive enzymes.

We receive our energy from two sources. First we receive energy from the food-digestive principle above and secondly from the atmosphere in which we live and breathe. Approximately 20% of the mineral energy comes from the food and 80% from the atmosphere. The more efficient the digestion process, the more efficient the body is in extracting mineral energy from the air.

Chapter Five

THE GENERALIZED CELL

I Corinthians 12:23, 24

"And those [members] of the body, which we think to be less honorable, upon these we bestow more abundant honor; and our uncomely [parts] have more abundant comeliness. For our comely [parts] have no need: but God hath tempered the body together, having given more abundant honor to that [part] which lacked."

The cell is the fundamental building block of all living matter. Everyone needs to understand its structure and function.

The activities of cells constitute and promote all of the life processes. These life processes include digestion (the breakdown of food for nourishment), assimilation, and excretion of the food residue, respiration (the interchange of gases—taking in of oxygen, oxygen utilization in the tissues and the giving off of carbon dioxide), synthesis and degradation of materials, movement, and the excitability or response to stimuli. The impairment or cessation of these activities in normal cells, whether caused by trauma, infection, tumors, degeneration, or congenital defects, is the basis of a disorder or disease process.

THE CHEMICAL COMPOSITION OF A CELL

A cell is made up of about 15% protein, 3% lipids (fats or fatlike substances characterized by their insolubility in water), 1% carbohydrates, and 1% nucleic acids. Nucleic acids are found in cells that have a complex chemical structure formed from sugars, pentoses, phosphoric acid, and nitrogen which is in the form of purines and pyrimidines. The most important chemical composition is deoxyribonucleic acid (DNA) and ribonucleic acid (RNA). A cell is also composed of other minerals and 80% water by volume.

CELLULAR STRUCTURE

Let's take a look at the diagram of a cell.

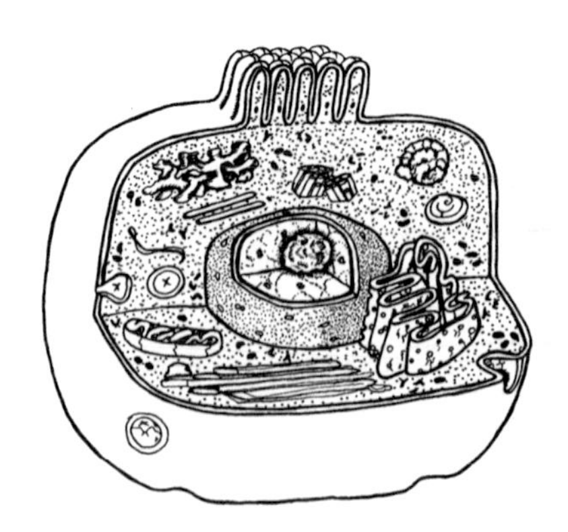

1. **The cell membrane**—The limiting membrane of the cell retains the internal structure. It also permits exportation and importation of materials. It is composed primarily of lipids and proteins and a smaller amount of carbohydrates.
2. **Microvilli**—These are finger-like extensions of the cell membrane covering the free surface of certain epithelial cells. They increase the surface area of the cell enhancing secretion and absorption.
3. **Nuclear membrane**—This is a porous membrane of similar construction to the cell membrane. It is the limiting membrane of the nucleus separating it from the cytoplasm. It also regulates passage of molecules.
4. **Nucleoplasm**—This is the ground substance of the nucleus. It contains the chromatin or thin threads of genetic material (DNA and related protein). During cell division the chromatin transforms into chromosomes.
5. **Nucleolus**—This is a mass of largely RNA (and some DNA and protein). It is found in the nucleus producing units of RNA which combine in the cytoplasm to form ribosomes. These ribosomes receive genetic information and translate those instructions into proteins.
6. **Cytoplasm**—This is the ground substance of the cell less the nucleus.
7. **Endoplasmic Reticulum**—This is the smooth or rough membrane—tubules to which ribosomes may be attached. Rough ER transports protein synthesized from the ribosomes. Smooth ER synthesizes complex molecules called steroids in some cells. It also stores calcium ions in the muscles and breaks down toxins in the liver.
8. **Ribosome**—This is the site of protein synthesis where amino acids are strung in sequence as directed by messenger RNA from the nucleus.
9. **Golgi complexes**—These are flattened membrane-lined sacs which bud off small vesicles from the edges. They collect secretory products and package them for export or cell use.
10. **Mitochondrion**—This is a membranous oblong structure in which the inner membrane is convoluted like a maze. The energy for cell operations is generated here through a complex series of reactions between oxygen and products of digestion.
11. **Vacuoles / pinocytotic**—vesicles—These are membraned lined containers which can merge with one another or other membraned lined structures such as the cell membrane. They function as transport vehicles.
12. **Lysosome**—This is a membraned lined container of enzymes with great capacity to break down structure especially ingested foreign substances.
13. **Centriole**—This is a bundle of microtubules in the shape of a short barrel usually seen-paired perpendicular (vertical) to one another. They give rise to spindles used by migrating chromatids. This is a substance present in the nucleus of a cell that contains the genetic material during cell division.
14. **Microtubules**—These are formed of protein and provide structural support for the cell.
15. **Microfilaments**—These are support structures formed from protein and are different from the microtubules. In skeletal muscles, the proteins called *actin* and *myosin* are examples of thin and thick microfilaments.
16. **Cell Inclusion**—This is the aggregation of material within the cell that is not a functional part of the cell, e.g. glycogen, fat, and etc.

CELLS

The adult human body consists of more than 50 trillion cells. There are many kinds of cells. Just like the billions of people on the Earth, cells are different. Most of these cells are specialized in structure and function; each having its own place and its own job. Just as we each have a place to work for our employment each cell has a working place designated for it by our Creator. We begin life as a fusion of two highly specialized cells. All people living on Earth today came from two ancestors, Adam and Eve.

All cells have things in common. They need oxygen and food in order to metabolize. Except for the red blood cells, each cell has a nucleus. It is here that the work of the cell is determined by a substance known as DNA or deoxyribonucleic acid. When a cell divides, through a very complicated process of chemistry, this double strand of DNA unzips to form two exact double strands. Basically, the enzymes break up the hydrogen bonds connecting the bases of the two strands. The enzyme moves up the base selecting material and forming new bonds. This is called complementary bonding. Each strand of the original DNA molecule acts as a pattern to make an exact copy of the other strand of the unzipped molecule. As soon as the new bonds are formed, the original strand and the newly formed strand begin to twist together. When this process is finished, each of the two newly formed DNA molecules will contain one strand from the original molecule and one new strand.

Just as there are 12 systems that make up the human body, there are 12 basic systems that charge and run the cell. Every day scientists are finding out more and more about how complicated this tiny cell structure is.

The cell membrane separates the cytoplasm inside a cell from the extra-cellular fluid outside. It also separates cells from one another and regulates the passage of materials into and out of cells. Cells have *selective permeability* which allows some things to be let in and some things to be kept out of the cell.

Some cells have **microvilli** in the membrane. These slender extensions of the membrane provide a more abundant surface on which the chemical reactions can occur in either the absorption or secretion of molecules.

Cilia (Sill-ee-a; *eyelashes*) and **flagella** (fluh-Jell-a; *small whips*) are microtubules that are threadlike appendages of some cells. They are anchored by a *basal body* which protrudes into the cytoplasm of the cell to which it is attached. Cilia beat in unison creating a rhythmic wavelike movement in only one direction. They are found in the upper portions of the respiratory system and the female reproductive system.

Flagella only exist on male sperm cells and form the tail that enables the free-swimming cell to reach its destination.

RNA—ribonucleic aid consists of 3 types of nucleic acid which are involved in transcription and translation of genetic code. (1) **Messenger** RNA carries genetic information from DNA; (2) **transfer** RNA is involved in amino-acid activation during protein synthesis; (3) **ribosomal** RNA is involved in ribosome structure. (Ribosomes are sites of protein synthesis.)

The control center of the cell is the nucleus (Noo-klee-uhss; *a kernel.)* It transcribes and translates genetic information in DNA into specific enzymes that determine the cell's metabolic activities. It contains hereditary information in the DNA which is transferred during cell division. It is enclosed in two distinct membranes with a space between them. This is called an *envelope* and has many openings called *nuclear pores* which make it look like a wiffle ball. This allows larger materials to enter and leave the nucleus and for the nucleus to be in direct contact with the membranes of the endoplasmic reticulum. (NOTE: The red blood cell does not have nuclei or DNA.)

Cells usually contain a nucleolus (new-Klee-oh-luhs or *little nucleus).* This is a dark mass inside the nucleus that contains the DNA the RNA, and the protein. It is a pre-assembly point for ribosomes. Ribosomes pass into the cytoplasm and some of them become attached to the endoplasmic reticulum.

The human body functions by chemical and electrical activity. Without this chemical and electrical activity, the nervous system would be like an unplugged television set. The muscles would not work, and oxygen could not be utilized.

The energy for life is generated from chemicals and their reactions. The immediate source of energy for most biological activities in the cell is a small organic molecule known as **ATP** (adenosine triphosphate) Smaller units of adenine (one of the bases in nucleic acids), ribose (a 5 carbon sugar) and 3 phosphate groups make up the ATP ***molecule***. When energy is needed for cellular activity, ATP is broken by water, and the last of the three phosphate bonds is broken. Because one of the phosphates is gone, the compound that is left is ADP (adenosine diphosphate). When energy needs to be stored the last phosphate is *added* to ADP. It is stored in the cell as ATP. The cells are so tiny that they use only a little ATP at a time. By these chemical reactions, energy is converted when it is needed. We are fearfully and wonderfully made!

The center of the life of the Israelite was the sanctuary and its services. Just as the Israelites depended upon the Levites to keep the sanctuary functioning, the body depends upon ATP chemical reactions for its energy. The Hebrews were to support the Levites. Just as the Levites inherited no land of their own, the ATP has no system of its own, but is present in every cell in the body. When there is no energy on a cellular level, the body can have no energy.

Chapter Six

TISSUES

I John 4:8-10

"He that loveth not, knoweth not God; for God is love. In this was manifested the love of God toward us, because that God sent His only begotten Son into the world, that we might live through Him. Herein is love, not that we loved God, but that He loved us, and sent His Son to be the propitiation for our sins."

Cells differentiate in the human body during development to perform their special jobs. These specialized cells are grouped together as tissues. The structures in the body have cellular arrangements that are closely related to their functions showing again the hand of the Master Designer. The body is a demonstration of love demonstrated from the very moment God began to create Adam dust particle by dust particle. So it is to be expected that the lessons on the tissues would be lessons of love.

There are four major tissue types: **epithelial tissue**—*covering love*, **connective tissue**—*supporting love*, **muscle tissue**—*judging love*, and **nervous tissue**—*communicating love*. 02.21.17

EPITHELIA

Epithelia (epp-uh-THEE-lee-uh, on, over, upon, GR. nipple) is a word that denotes something that is being covered or lined. The epithelia is arranged in 2 different ways depending on their function. Some are arranged in layers and cover surfaces or line body cavities or ducts that often connect with the surface of the body. If these cells are lining one of the internal organs, they are shaped to carry out absorption and protection. Glandular epithelial cells are adapted for secretion and are organized into glands.

Epithelial tissue (1) ***absorbs*** in the lining of the small intestine, (2) ***secretes*** as glands, (3) ***transports*** as kidney tubules, (4) ***excretes*** as sweat gland, (5) ***protects*** like skin, and (6) ***receives*** sensory signals such as, taste buds.

Spiritual Application

1. **Absorbs:** *"Christ was treated as we deserve, that we might be treated as He deserves. He was condemned for our sins, in which He had no share, that we might be justified by His righteousness, in which we had no share. He suffered the death which was ours, that we might receive the life which was His."* E. G. White, Desire of Ages, p.25 *"He was wounded for our transgressions, He was bruised for our iniquities: the chastisement of our peace was upon Him; and with His stripes we are healed."* Isaiah 53:5.
2. **Secretes:** *"For God so loved the world that He gave His only begotten Son, that whosoever believeth in Him should not perish, but have everlasting life."* John 3:16. *"He that spared not His own Son, but delivered Him up for us all, how shall He not with Him also freely give us all things?"* Romans 8:32.
3. **Transports:** *"But God, who is rich in mercy, for His great love wherewith He loved us, even when we were dead in sins, hath quickened us together with Christ, (by grace are ye saved;) and hath raised up together and made us sit together in heavenly places in Christ Jesus."* Ephesians 2:4-6. *"In all their affliction He was afflicted, and the angel of His presence saved them: in His love and in His pity He redeemed them: and He bare them, and carried them all the days of old."* Isaiah 63:9.

4. **Excretes:** *"There is no fear in love; but perfect love* *casteth* *out fear: because fear hath torment. He that feareth is not made perfect in love."*—1 John 4:17.
5. **Protects:** *"He that dwelleth in the secret place of the most High shall abide under the shadow of the Almighty. I will say of the LORD, He is my* *refuge* *and my* *fortress**: my God; in Him will I trust."*—Psalm 91:1,2.
6. **Receives sensory signals:** *"I love the LORD, because He hath* *heard* *my voice and my supplications. Because He hath inclined His ear unto me, therefore will I call upon Him as long as I live."*—Psalm 116:1,2.

CONNECTIVE TISSUE

Biological and Spiritual Application: Connective tissue *connects, binds, and supports* body structure. *"Fear thou not; for I am with thee: be not dismayed; for I am thy God: I will* *strengthen* *thee; yea, I will* *help* *thee; yea, I will* *uphold* *thee with the right hand of my righteousness."*—Isaiah 41:10.

These tissues have a great deal of extra-cellular fibrous material that helps support the cells of other tissue. They also form protective sheaths around hollow organs and are involved in storage, transport, and repair. They vary in structure and function. There are three basic types*: loose, dense and elastic.*

Spiritual Application

Loose Tissue: The loose connective tissue represents the support of God even for the ungodly. For if He did not give them breath, they could not continue to exist. He is not able to give them the dense support that He can give His followers, because they do not ask for it. *"But I say unto you, Love your enemies, bless them that curse you, do good to them that hate you, and pray for them which despitefully use you. That ye may be the children of your Father which is in heaven: for He maketh His sun to rise on the evil and on the good, and sendeth rain on the just and on the unjust."*—Matthew 5:44,45.

Dense Tissue: The dense connective tissue represents the firm support that our heavenly Father gives to those that are members of His family. *"Let your conversation be without covetousness, and be content with such things as ye have: for He hath said, I will never leave thee, nor forsake thee. So that we may boldly say, The Lord is my* *helper**, and I will not fear what man shall do unto me."*—Hebrews 13:5,6. *"Now unto him that is able to keep you from falling, and to present you faultless before the presence of his glory with exceeding joy."*—Jude 24.

Elastic Tissue: Like rubber bands, elastic fibers stretch easily when they are pulled and return to their original shape when the force is removed. Sometimes the Lord has to withdraw some of His support in order for us to learn our lessons. *"Because they rebelled against the words of God, and despised the counsel of the most High: Therefore he brought down their heart with labour; they* *fell* *down, and there was none to help. Then they cried unto the LORD in their trouble, and he saved them out of their distress."*—Psalm 107:11-13.

MATRIX TISSUE

The extracellular component of connective tissue is called *matrix.* Included are (1) *collagenous fibers* which support and protect organs, connect muscles to bones and bones to bones; (2) *reticular fibers* which support fat cells, capillaries, nerves, muscle fibers, and secretory liver cells. These form reticular framework of the spleen, lymph nodes, and bone marrow; (3) *elastic fibers* which allow hollow organs and other structures to stretch and recoil and provide support and suspension; and (4) *ground substance* which provide a medium for passage of nutrients and wastes between cells and bloodstream and lubricates and functions as a shock absorber.

Collagenous fibers

Collagen (Gr. *kolla*, glue + *genes*, born, produced) is a protein of which collagenous fiber is composed. This fiber can be arranged in various ways from loose and pliable, as the connective tissue that supports most of the organs, to tightly packed and stretch-resistant as in tendons.

Connective tissue cells

There are two types of connective tissue cells called *fixed* and *wandering* cells. **Fixed cells** have a permanent site and are usually concerned with long-term functions such as synthesis, maintenance, and storage. They include (1) *fibroblasts* which manufacture matrix materials and assist in wound healing; (2) *adipose* cells which synthesize and store lipids or fats; (3) *macrophages* which engulf and destroy foreign bodies in bloodstream, lymph, and tissues; (4) *reticular* cells are involved with the immune response. The **wandering cells** are usually involved with short-term activities such as protection and repair. They include:

1. *Plasma* cells, which are the main producers of antibodies which help protect the body against microbial infection,
2. *Mast* cells, which produce heparin which reduces blood clotting and histamine, and
3. *Leukocytes,* which are white blood cells which destroy harmful microorganisms and destroy garbage.

There are a total of seven (7) classified cells in this support system.

CARTILAGE TISSUE

This specialized connective tissue provides support and aids in the movement of joints. Cartilage cells are called chondrocytes (KON-droh-sites; Gr. khondros, cartilage). These cells are embedded in small cavities called lacunae (la-KYOO-nee; L. cavities).

Biological and Spiritual Application

Hyaline cartilage

Hyaline cartilage has a translucent pearly blue-white color. Blue is the color of obedience. (Numbers 15:38,39.) The main location of hyaline cartilage is in the respiratory passages, the ends of long bones, and the rib cartilages that connect the ribs (the 10 commandments) to the sternum (the righteousness of God). This gives some flexibility to the law. (Luke 13:14-16; John 8:3-11.)

Fibro cartilage

Fibro cartilage is located in the intervertebral disks and cushions pressure forces.

Elastic cartilage

Elastic cartilage is found in areas that require lightweight support and flexibility.

Chapter Seven

THE BODY SYSTEMS

I Corinthians 6:19, 20

"What? know ye not that your body is the temple of the Holy Ghost [which is] in you, which ye have of God, and ye are not your own? For ye are bought with a price: therefore glorify God in your body, and in your spirit, which are God's."

Biological and Spiritual Application

The **SKELETAL SYSTEM** refers chiefly to the bones that support and protect the body. Bones represent **principles and laws**, which provide the support for the government of God just as bones provide the support for the body. They are never seen unless they are broken. *"And the LORD commanded us to do all these statutes, to fear the LORD our God, for our good always, that He might preserve us alive, as it is at this day. And it shall be our righteousness, if we observe to do all these commandments before the LORD our God, as He hath commanded us."*—Deuteronomy 6:24,25.

The **IMMUNE and LYMPHATIC SYSTEMS** work closely with each other—the lymphatic system quietly ***cleans up;*** the immune system ***goes to war***! The plan of salvation must of necessity include not only **forgiveness of sin** but complete **victory.** Sin, like some diseases, leaves man in a deplorable condition—weak, despondent and disheartened. The sinful soul has little control of his mind, his will fails him, and with the best of intentions he is unable to do what he knows to be right. He knows that he has himself to blame for his weakness, and remorse fills his soul. As a disabled ship towed to port is safe but not sound, so the man is "saved" through forgiveness but not sound. Repairs need to be made on the ship before it is pronounced seaworthy, and the sin wrecked man needs reconstruction before he is fully restored. This process of ***restoration*** is called sanctification. This includes in its finished product restoration of body, soul, and spirit. The restoration of the physical body is not complete until the coming of Christ when "this mortal puts on immortality."—I Cor. 15:53. When the work is finished, the man is "holy," completely sanctified, and restored to the image of God.

The **MUSCULAR SYSTEM** operates by contracting a muscle and relaxing an opposing muscle **representing judgment**. There are three types of muscle tissue: skeletal, cardiac, and smooth. Each muscle type is designed to perform the function it performs. There are three classes of judgment; human judgment, God's judgment and the judgment of the universe. There are three basic works of muscle tissue—movement, posture, and heat production. There are three phases to judgment: investigation, declaration and execution. *"Let us hear the conclusion of the whole matter: Fear God, and keep his commandments: for this is the whole duty of man. For God shall bring every work into judgment, with every secret thing, whether it be good or whether it be evil."*—Ecclesiastes 12:13,14.

The CIRCULATORY or CARDIOVASCULAR SYSTEM is composed of the heart and the blood vessels. It is the main transport system of the body and carries the blood to every cell. The job of the church is to provide the "Word" to all of the world; the job of the **heart** is to provide the **blood** to all parts of the body. The Bible uses the word for heart to describe the ***innermost feeling and desires*** because this organ is located in an innermost protected place, surrounded by ribs, the ***commandments* of God**, and covered by the breastbone, the ***righteousness* of God**. When you see the role of the heart in the body, then the reason of this becomes clear.

The **RESPIRATORY SYSTEM** is the story of the **Holy Spirit's help** symbolized by the role of oxygen in the body. Breath in the Bible is synonymous with life. The blood has to pick up the oxygen in the lungs. Without oxygen the body does not function. We cannot see the wind, but we know it is present because we can hear it, we can feel it, and we can see the results of its contact with the earth. We cannot explain it nor predict what it will do. It is the same with the Holy Spirit.

The **NERVOUS SYSTEM** includes the brain (the master control), the spinal cord, and all of the nerves. The spiritual lessons in the nervous system are to teach us about the **activities that are going on in Heaven**. As the brain works in behalf of the body, Christ is working in our behalf in the heavenly courts sending out His messengers to all parts of the globe. He sends assistance to every suffering one who looks to Him for relief, for spiritual life, and knowledge.

The **DIGESTIVE SYSTEM** is a lesson for our **mental and spiritual nourishment**. What food is to the body, Christ must be to the soul. Food cannot benefit us unless we eat it, unless it becomes a part of our being. So Christ is of no value to us if we do not know Him as a personal Saviour. A theoretical knowledge will do us no good. We must feed upon Him, receive Him into our heart, so that His life becomes our life. His love, and His grace, must be assimilated.

The mechanical and chemical processes that break down large food molecules into smaller ones is called digestion. Unless the small molecules can enter into the cells, they are useless. They must pass through the cells of the intestine into the bloodstream and lymphatic system. This is called absorption. The same is true of spiritual nourishment. It does not do any good to study the Word of God if it does not become part of us.

The **URINARY SYSTEM**, also known as the *renal system,* removes waste products through the production of urine and regulates the water content of the body. It consists of 6 structures: 2 kidneys, 2 ureters, the urinary bladder, and the urethra. Our blood is our life and needs to be cleansed, balanced, and purified in order to keep us alive. The kidneys are used by the body as the means of blood purification. **They represent works**. They have to help the liver, which represents faith, to keep the body clean. Satan has always sought to shut out from men a knowledge of God, to turn their attention from the Temple of God, and to establish his own kingdom. The principle that man can save himself by his own works lies at the foundation of every heathen religion. Satan implanted this principle, and wherever it is held, men have no barrier against sin. In ourselves we are incapable of doing any good thing, but that which we cannot do will be wrought by the power of God in every submissive and believing soul. It is through faith that spiritual life is begotten. We are then enabled to do the works of righteousness. If it were not for the help of the liver, the kidneys could not do their job.

The **REPRODUCTIVE SYSTEM** tells the story of **creation and recreation**, of spiritual oneness with God, the purpose of marriage, the unity of male and female, and the process of spiritual rebirth. After God created man He took a rib and created the woman. In this process we see God actually taking part of the human—dividing them into two—and the result was man and woman each with half of the attributes of God. They were perfectly equal with different roles to perform. The act of two becoming one is a symbol of the spiritual togetherness that Christ wishes with His people.

The organs of the reproductive system create human offspring by combining genes from the male and female, from which a fetus or human baby will develop.

The **ENDOCRINE SYSTEM** represents the **promises of God**, which are only applicable **IF** we conform to His conditions. To every promise of God there are conditions. If we are willing to do His will, His strength is ours.

Whatever gift He promises is in the promise itself. *"Whereby are given unto us exceeding precious promises that by these ye might be partakers of the divine nature, having escaped the corruption that is in the world through lust"* —II Peter 1:4.

The principal function of the endocrine system is to secrete hormones and chemical agents that feed back and activate homeostasis, the metabolic stability of the body.

The **INTEGEMENTARY SYSTEM**—*(Skin)* is the **covering of love** that holds in the fluids, flesh, and bones of the body. The skin is the largest organ of the body. It occupies about 21 square feet of area, depending, of course, upon the size of the person. It can vary in thickness from less than 0.5mm on the eyelids to more than 5mm in the middle of the upper back. It is multi-layered like love and has many facets. There is love that protects, love that provides, love that comforts, and love that disciplines. *"(Love) suffereth long, [and] is kind; (love) envieth not; (love) vaunteth not itself, is not puffed up, Doth not behave itself unseemly, seeketh not her own, is not easily provoked, thinketh no evil; Rejoiceth not in iniquity, but rejoiceth in the truth; Beareth all things, believeth all things, hopeth all things, endureth all things: (love) never faileth: but whether [there be] prophecies, they shall fail; whether [there be] tongues, they shall cease; whether [there be] knowledge, it shall vanish away."*—**I** Corinthians 13:4-8. (Charity = Love)

Chapter Eight

BONE

Psalms 34: 19, 20

"Many [are] the afflictions of the righteous: but the LORD delivereth him out of them all. He keepeth all his bones: not one of them is broken."

Biological and Spiritual Application

Bones give the human body support. Bones represent laws and principles. This is the highest expression of love in the connective tissue. *". . . love is the fulfilling of the law."* Romans 13:10. Bone construction begins with cartilage which has no blood. The blood carries cells called osteoblasts. These cells secrete material that makes the cartilage that becomes bone.

It is interesting to note that the 10 commandments have a 40% - 60% division. Four commandments cover our relationship with God and six cover our relationship with each other. The bones are divided the same way: 40% in the axial skeleton (80) and 60% in the appendicular skeleton (120) with six bones that yoke it all together.

What is added to make the law firm? *"Do we then make void law through faith? God forbid: yea, we establish the law."* Romans 3:31. When we add faith in God's ability to provide us with the character of Christ, we then establish the law in our hearts and do it ***willingly***. It becomes a great support and comfort to us. We shall study much more about the role bones play in our lives in future lessons.

Chapter Nine

HUMAN ANATOMY AND THE SANCTUARY

I Corinthians 3: 16, 17

"Know ye not that ye are the temple of God, and that the Spirit of God dwelleth in you? If any man defile the temple of God, him shall God destroy; for the temple of God is holy, which temple ye are."

Biological and Spiritual Application

The Sanctuary that was built by Moses and the Israelites in the wilderness was not the original plan. God always wanted to dwell in our hearts through His Holy Spirit. Before sin entered the world, God dwelt in the hearts of Adam and Eve. Daily they walked and talked with Him. We now have a choice. We can let Satan rule our lives or we can let God rule our lives. God wanted His presence known in the camp of the Israelites, so that His children could make the *right* choice. Therefore; God instructed Moses to build a sanctuary so that the children of Israel could come before their Redeemer, confess their sins, obtain forgiveness, and be a vessel for His dwelling.

Man was made in the image of God. (Gen. 1:27) As the Sanctuary is a miniature representation of God's Throne Room, so our bodies reflect the Sanctuary that was constructed by Moses in the wilderness.

In the book of Revelation, God gives us a choice. We can either remain in the "lukewarm" state, or we can buy the eyesalve, the white raiment and become as gold tried in the fire. Jesus is ever ready to speak peace to souls that are burdened with doubts and fears. He waits for us to open the door of the heart to Him, and say, "abide with us." He says, *"As many as I love, I rebuke and chasten: be zealous therefore, and repent. Behold, I stand at the door, and knock: if any man hear my voice, and open the door, I will come in to him, and will sup with him, and he with Me. To him that overcometh will I grant to sit with Me in my throne, even as I also overcame, and am set down with My Father in His throne."*—Revelation 3:19-21.

"Then opened He their understanding, that they might understand the Scriptures."—Luke 24:45.

In the next few pages we shall look at the sanctuary constructed in the wilderness and compare its construction in correlation with the human body. God gave Moses specific details as to how to build the sanctuary so that it would reflect the divine plan of redemption. Our bodies are also constructed to reveal God's plan of restoration to perfection.

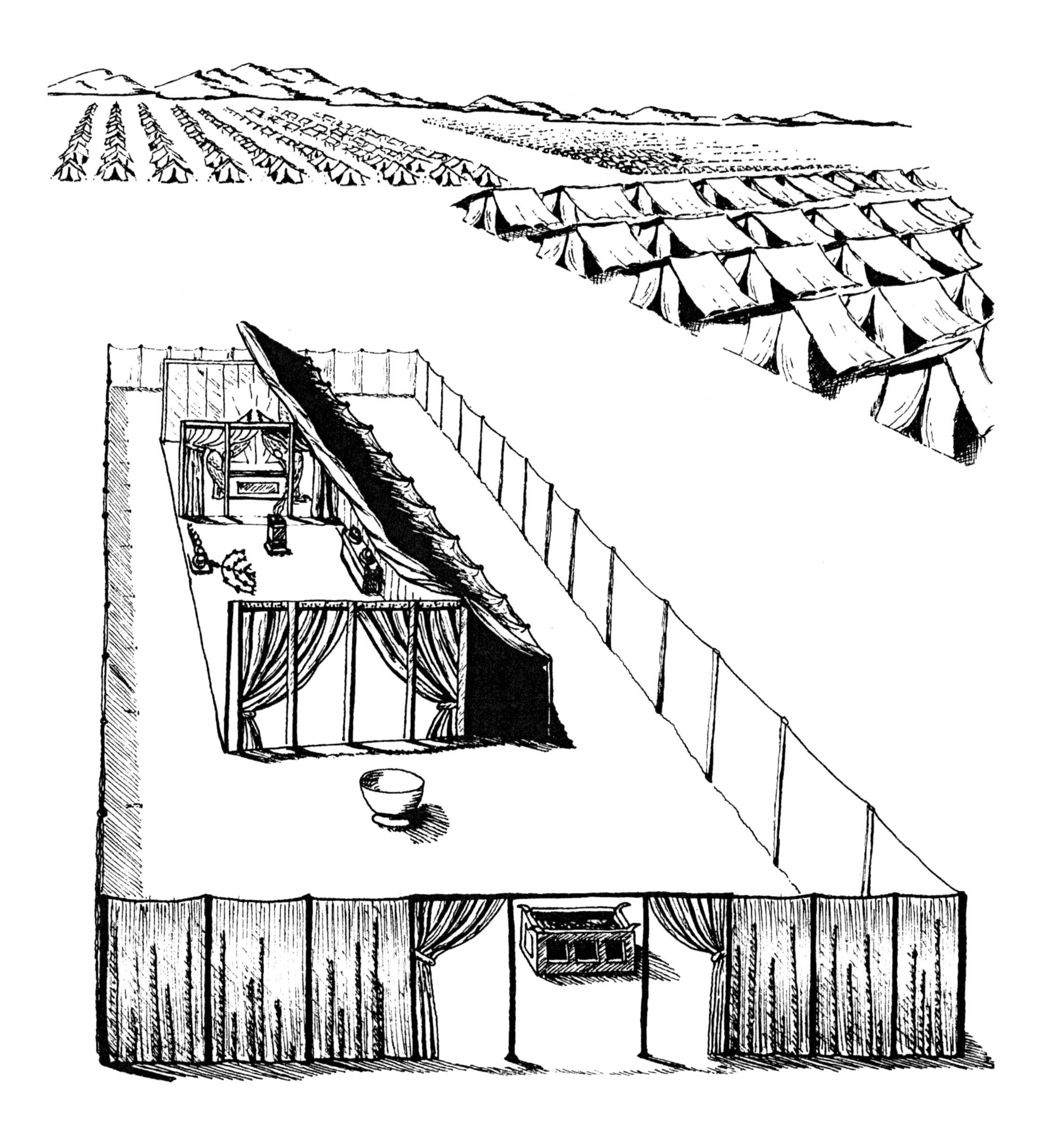

In Exodus 26, we read that there were four layers on the roof of the sanctuary building. The ***inner one*** was woven of blue, red, and purple with embroidered gold angels. It had five parts on each side coupled together in the middle. The ***second*** was white, six and five parts coupled together; the sixth curtain to be doubled in the forefront. The ***third*** was red with no number. The ***fourth*** was badger skin.

God was very particular about the numbers of the Sanctuary so that it would truly represent the coverings on the human head

The head of our body temple has the same four coverings as the sanctuary. The ***first*** has five outer lobes of the brain on each side, coupled in the middle. This layer has 3 parts representing the three colors in the veil. The ***second*** layer is white bone that protects the brain with the frontal bone that doubled at birth. The ***third*** is a solid piece of scalp with no couplings which contains many blood vessels. The ***fourth*** is the hair and is the outside covering.

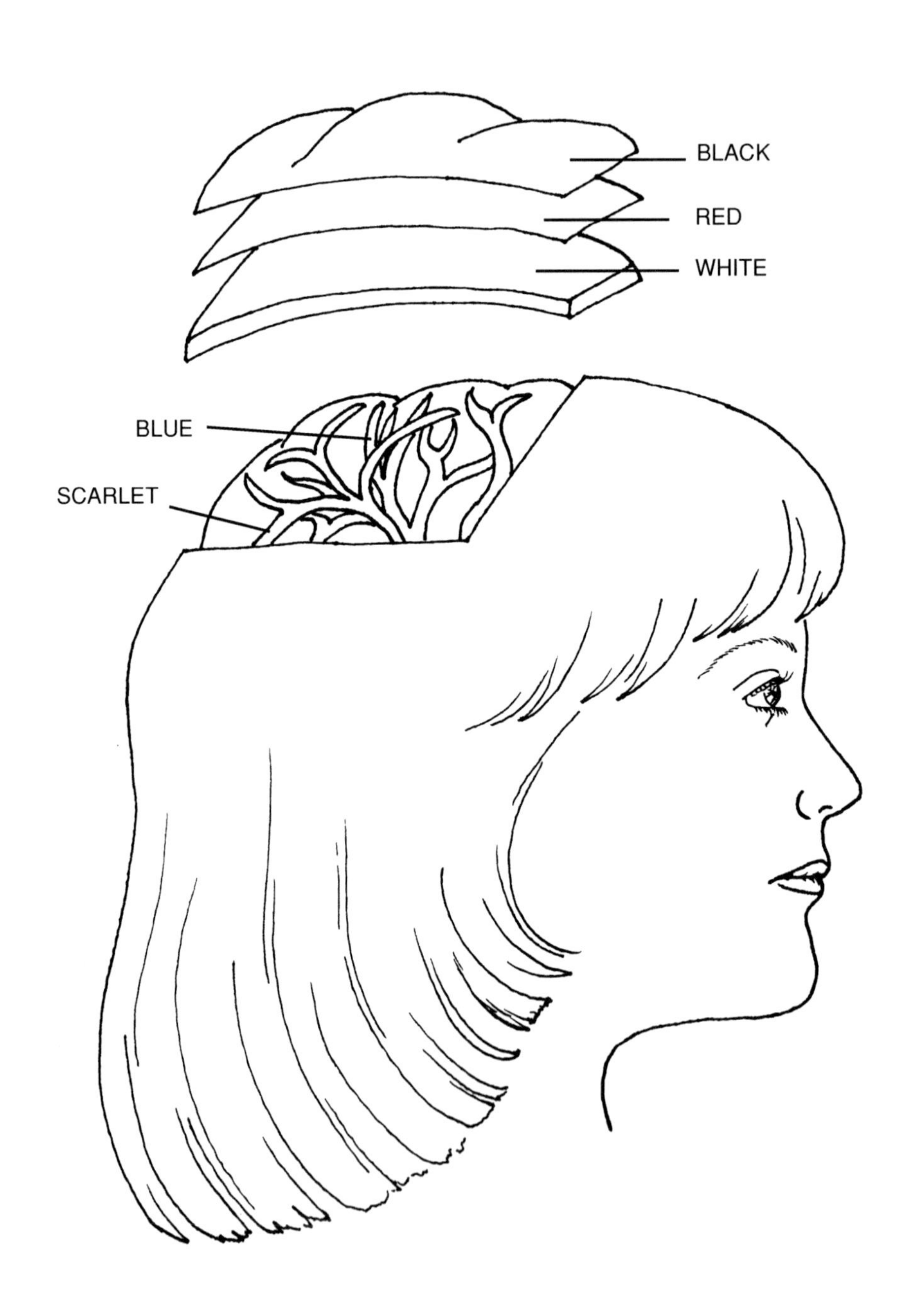

The sanctuary had three parts: The courtyard, the Holy Place, and the Most Holy Place. These represented the different parts of the head. (The Ark of the Covenant was behind the curtain, but is viewed in front in this picture so that it may be seen.)

There were 48 boards in the building. In the body there are 24 vertebrae, plus 24 ribs coupled together in the central part of the axial skeleton. The design matches as well as you can make a rectangular structure to represent a many sided body.

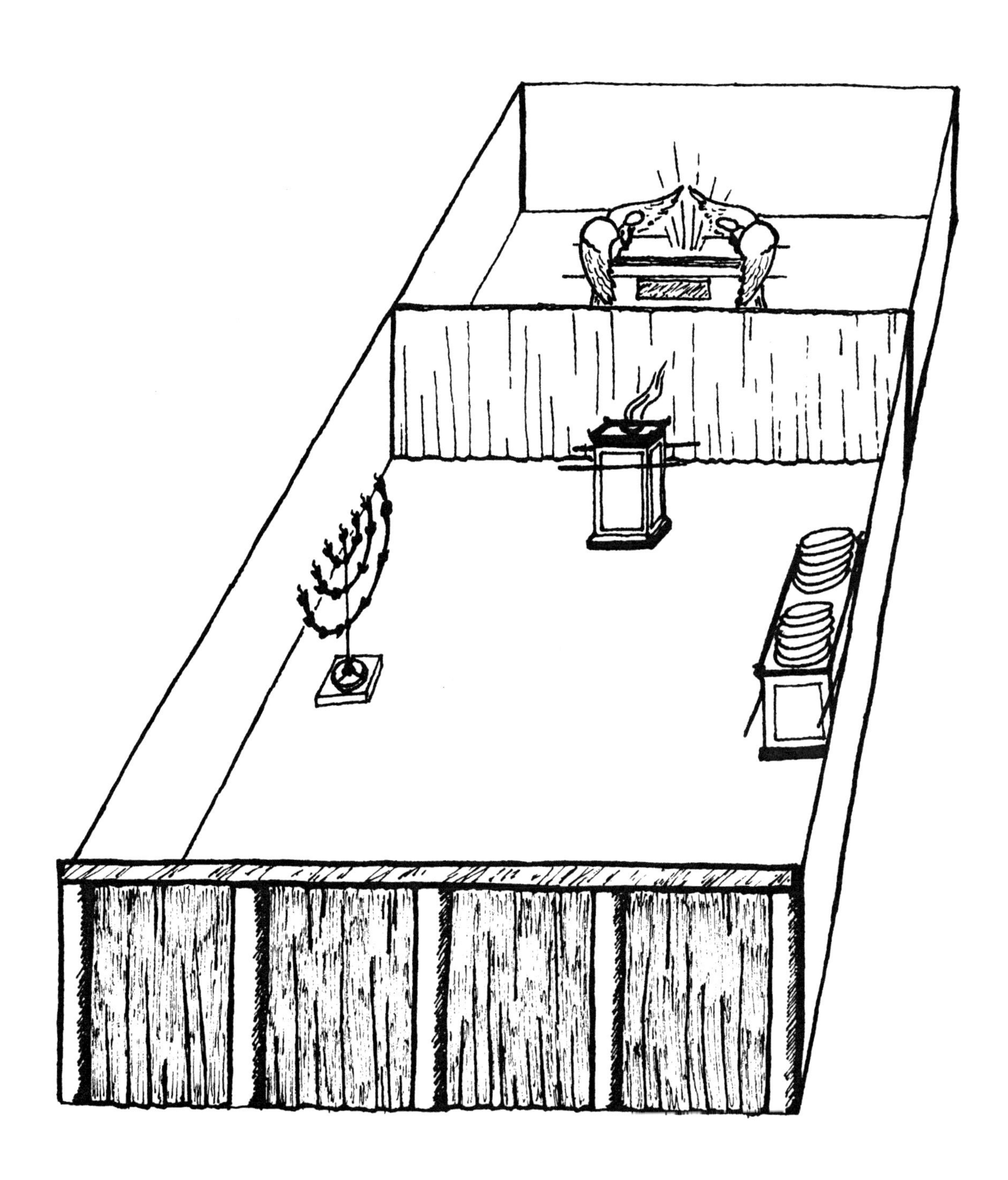

As we saw earlier, there were five pillars holding the entrance curtain to the sanctuary. They represent the five senses by which information is reported to the brain to be processed and become part of our character. These are avenues to the soul. Be careful what you see, hear, smell, touch, and taste because from these sensations decisions are made that decide eternal destiny. There are only two sources of information: Our Heavenly Father or the devil. Jesus told some people that they were of their father, the devil. (John 8:42-47.) Let us choose the right father to instruct us!

THE FIVE SENSES

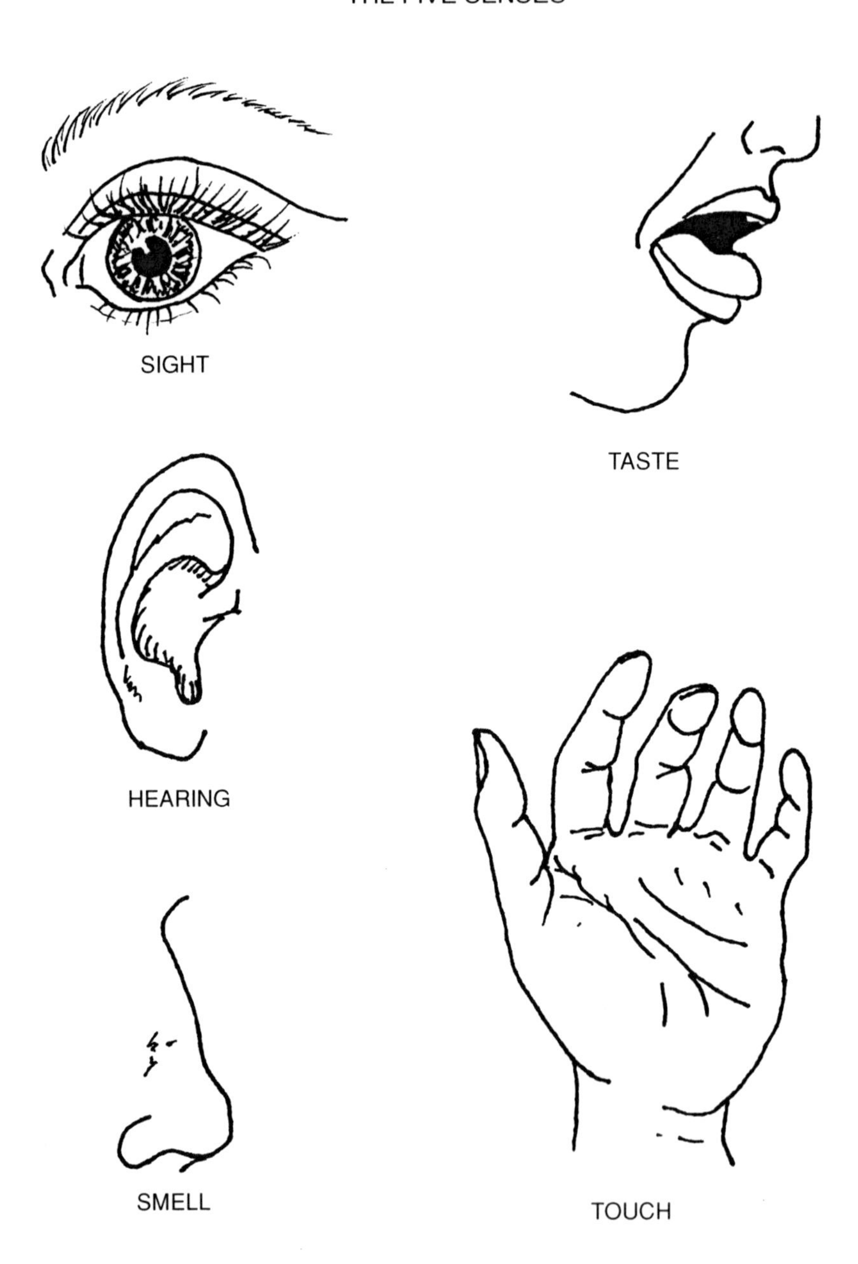

There were angels embroidered on the first covering of the sanctuary to impress the fact that angels are ever present to help us. (Hebrews 1:14.) Angels are a special order of beings that were created before man. (Genesis 3:24.) In this coronal view of the brain from Gray's Anatomy, the form of angel wings is apparent. This is the throne room area of the brain.

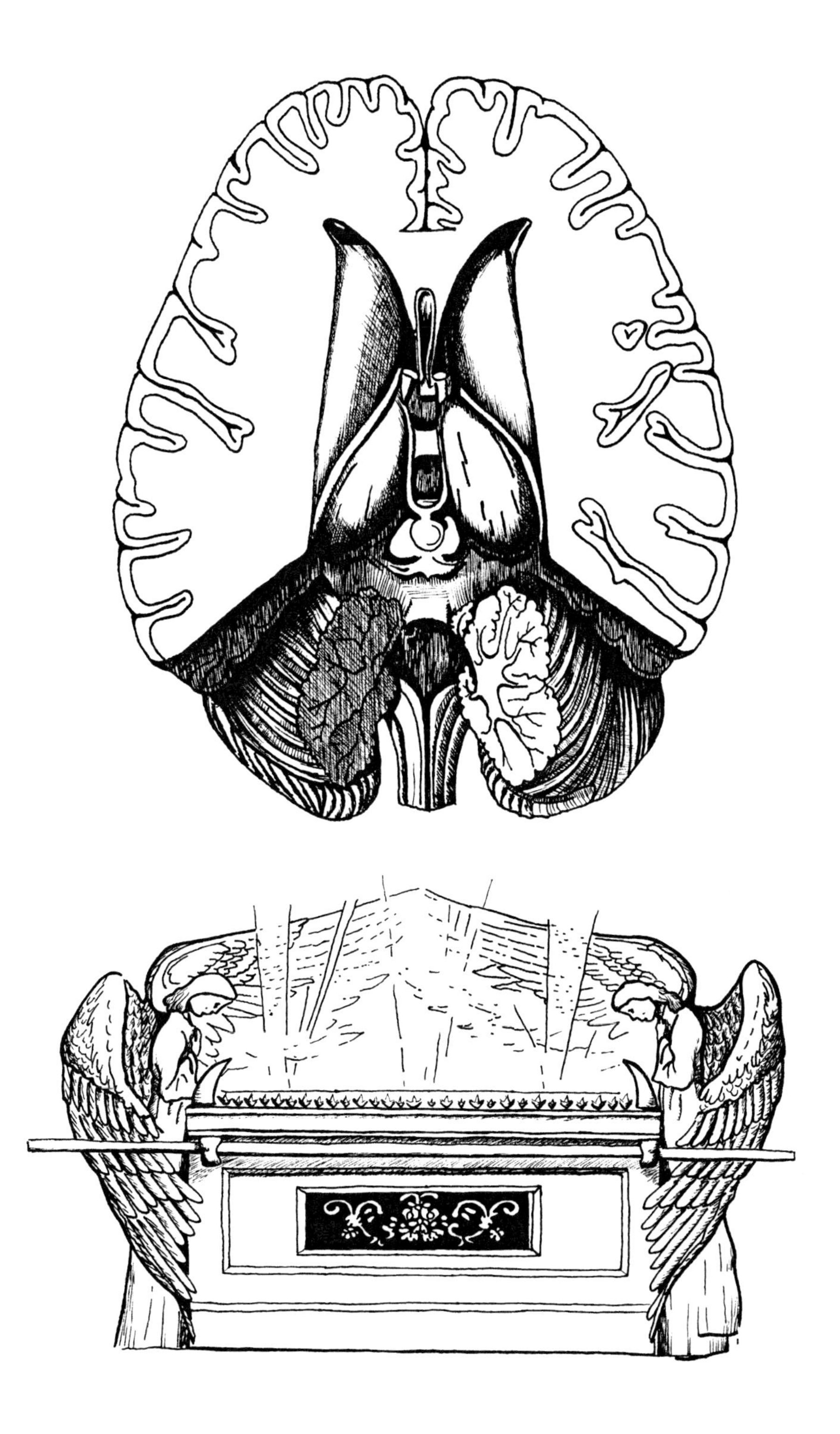

The sphenoid bone is situated at the anterior part of the base of the skull, articulate with all the other cranial bones which it binds firmly and solidly together. In its form it resembles angel wings with two greater and two lesser wings extending outward on each side of the body. In the head, the spenoid bone represents the Ark of the Covenant. Notice the box called the "Sella Tureica." It is the space where the pituitary gland sits. Over it is a membrane called the *Olivary process* that represents the mercy seat. The pituitary has two working parts which represents the two tablets of the Ten Commandments. Notice the wings in this picture.

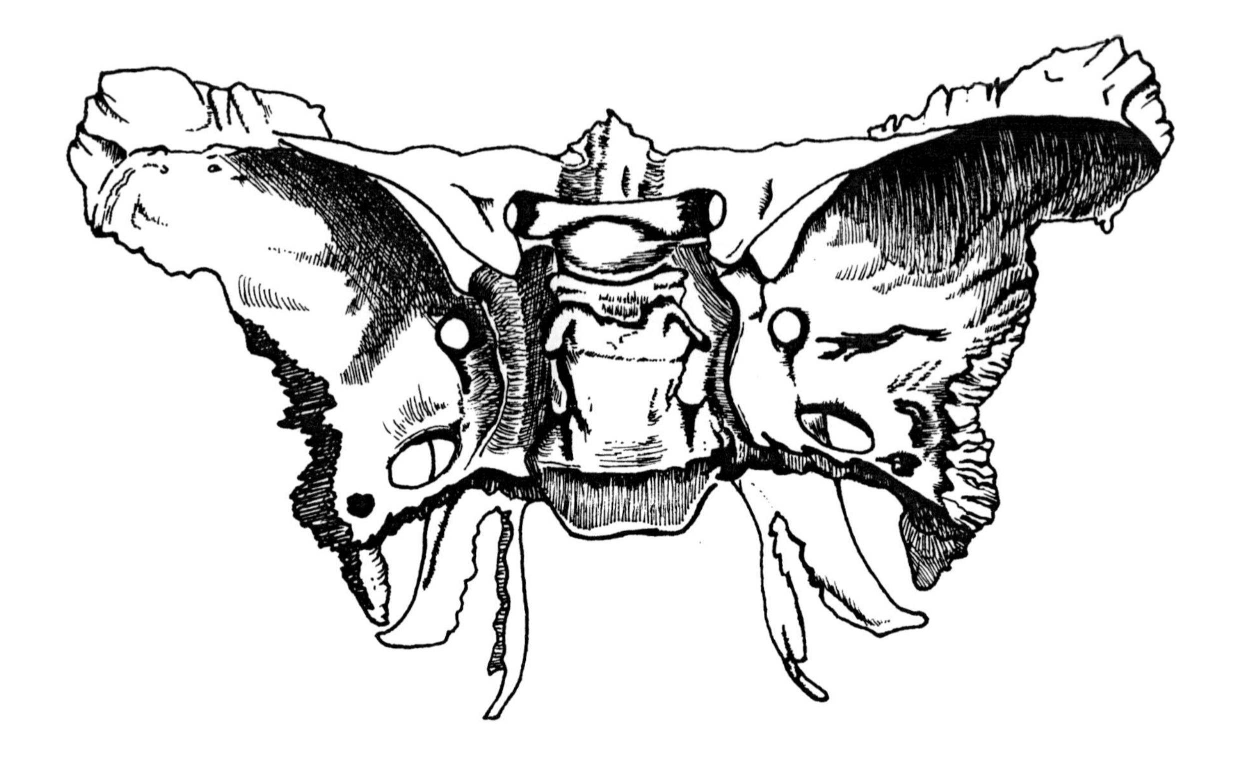

In the sanctuary the glory of God was manifested above the ark of the Covenant in the Most Holy Place. There are eight nuclei in the hypothalamus. Eight is the special number of God's Holy Spirit which represents regeneration. There is one thalamus on each side. *"Behold, I see the Heavens opened, and the Son of Man standing on the right hand of God."* (Acts 7:56.)

The thalamus interprets and defines messages and passes instruction to the hypothalamus. The nerves of the sense of smell travel through the ethmoid bone to the olfactory bulbs. This is the only one of the five senses that is not routed through the thalamus. It represents our prayers. It is left up to us to pray. We have an atmosphere around us that smells according to who is in charge of the soul, Christ or self. We make this decision.

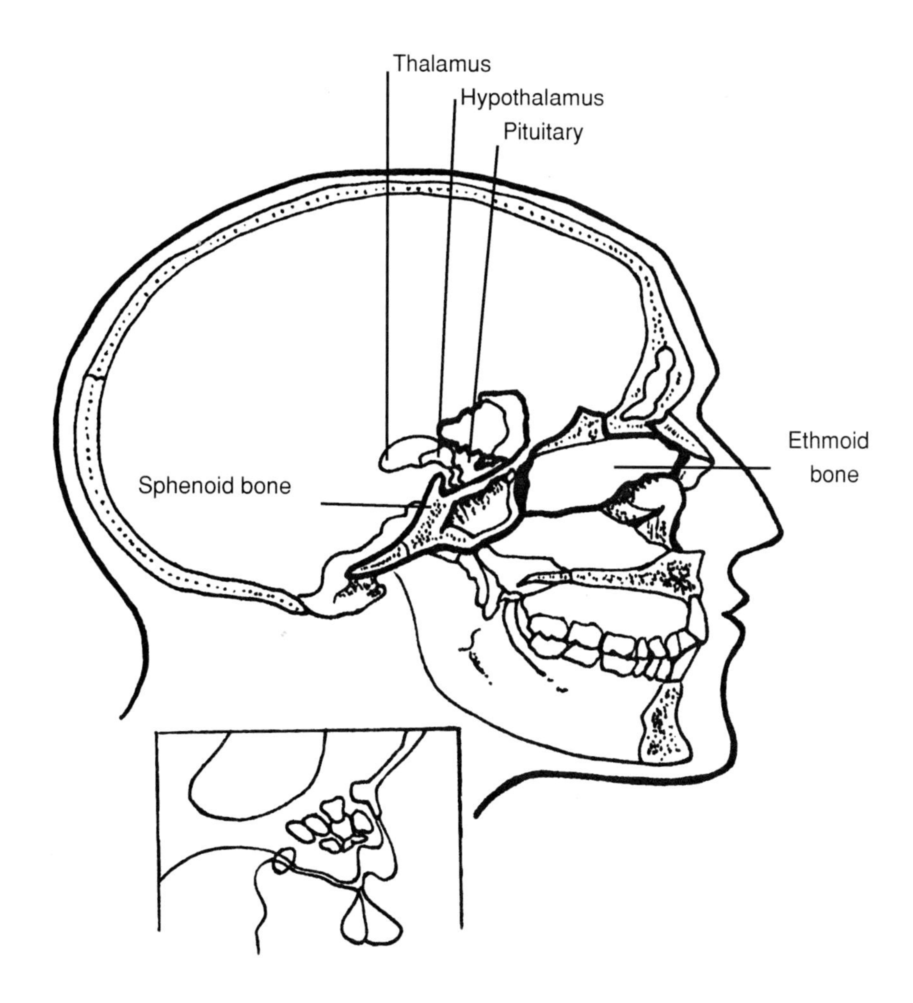

The Altar of Incense was the place of prayer. The incense had four spices plus salt, making five ingredients in all. Five is the number of grace and redemption. The incense represented the merits of Jesus, which went up with the prayers of the people. The incense was "beaten." *"But he was wounded for our transgressions, He was bruised for our iniquities; the chastisement of our peace was upon Him; and with His stripes we are healed."*—Isaiah 53:5. The odor of the sweet incense covered up the smell in the courtyard where the sacrifices were burned.

The candlestick was to give light in the sanctuary. It was never to go out. Vision represents understanding; when we understand something we say, "I can see that." *"The light of the body is the eye: therefore when thine eye is single, thy whole body also is full of light; but when thine eye is evil, thy body also is full of darkness."*—Luke 11:34.

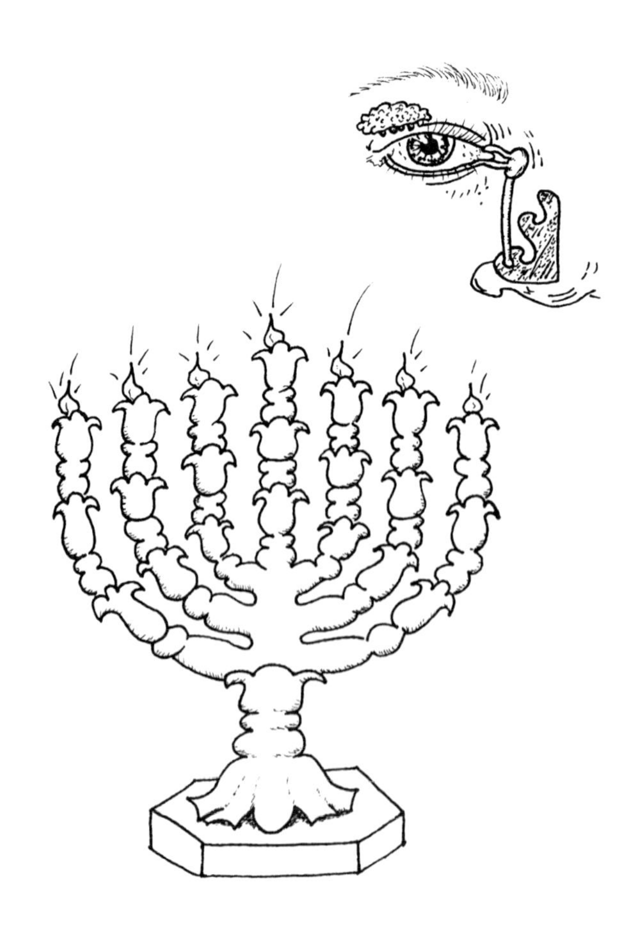

It was commanded that the priests wash their hands and feet in the laver before they had any part in the services, or they would die. (Exodus 30: 19,20.) Because it is the person's part to do the washing, there were no instructions on how to build the laver or who was to cover it or to carry it. It represents our part in the plan of salvation, or our free choice.

Buried under all the other lobes is the insula, and this is especially protected. Science has not discovered the function of this part of the brain. It is shaped like a bowl. It represents the laver which was a symbol of man's part in the plan of salvation—his free choice. Our free will is so important to God that He died to protect it for us to give us our choice back. He frees us from bondage and asks us to choose Him. *"Choose you this day whom ye will serve."* (Joshua 24:15.)

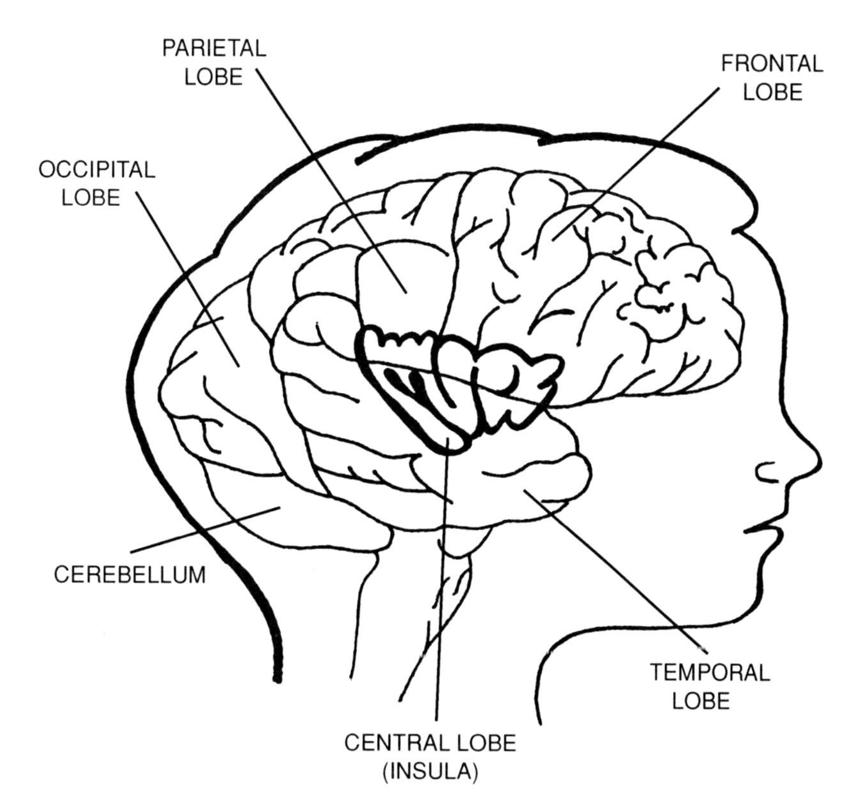

The altar of sacrifice was the first thing seen when entering the courtyard. It represents the sacrifice of the Lamb of God upon the cross. "All this was done that the Scriptures of the prophets might be fulfilled." (Matt. 25:56.) Sin requires a punishment or there will be anarchy; Christ paid the penalty, so that we need not die. "*. . . Behold the Lamb of God which taketh away the sins of the world.*" (John 1:29.)

The first structure upon entering the foramen magnum is the brainstem. When you slice the medulla, it looks like a grate. The grate on the altar and the table of shewbread were the same height as are the mouth and the medulla. This part of the brain is involved in the involuntary functions such as respiration, heart rate, etc. that occur without conscious thought. Ever since the fall of man, sin is practiced naturally, mostly without conscious thought. The sacrifice of Christ was to change that process and make us a new creature, if we so choose.

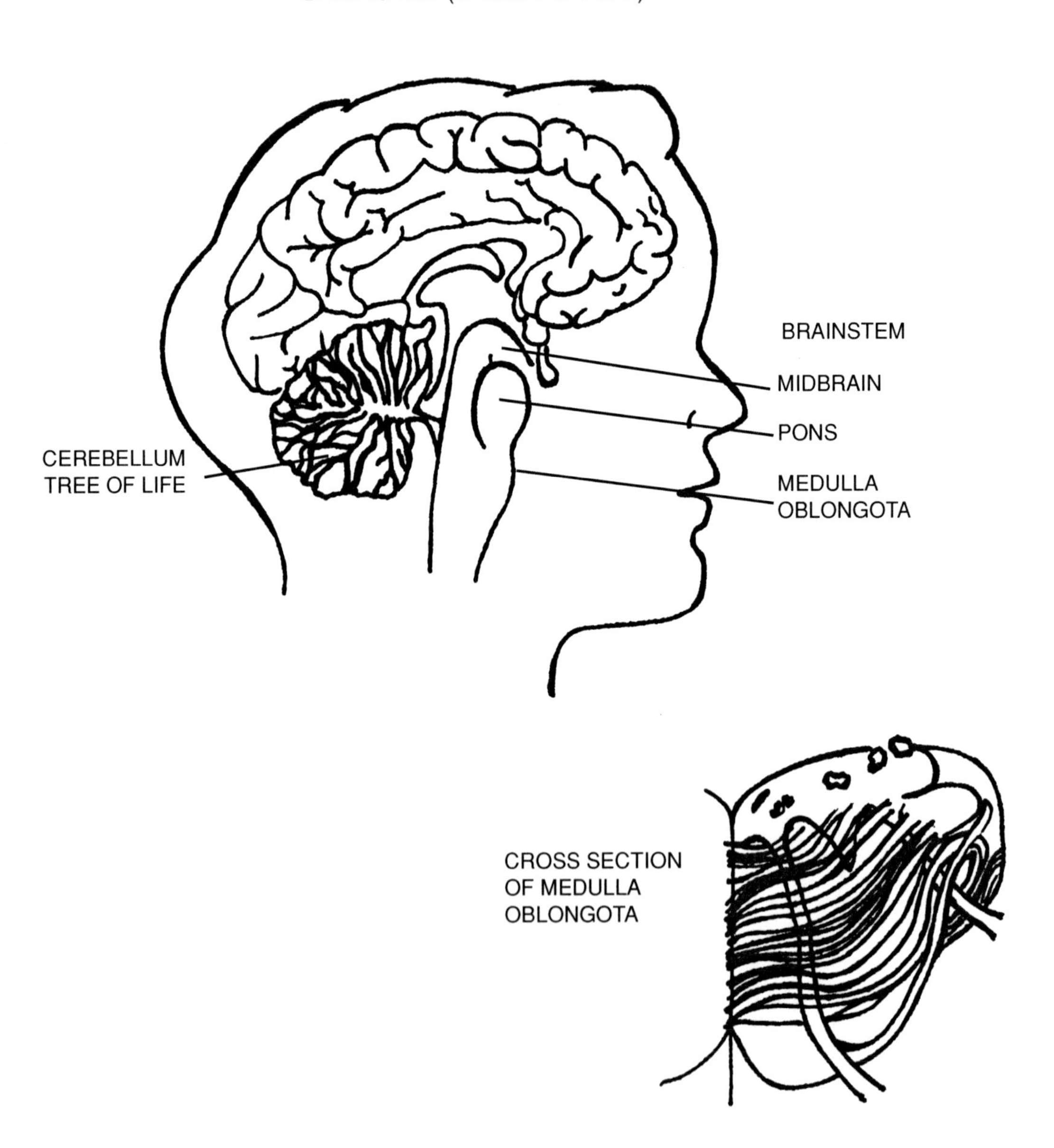

The table of shewbread had a shelf with a crown, and a border of a hand's breadth with another crown. God was very specific in this instruction.

THE TABLE OF SHEWBREAD

Bread is ingested through the mouth, where we have two sets of crowns. One crown is on a wide bone, the other one on a narrow bone. Eating represents taking in the Word of God and absorbing it. In the sixth chapter of the book of John, Jesus talked of eating His flesh and drinking His blood. He said, *"the words that I speak unto you, they are spirit and they are life."* (John 6:63.) There were 12 loaves of bread, one for each tribe. There is something special in God's word for each personality represented by a tribe of Israel. (Jeremiah 15:16.)

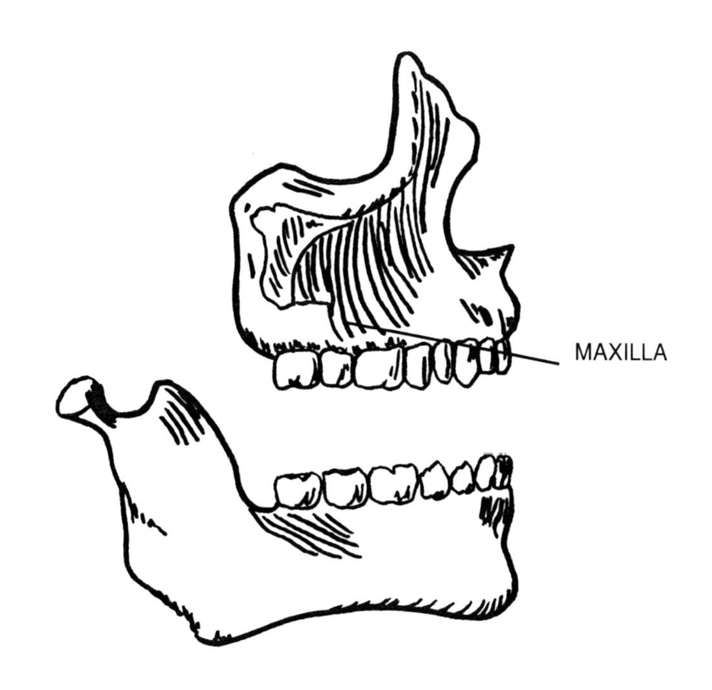

Note the picture from Gray's Anatomy showing the parts of the middle brain. The trumpet that called the people to the sanctuary was the *shofar* or a ram's horn. It was to instruct and bring to remembrance. It was a call to worship, a call to war, a call to march, or a warning, etc. This part of the brain is involved in memory and emotions. The hearer had to remember the different signals. Imagine the emotion when Jesus comes and we hear the trump of God. (I Thessalonians 4:13-17.)

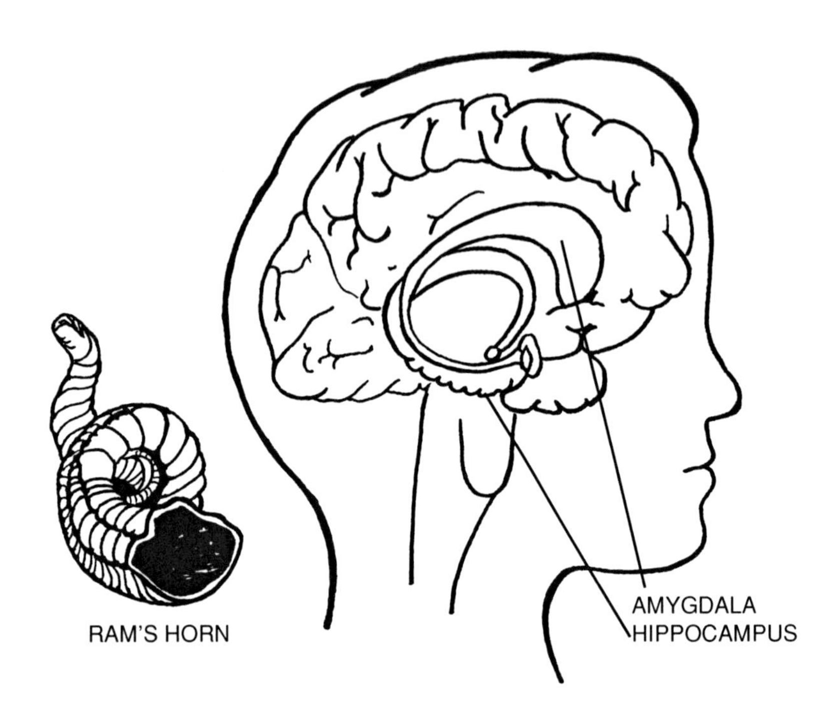

There are 12 cranial nerves on each side of the brain to govern the body. They represent the 24 elders in Revelation chapters 4 & 5. These people went to Heaven with Jesus at His ascension.

CRANIAL NERVES

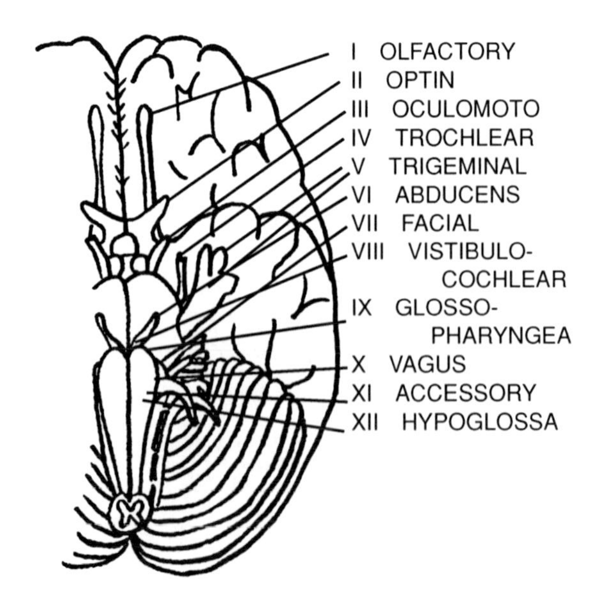

The twenty-four elders represent the vessels of the temple which were necessary to help perform the temple services. *"If any man therefore purge himself from these, he shall be a vessel unto honor, sanctified, and meet for the master's use, and prepared unto every good work."*—II Timothy 2:21.

"And immediately I was in the spirit; and, behold, a throne was set in heaven, and [one] sat on the throne. And he that sat was to look upon like a jasper and a sardine stone: and [there was] a rainbow round about the throne, in sight like unto an emerald. And round about the throne [were] four and twenty seats: and upon the seats I saw four and twenty elders sitting, clothed in white raiment; and they had on their heads crowns of gold.

"And out of the throne proceeded lightnings and thunderings and voices: and [there were] seven lamps of fire burning before the throne, which are the seven Spirits of God. And before the throne [there was] a sea of glass like unto crystal: and in the midst of the throne, and round about the throne, [were] four beasts full of eyes before and behind. And the first beast [was] like a lion, and the second beast like a calf, and the third beast had a face as a man, and the fourth beast [was] like a flying eagle. And the four beasts had each of them six wings about [him]; and [they were] full of eyes within: and they rest not day and night, saying, Holy, holy, holy, Lord God Almighty, which was, and is, and is to come. And when those beasts give glory and honour and thanks to him that sat on the throne, who liveth for ever and ever.

"The four and twenty elders fall down before him that sat on the throne, and worship him that liveth for ever and ever, and cast their crowns before the throne, saying, Thou art worthy, O Lord, to receive glory and honour and power: for thou hast created all things, and for thy pleasure they are and were created."—Revelation 4:2-11.

Chapter Ten

THE GREATEST WORK IN THE WORLD

Psalms 33:12

"Blessed [is] the nation whose God [is] the LORD; [and] the people [whom] he hath chosen for his own inheritance."

Everyone who has been chosen by God to be alive at this time, at the end of the world, must understand that they are charged by God, to share the "truth" to everyone they come in contact with and to minister to the needs of mankind. The servants of Christ are, in a large measure, responsible for the salvation of the world! This is a hard thing to say and an even harder thing to do! To be a "brother's keeper," so to speak, might be a great burden for some. Nevertheless God has charged His servants to be co-laborers with Him in the work of winning souls to Christ. To all, great and small, learned or unlearned, old and young the command is given: *"Go ye therefore and teach all nations, baptizing them in the name of the Father, and of the Son, and of the Holy Ghost. Teaching them to observe all things whatsoever I have commanded you, and lo, I am with you always, even unto the end of the world."*—Matthew 28:20.

It is the duty and privilege of every Christian to prepare himself and help others prepare themselves for eternity. It is also our sacred duty and privilege to hasten the coming of our Lord, Jesus Christ. We can help accomplish this only by submitting ourselves to God's will every second of every day. Those of us that are willing to do this will hasten to enlist in God's service. Those whose faces are turned toward Jerusalem and whose eyes are continually fixed on the Lamb of God will work in sacrifice, leaving all worldliness behind, and will die daily to selfishness. Quickly then the last harvest will be ripened, and Christ will come to gather His precious grain.

Everyday we are "cadencing" time marching toward the midnight hour. Even worldlings know that soon a calamity *"such as never was"* (Daniel 12:1) will come upon the Earth. In this fearful and exciting time in which we live, God has called His people and given them a message to bear, Isaiah 58, of the 'Right Arm'. The right arm hedge is the protection canopy given by God for restoration and sanctification of the redeemed.

The last great conflict between truth and error is almost upon us. Great decadence and confusion is soon to break upon the world in terror and as an overwhelming surprise. Everyone will be tested in this conflict. Only those who have been diligent students (and doers) of the Scriptures and who have been lovers of the truth will be shielded from the terrors by night as outlined so graphically in the ninety-first psalm. Precious time is rapidly passing, and most of the world is in a lost and dying state.

Who will go and testify of God's love and healing laws to a despotic world? Who will share meager means or, if necessary, all available to his dying brother or sister? Who will train missionaries, technicians, and doctors to reach out to a fallen and lost world? Only those who place themselves under God's control to be led and guided by Him will do this holy mandated work. God's servants will respond imbued with the Spirit of Him who gave His life for the life of the World. They will put on their armor of Heaven and go forth to war in the Army of our Lord and Savior, Jesus Christ. They will be willing to do whatever is required knowing that God's omnipotence will supply their every need. This is our quest, this is our hope, this is our mission and goal at the International Institute of Original Medicine. And this is the ***greatest need and work*** of the world. (A speech delivered by Dr. Nancy McEndree at campmeeting, October 2001).

STUDY QUESTIONS

STUDY QUESTIONS FOR CHAPTER ONE

1. Discuss in written format the following quotation. "So God created man in His own image, in the image of God created He him; male and female created He them."
2. Where is the best place to turn when sickness comes?
3. What are the results of obedience? Of Disobedience?
4. "A practical knowledge of science of human life is ____________ in order to glorify God in our bodies."
5. In your own words give the definition of the study of anatomy.
6. In your own words give the definition of the study of physiology.
7. What governs all the body functions?

STUDY QUESTIONS FOR CHAPTER TWO

1. Essay question. What manner of people should we be in light of the days in which we are living?
2. Anything that occupies space and has mass is called ____________.
3. We are "builded together for an ____________ of _________."
4. The human body is made up of how many elements?
5. True or False? "Relative to the difference in size, electrons are as far away from the nucleus as the planets are from the sun."
6. The hydrogen atom is the only atom without any _________ in the nucleus.
7. When two or more elements are joined together they form what?
8. Define the word "frequency."
9. Give the two forms of energy and discuss them in your own words.
10. Illness is defined as a _________________________.

STUDY QUESTIONS FOR CHAPTER THREE

1. Our bodies are made to be a ______________ of God through His Spirit.
2. Define "electromagnetic structure."
3. Electricity will always follow the line of _________________.
4. What is a cation? Define "anion."
5. What are the three ways that elements are bonded together?
6. True or False? When chemical bonds are broken they require energy.
7. Define "homeostasis."
8. True or False? When a molecule of water breaks up the bonds in a compound into smaller and different molecules it is called hydrolysis.
9. Define "dehydration synthesis."

STUDY QUESTIONS FOR CHAPTER FOUR

1. What is the source of all power? Write an essay discussing this.
2. Define "energy" in your own words.
3. What is the "Law of Conservation of Energy"? Explain
4. True or False? "Often the movement of digested material to cells and tissues must be done against a resistance to the flow."

5. List the five types of energy and define how they differ.
6. Each one of the atoms in our bodies is an ____________ generator.
7. True or False? Those who make great exertions to accomplish a great deal in a short period of time and continue to labor when their judgment tells them they should rest, will succeed in attaining life's goals.
8. We cannot receive energy from ________ food. Define this type of food.
9. What percent of our energy comes from the atmosphere?

STUDY QUESTIONS FOR CHAPTER FIVE

1. True or False? The cell is the fundamental building block of all living matter.
2. The most important chemical composition in our bodies is called __________________ acid. Define your answer.
3. The cell membrane is composed primarily of what?
4. What function does the "nucleolus" play in the cell?
5. Where is the energy for cell operations generated?
6. Define "actin" and "myosin."
7. True or False? All people living on Earth today came from two ancestors, Noah and his wife. Express your answer in essay format.
8. Define "complementary bonding."
9. What is "selective permeability"?
10. Name the two places where you can find "cilia" in the body.
11. What is the function of flagella?
12. Define "Deoxyribonucleic acid."
13. What allows the nucleus to be in direct contact with the membranes of the endoplasmic reticulum?
14. The immediate source of energy for most biological activities in the cell is a small organic molecule known as __________.
15. How is ADP formed?

STUDY QUESTIONS FOR CHAPTER SIX

1. True or False? It is impossible for the person that does not love others to know God.
2. Name the four major tissue types.
3. What is the function of epithelial tissue?
4. Why are we not to fear or be afraid of anything in the world today?
5. What does the loose connective tissue represent?
6. Define "matrix."
7. What is the function of collagen?
8. There are two types of connective tissue cells called ________ and ______________.
9. Name the three types of cartilage and their functions.

STUDY QUESTIONS FOR CHAPTER SEVEN

1. How do we glorify God?
2. How many body systems are there? Write a brief description for each.
3. What do the bones represent in our body? (Spiritual application.)
4. Define "sanctification."
5. How does the muscular system operate? (Spiritual application.)
6. What is the whole duty of man?

7. "The cardiovascular system is the __________________ system of the body."
8. What system represents God's Holy Spirit? What is the master control?
9. True or False? The nervous system contains the master control.
10. What __________ is to the body ___________ must be to the soul.
11. The urinary system is also known as the ___________ system.
12. The reproductive system tells the story of what?
13. Define "integementary."

STUDY QUESTIONS FOR CHAPTER EIGHT

1. The strongest support from the ____________ is found in the human body.
2. Research question? What is the relationship between the bones and the immune system?

STUDY QUESTIONS FOR CHAPTER NINE

1. What will happen to those that defile the body temple?
2. What must happen before Christ can enter and make His abode with us?
3. Our bodies are constructed to reveal God's plan of ____________ to perfection.
4. How many layers were on the roof of the sanctuary building?
5. What do the five pillars holding the entrance curtain represent?
6. Where is "God's Throne room" located in the body?
7. Define "articulating."
8. The _______________ interprets and defines ____________ and passes ______________ to the __________________.
9. The light of the body is the ___________.
10. Why was there no instructions on how the build the laver?
11. What is man's part in the plan of salvation?
12. What was the first thing seen upon entering the courtyard?
13. Thought question. Do we have the ability to control our subconscious thoughts such as our dreams?
14. What part of the brain is involved with memory and emotions?
15. List the 12 cranial nerves.
16. Who are the 24 elders that are seated around God's throne?

STUDY QUESTIONS FOR CHAPTER TEN

1. What charge have we been given by God?
3. How do we go about preparing for the above charge?
4. The right arm hedge is the __________________ given by God for the _______________ and ______________ of the redeemed.
5. What are some of the "terrors by night" listed in Psalms 91?
6. What will supply our every need during the last days on Earth?

Chapter Eleven

THE OVERCOMER

Genesis 32:28

"And he said, Thy name shall be called no more Jacob, but Israel: for as a prince hast thou power with God and with men and hast prevailed."

The name Israel means "A prince with God," or a more accurate translation would be "ruling with God." The Hebrew man Jacob became Israel when God changed his name. It was God who wrestled with Jacob and blessed him and then changed his name, from Jacob, which means "deceiver" or "supplanter," to Israel, which signifies "overcomer." Names indicated significant data for Biblical characters. Some name changes indicate a character or social or political change. If we, the last generation, are faithful, we will be receiving a new name when the great controversy is over and we are translated into the Kingdom of God.

"He that hath an ear, let him hear what the Spirit saith unto the churches; To him that overcometh will I give to eat of the hidden manna, and will give him a white stone, and in the stone a new name written which no man knoweth saving he that receiveth it."—Revelation 2:17.

The book of Numbers, Chapter 10 and verses 13-28 tells of the marching order of the Tribes of Israel. They marched in the order that they camped. The Levites were charged with the sacred duty of carrying the Sanctuary.

The Gershonites carried *"the tent, the covering thereof, and the hanging for the door of the tabernacle and the hangings of the court and the curtain for the door of the court, which is by the tabernacle, and by the altar round about, and the cords of it for all the service thereof."*—Numbers 3:25-28.

The family of Merari carried *"the boards of the tabernacle, and the bars thereof, and the pillars thereof, and the sockets thereof, and all the vessels thereof, and all that serveth thereto. And the pillars of the court round about, and their sockets and their pins, and their cords."*—Numbers 3:36,37. The family of Kohath with *"the ark, and the table, and the candlestick, and the altars, and the vessels of the sanctuary wherewith they minister, and the hanging, and all the service thereof."* (Numbers 3:31.) Numbers 4:5-15 tells how Aaron and his sons covered the furniture before it was carried. The Bible indicates that next in the marching order was the tribes of Judah, Issachar, and Zebulun. The tribes of Reuben, Simeon, and Gad followed. Next in procession came the tribes of Ephraim, Manasseh, and Benjamin. And finally the tribes of Dan, Asher, and Naphtali.

The following three charts show the order in which they camped around the sanctuary and an overview of each tribe. The systems represent the circle of service, the circuit of beneficence with every system serving all other systems. The student should now turn to Genesis 49 and read verses 1-28 and Deuteronomy 33:6-29. With these charts you will receive a condensed overview of each tribe, the meaning of the name, the strengths and weakness of the tribe, the name of the leader and his father in the wilderness, the disciple that represented that tribe, the system it represents, the job done by the system, and the product of that system. This is done in the order in which they camped around the sanctuary according to Numbers 3. The name of the leader chosen from each tribe and the father of each leader is taken from Numbers 1:5-15.

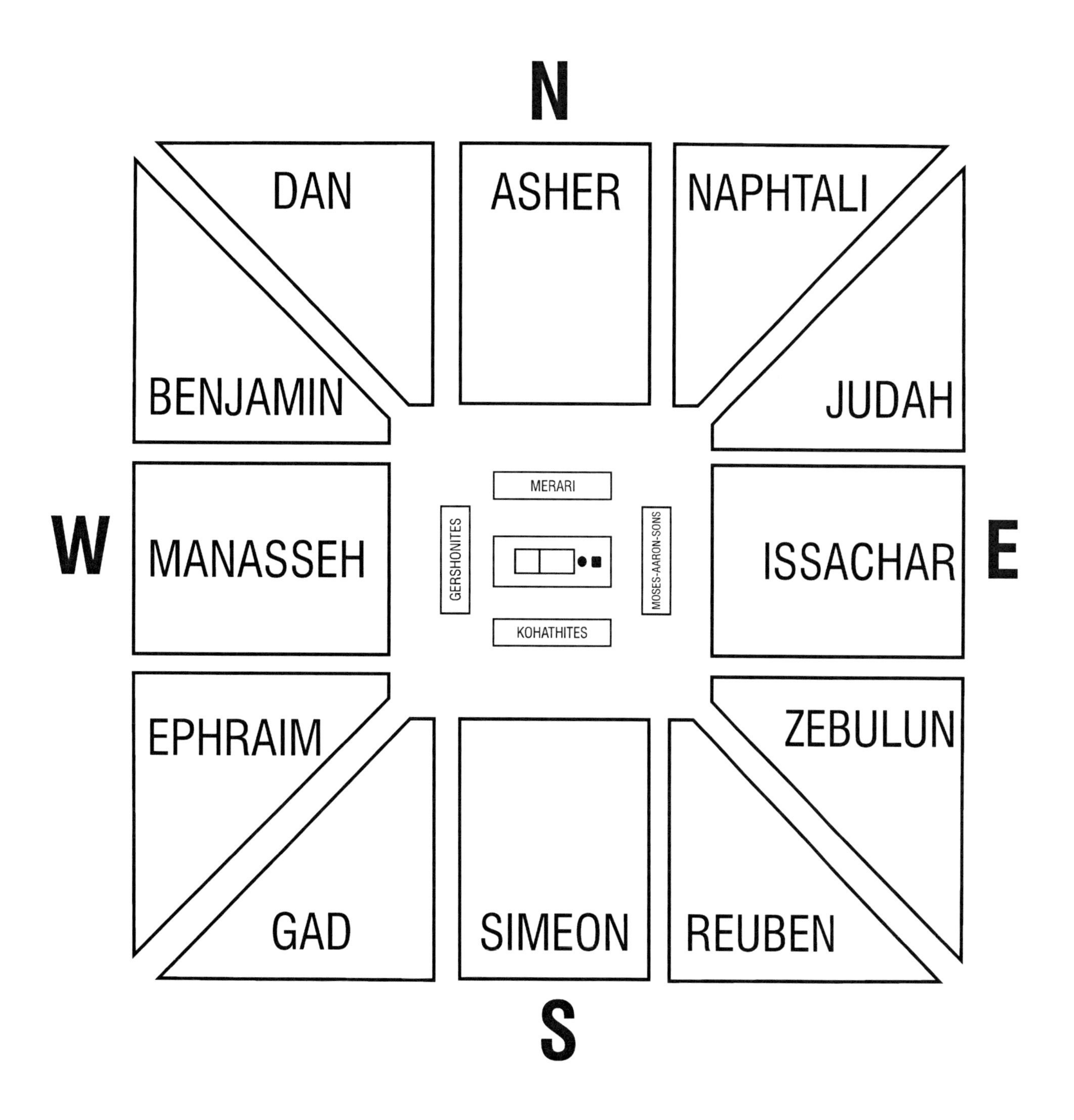
N
DAN
ASHER
NAPHTALI
BENJAMIN
JUDAH
MERARI
W
MANASSEH
GERSHONITES
MOSES-AARON-SONS
ISSACHAR
E
KOHATHITES
EPHRAIM
ZEBULUN
GAD
SIMEON
REUBEN
S

Tribe	Ephraim	Manasseh	Benjamin	Dan	Asher	Naphtali
Meaning	Doubly fruitful	Causing to forget	Son of the right hand	Judge	Happy	Wrestling
Strength	Energetic, enthusiastic	Competent	Protective, daring	Judgment	Diplomacy	Eloquent
Weakness	Impulsive	Reluctant	Fierce, headstrong	Critical, backbiting	Crafty	Impractical
Disciple	Peter	Andrew	Paul	Judas Iscariot	Nathanel	Phillip
System	Reproductive	Urinary	Digestive	Muscular	Endocrine	Integementary
Duty	Increases	Filters	Nourishes	Moves	Regulates	Covers and connects
Product	New being	Mineral balance	Fuel	Mobility	Hormones	Protection and adhesive
Leader in wilderness	Elishama	Gamaliel	Abidan	Ahiezer	Pagiel	Ahira
Meaning of name	God is hearer	God is recompenser	Father of liberality	Helping brother	God meets	Brothers of wrong (distress)
Father of leader	Ammihud	Pedahzur	Gideoni	Ammishaddai	Ocran	Enan
Meaning of name	My people is honorable	The Rock delivers	Warlike	My people is mighty	Troubler	Fountain

Tribe	Levi	Judah	Issachar	Zebulun	Reuben	Simeon	Gad
Meaning	Joined	Praise	Bearing a reward	Dwelling	Behold a son	Hearing	A troop
Strength	Loyal	Generous, Leader	Strong, Burden bearer	Thrifty, Discreet	Kind, gentle	Zealous	Perserver-ance
Weakness	Cruel	Independent	Slow, lazy	Stingy, subtle	Vacillating, weak	Excitable, bad-tempered	Intolerance
Disciple	John	James	James, son of Alpheus	Matthew	Thomas	Simon, the zealot	Jude
System	Life force	Nervous	Skeletal	Lymphatic	Cardio-vascular	Respiratory	Immune
Duty	Ignites, fires	Communi-cates, directs, commands	Supports	Cleanses, transports	Transports	Exchanges	Defends
Product	Electrical force	Information, direction	Blood cells	Plasma cells that secrete anti-bodies	Pressure	Breath	Protection
Leader in wilderness	Aaron	Nahshon	Netaneel	Eliab	Elizur	Shelumiel	Eliasaph
Meaning of name	Illuminated	Oracle	God gives	God is Father	God is a Rock	God is peace	God is gatherer
Father of leader		Amminadab	Zuar	Helon	Shedeur	Zurishaddai	Deuel
Meaning of name		My people is willing	Little	Strong	Shedder of light	Almighty is a Rock	God is knowing

Levi. Their leader was Aaron and they do not give the name of his father. This tribe was special, set aside as sons of God to serve the Lord in the tabernacle and tend to the holy things. Joseph is not included in this chart because his two sons took the place of his name; Also he was a type of Jesus, our Saviour, as the story of Joseph explains. Jacob was the father of sons that are the patriarchs of the tribes of Israel. In this volume of the Body Temple we will be studying and examining the Tribes of Israel. Our study makes an analogy between the characteristics of the body.

Chapter Twelve

The Twelve Tribes

<u>**Genesis 49:5-7**</u>

"Simeon and Levi are brethren; instruments of cruelty are in their habitations. O my soul, come not thou into their secret; unto their assembly, mine honour, be not thou united: for in their anger they slew a man, and in their self-will they digged down a wall. Cursed be their anger, for it was fierce; and their wrath, for it was cruel: I will divide them in Jacob, and scatter them in Israel."

Levi was the third son of Jacob and Leah; his name meant "joined." His mother probably named him this because she thought maybe his birth would join her to Jacob's affection. His name was a prophecy because it was the ministration of this tribe in the sanctuary that reconciled the spiritual bride, and the people of the covenant to their Maker Husband. (Isaiah 54:5.)

Genesis 34 records the horrible deed of revenge done by the brothers in Scechem. When Jacob blessed his sons, there was no indication that Levi would come to any good. In no other tribe are the words concerning them more opposite in the above passage and in the words of Moses:

"And of Levi he said, Let thy Thummim and they Urim be with thy holy one, whom thou didst prove at Massah, and with whom thou didst strive at the waters of Meribah: Who said unto his father and to his mother, I have not seen him; neither did he acknowledge his brethren, nor knew his own children: for they have observed thy word, and kept they covenant. They shall teach Jacob thy judgments, and Israel thy law: they shall put incense before thee, and whole burnt sacrifice upon thine altar. Bless, LORD, his substance, and accept the work of his hands: smite through the loins of them that rise against him, and of them that hate him, that they rise not again." —Deuteronomy 33:8-11.

In his younger years, Levi was hot-tempered and wicked just like his brothers. He took part of the sale of Joseph. Even so, this tribe was honored with the priesthood because of the singleness of purpose and the role they played in the great apostasy of the golden calf in Exodus 32. Personal feelings were set aside to comply with God's requirements, no matter how painful it was. Levi does not represent any particular body system, but represents the electrical life force that runs the body. The Levites camped directly around the sanctuary, and without the tribe of Levi, there was no life in the sanctuary. The Levites did not have an inheritance in the land, but were scattered throughout all Israel to minister to the people. The Levites were given forty-eight (12x4) cities, and they were dependent upon the faithfulness of the other tribes for their livelihood. The Levi character was revealed in crisis. Like him, we should be developing our characters by God's grace every day, so that when the great crisis comes, we will be ready to stand with unwavering obedience to God.

JUDAH

Genesis 49:8-12

"Judah, thou [art he] whom thy brethren shall praise: thy hand [shall be] in the neck of thine enemies; thy father's children shall bow down before thee. Judah [is] a lion's whelp: from the prey, my son, thou art gone up: he stooped down, he couched as a lion, and as an old lion; who shall rouse him up? The scepter shall not depart from Judah, nor a lawgiver from between his feet, until Shiloh come; and unto him [shall] the gathering of the people [be]. Binding his foal unto the vine, and his ass's colt unto the choice vine; he washed his garments in wine, and his clothes in the blood of grapes: His eyes [shall be] red with wine, and his teeth white with milk."

Judah is the representative of the *nervous system,* which is the ruler of the body. Psalm 60:7: ". . . Judah is my Lawgiver." You will notice that he is camped on the eastern side of the sanctuary. He was the fourth son of Jacob, and his mother was Leah. He became the prominent leader in his family, and his standard was the lion. The positive trait of Judah's tribe is generosity with independence becoming its weakness. Independence is always a weak characteristic if it is not controlled by the Holy Spirit. Jesus and all the kings after David were descended from this tribe.

"For Judah prevailed above his brethren, and of him came the chief ruler..."—I Chronicles 5:2.

Jacob accepted Judah's assurance that he would be responsible for Benjamin on their trip to Egypt, because Judah was a leader and dependable. He would not accept Reuben's offer. (Genesis 42:37, 38; 43:1-13.)

In Genesis 44 is the story of the cup that was found in Benjamin's bag. In verse 14 we find it was Judah who was the leader and spokesman when the family was in trouble. This was true also in time of joy when the family arrived in Egypt with Jacob. (Genesis 46:28.)

Judah's character was one of being consistent and dependable. By his integrity Judah won the respect of his father and all of his brothers.

The tribe of Judah was independent. They were the first to acknowledge David as king. (II Kings 4.) This tribe maintained their own land 142 years after the northern tribes had lost theirs. (II Kings 17:6; II Chronicles 36.) When independence is carried to extremes, it becomes a weakness. This is true of any of the strengths without the converting power of the Holy Spirit. Genesis 38 tells the story of Judah's early life among the heathen. He had left his brothers in disgust after the selling of Joseph, even though he was a part of it. After marrying a heathen woman, he eventually lost two of five sons to wickedness and then returned to the family.

The greatest member of the tribe of Judah was Jesus, the true Lawgiver, the true Lion, and the true Lamb. Daniel and the three worthies, who were thrown into the furnace, were also of this tribe along with faithful Caleb. All the kings of Judah were of this tribe, King David being the type of Christ. Some were good and faithful kings, but some were wicked. Whichever they were, they did their deeds wholeheartedly. The wicked kings established pagan worship and the faithful kings tore down the idols and places of worship. And so the struggle goes on in each one of us today. Solomon was the wisest king at one time and the most wicked at another time. None of us can be safe without the constant guidance of God.

James, one of the sons of thunder, was of this tribe. He was the first of the apostles to be martyred.

ISSACHAR

Genesis 49:14,15

"Issachar [is] a strong ass couching down between two burdens: And he saw that rest [was] good, and the land that [it was] pleasant; and bowed his shoulder to bear, and became a servant unto tribute."

Issachar is the representative of the skeletal system. This tribe camped next to Judah on the eastern side of the sanctuary. Issachar was the ninth son of Jacob. Leah was his mother. He is represented as a burden-bearer. Issachar understood the ultimate reward and, because of this, he was willing to bear heavy burdens patiently without complaining. In I Chronicles 7:5, we are told that the men of Issachar were valiant men of might. They are called men of *"understanding of he times, to know what Israel ought to do"* in I Chronicles 12:32. This indicates the gift of discernment.

In Moses' evaluation of the tribes he said, *"Rejoice, Zebulun, in they going out; and Issachar, in thy tents."* —Deuteronomy 33:18. This indicates that those of this tribe enjoyed staying at home and working in the background. Their weakness was being slow to move and unwilling to volunteer; their strengths were great understanding and that quiet strength that made them good dependable friends. There is no mention of the life of Issachar after his birth until the blessing given by his father. The skeletal system is in the background bearing the weight of the body. It is never openly exposed, but when a break occurs, the results are seen in the form of swelling or a cast. Bones represent laws or principles. By principle we have to make our decision of whether God or self will reign in our lives.

Whether right or wrong principles guide us will be manifested in our lives. The valley of Megiddo, a place of final decision and battle, was located in the land of Issachar. (Joshua 17:11.) Jezebel's stronghold was in the land of Issachar, and from there all Israel was led astray. It was in the land of Issachar that Saul consulted the witch of Endor, thus sealing his total separation from God. It is in the realm of principle that our decisions for life and death are made. The battle wages between selfishness and self-sacrifice. May we be true to the principles of God.

ZEBULUN

Genesis 49:13

"Zebulun shall dwell at the haven of the sea; and he [shall be] for an haven of ships; and his border [shall be] unto Zidon."

Zebulun represents the lymphatic system. This tribe was camped on the eastern side of the tabernacle next to Issachar and Judah. Zebulun was the youngest son of Leah. Zebulun's tribe was a sanctuary for their family & friends. They were people gifted in social graces as depicted by being good listeners.

Deuteronomy 33:18 pictures Zebulun as one who loves to travel.

During the battle of Megiddo, which is a type of the last great battle of Armageddon, the tribe of Zebulun is portrayed as "a people who jeoparded (margin: exposed to reproach) their lives unto the death in the high places of the field." (Judges 5:18.) This shows their courage in times of crises. In the margin of I Chronicles 12:33, they are portrayed as "rangers" of battle, or experts in war. *"Of Zebulun, such as went forth to battle, expert in war, with all instruments of war, fifty thousand, which could keep rank: they were not of double heart."*

The people who are among this tribe are neat and orderly and will not be found among the hypocrites. They are those who *"handled the pen (margin, draw with the pen) of the writer."* (Judges 5:14.)

The people of this tribe are found thrifty with their goods and talents. If not under the control of he Holy Spirit they could become stingy and self-centered. When a lymph system as with the Zebulunites ceases to work, all the toxins of the body are retained and poisons the whole system.

The land of Zebulun was the childhood of Jesus. His first miracle was done here in the city of Can. Isaiah prophesied that the land of Zebulun would see a great light. (Isaiah 9:1,2.)

REUBEN

Genesis 49: 3,4

"Reuben, thou [art] my firstborn, my might, and the beginning of my strength, the excellency of dignity, and the excellency of power: Unstable as water, thou shalt not excel; because thou wentest up to thy father's bed; then defiledst thou [it]: he went up to my couch."

Reuben represents the cardiovascular (circulatory) system. He was the firstborn of Jacob. This tribe camped on the south side of the sanctuary; their banner had a man on it. The four tribes on the four corners led with their banners and represented the four beasts around the throne. (Revelation 4:7.)

Reuben was a leader, strong with natural dignity, and he had great influence with men. As the firstborn, he should have had a threefold blessing. He should have received a double portion of his father's possessions, the priesthood of God's people, and the honor of becoming the progenitor of the Messiah. But this blessing was not bequeathed to him because he was unstable as water.

Water is one of the great forces of nature. It cleanses and gives life to all living creatures. It can also be one of the greatest destroyers and ruinous forces in the form of a flood. Water naturally takes the path of least resistance as it flows along, always downward, and this well described the life of Reuben. His many talents were wasted, and he never seemed to stand for what he knew to be right. All the good he might have done seemed to be destroyed, because he could not control his impulsiveness and passion.

Reuben did use his influence to keep Joseph from being destroyed, but lacked the courage to set him free. Then he joined in the deception that caused his father years of grief. (Genesis 37.) Reuben displayed kindness when he brought his mother mandrakes, but displayed selfishness in his disregard for Joseph's plight.

There are people everywhere that live without fixed principles. It is hard for them to resist temptation no matter from which direction it comes and in what form the temptation may take. Every precaution must be taken to surround them with influences that will strengthen their moral power. Let them be separated from these helpful influences and association and be thrown in with a class who are irreligious, and they will soon show that they have no real hold from above. They only trusted in their own strength. They have been praised and exalted when their feet were standing in sliding sand. They are like Reuben, unstable as water, having no inward rectitude. And like Reuben, without help from the Lord, they will never excel. That kind of easy good nature is usually under the control of Satan far more than under the control of the Spirit of God. These people are led into evil very easily because they have a very accommodating disposition. It hurts them to give a square "No" to any situation. If invited to do something wrong, they have no interior strength to fall back on. They do not make God their trust.

We need to see our dependence upon God, and to have a resolute heart. We need to show a Christlike strength of character. We need to be able to say, *"No, I will not do this great wickedness and sin against God."* The natural hearts of Reubens have no high principles of duty. They demonstrate no power of God to redeem and restore. Reuben's name is on one of the gates in the holy city and 12,000 of the 144,000 will be from this tribe. When submitted to the Holy Spirit, weak, vacillating characters can become a mighty force for God, just as water under heat and pressure becomes steam and can move mountains. It was when the tribe of Reuben had "great thoughts of heart" and "great searchings of heart" that they became a mighty force for God. (Judges 5:15,16.) "The Moses" prayer for this tribe became a reality: "Let Reuben live, and not die; and let not his men be few." (Deuteronomy 33:6.)

The body of Christ needs converted Reubens today. A hurting world is dying for the kindness and gentleness of Reuben. The heart, just like Reuben, has a great potential to serve the body for many years. But if it finds an obstruction in its pathway, it decreases in its ability to supply blood. About 40 million Americans have some form of cardiovascular disease, and it is responsible for more deaths than all other causes of death combined. The cause of most of this trouble is plaque in the arteries which is caused, in a great measure, by fat. In the book of Leviticus, fat represents sin which was burned on the altar of sacrifice. When, like Reuben, we allow fat or sin to build up in our heart, we are in big trouble.

SIMEON

Genesis 49:5-7

"Simeon and Levi [are] brethren; instruments of cruelty [are in] their habitations. O my soul, come not thou into their secret; unto their assembly, mine honour, be not thou united: for in their anger they slew a man, and in their self-will they digged down a wall. Cursed [be] their anger, for [it was] fierce; and their wrath, for it was cruel: I will divide them in Jacob, and scatter them in Israel."

Simeon was the second son of Jacob and Leah. This tribe represents the respiratory system.

Members of this tribe are born leaders and headstrong. The Bible first mentions Simeon as a young man with a story of violence. This was prompted by misplaced family loyalty. Therein lies the key to the personality of this tribe. Zealous to protect his family from dishonor, resenting any intrusion of the family circle, his unconverted heart thought it was right to slaughter in order to make things right in his sight. He brought suffering to his family, and his enemies, and instead of honoring his family as he was trying to do, he brought disgrace by such a horrible act. Even though his motive was love for his sister, who was viciously attacked and dishonored by his actions were totally out of harmony with the will of God.

After committing murder to break up the marriage of his sister to a Canaanite, he himself married a heathen woman, totally disregarding God's command. Sometimes the first to be overcome by a weakness are those who criticize others with that weakness. Simeon was ruled by force. When zeal is uncontrolled, it becomes a weakness and unconverted passion is manifest. *"He that hath no rule over his own Spirit is like a city that is broken down, and without walls."* (Proverbs 25:28.)

Because of the violence at Shechem led by Simeon, shame was brought upon his father. Because of Simon's action, he was not given land but was allowed to occupy a portion of the inheritance of Judah. The Simeonites' numbers were greatly decreased between Sinai and the entrance into the Promised Land. It is believed that many were slain as punishment of Baalpeor because of their blatant disobedience. (Numbers 25)

In Genesis 42:24, it was Simeon that Joseph had bound and kept as a hostage in the land of Egypt. It may be that he was the instigator of Joseph's bondage (Genesis 37), although no names are mentioned as to who originated the idea for the sale of Joseph. *"He that is slow to anger is better than the mighty; and he that ruleth his spirit than he that taketh a city."*—Proverbs 16:32.

When Simeon's zeal was harnessed by the Holy Spirit, it created a fearless warrior for God, much like Judith in the book of the same name in the Apocrypha. She slew the leader of the enemies army and prayed to the "Lord God of my father, Simeon."

Simeon's name is over one of the gates in the Holy City and 12,000 of the 144,000 are of this tribe. Even though this tribe is famous for murder and sin, it demonstrates that God is able to change and use anyone who is willing to be converted.

This tribe represents the respiratory system that must have lots of zeal in order to supply oxygen to the body and remove the carbon dioxide. If the body is deprived of air for more than a few minutes, everything quits and the body dies. The Simeon people are always focused in on the job, are very productive, and are zealous workers for God. If, on the other hand, they are out of the control of the Holy Spirit, they can be pushy and intolerant of others' methods of working. When the respiratory system, like Simeon, becomes too active, hyperventilation takes place. An overabundance of oxygen is received, and it becomes toxic to the system.

GAD

Genesis 49:19

"Gad, a troop shall overcome him: but he shall overcome at the last."

Gad was the son of Jacob and Leah's maid Zilpah. His name, according to the marginal reading of Genesis 30:11, means "a troop or a company," and this suited him well. This tribe represents the immune system.

There is a battle indicated here, an invasion of temptation that must be fought and overcome. Gad wanted to have a good time. People from this tribe love the presence of a crowd. They are often the backsliders with good intentions, but they lack the will to swim against the current. A battle of double mindedness is exhibited today by God's people.

Members of this tribe are usually champions for the underdog who sometimes place loyalties in the wrong place. But they are fierce warriors, and as soon as they call upon Jesus, He gives them the strength they need to overcome.

The men of this tribe were brave enough to be on David's side when he was still a rebel. (I Chronicles 12:8-15.) They were "men of might, and men of war fit for the battle." The least of them could resist an hundred and the greatest a thousand. (margin, v.14). And of Gad he said, *"Blessed be he that enlargeth Gad: he dwelleth as a lion, and teareth the arm with the crown of the head. And he provided the first part for himself, because there in a portion of the lawgiver, was he seated (Hebrew: to hide by covering); and he came with the heads of the people, he executed the justice of the LORD, and his judgments with Israel."* (Deuteronomy 33:10.)

This is a perfect description of the workings of the immune system. Each warrior can fight many bacteria and viruses, UNLESS we weaken them by our wrong diet or lifestyle. We can fight off the many attacks of the enemy UNLESS we stop eating the Word and lose the connection with our leader.

Elijah was from this tribe. Read the story of his battle with the sun god (Satan) in I Kings 17-18. He took on the battle against 850 representatives of Baal. He did so fearlessly because he knew that when God is on your side, YOU ARE THE MAJORITY.

When the immune system gets too defensive, it attacks the body. Sometimes people of this tribe are very resistant to the standards of God. They do not like to be told what to do.

Without instruction, the immune system runs amok and it cannot identify the real enemy, as is the case of autoimmune diseases. Today in this world there are new viruses and temptations arising constantly. We need to keep our immune system strong by obeying the health laws, and we need the Holy Spirit abiding in our heart so that we may overcome and obey.

JOSEPH, EPHRAIM and MANASSEH

Genesis 49:22-26

"Joseph [is] a fruitful bough, [even] a fruitful bough by a well; [whose] branches run over the wall: The archers have sorely grieved him, and shot [at him], and hated him: But his bow abode in strength, and the arms of his hands were made strong by the hands of the mighty [God] of Jacob; (from thence [is] the shepherd, the stone of Israel:) [Even] by the God of thy father, who shall help thee; and by the Almighty, who shall bless thee with blessings of heaven above, blessings of the deep that lieth under, blessings of the breasts, and of the womb: The blessings of thy father have prevailed above the blessings of my progenitors unto the utmost bound of the everlasting hills: they shall be on the head of Joseph, and on the crown of the head of him that was separate from his brethren."

JOSEPH

Joseph was a type of Jesus. He was sold by his brethren. He was unwavering in principle, even though it meant imprisonment for him. He was Jacob's favorite son and received the blessing cited above. Ephraim and Manasseh were the sons of Joseph.

EPHRAIM

The tribe of Ephraim represents the reproductive system. Jacob adopted Joseph's sons and in the process crossed his hands and put Ephraim before Manasseh. Joseph said that Ephraim should be greater than Manasseh and *"his seed shall become a multitude of nations."* (Genesis 48.) The tribe of Ephraim was a lead tribe and their standard was a bull or ox. This tribe produced many great leaders including Joshua, Deborah, and Jeroboam. They had vigor and courage, but not all of them were as noble as Joshua. Like Jeroboam, they didn't always go about doing right. As the first King of the northern tribes of Israel, he saw what needed to be done, but he created havoc and war getting it done. (II Chronicles 13.)

Ephraim had inherited his father's wisdom. He was raised in the palace and consequently did not have Joseph's training in hardship. This tribe did not want to deal with the Canaanites' chariots of iron, but wanted Joshua to give them more land that was easily attained. (Joshua 17:14-17.)

In Gideon's campaign against the Midianites, there were no men from Ephraim's tribe. They joined the ranks at the end of the war, and then criticized the way it was done. (Judges 7 and 8:1-3.) When Gideon soothed their ruffled feathers by praising them, they were happy.

The same impulsiveness, selfishness, and arrogance was shown in Judges 12:1-6. It was this very attitude that caused this tribe to throw away its place, and their name will not be found on the gates of the Holy City or among the 144,000. People from this tribe will go in through the gate of Joseph and will have overcome these unlovely traits of character as Peter overcame them.

Ephraim represents the <u>reproductive system</u> and demonstrates the lesson of procreation. We may create our own unlovely character or we may allow the LORD to recreate us so that we may be born again a new creature in Christ.

The people from this tribe make mighty preachers like Peter. These people speak as a good seed (the word of God), and it comes flowing forth from their mouths. If they are not converted, then usually ugly or foolish words come flowing forth like a bad seed.

MANASSEH

Manasseh represents the urinary system and the tribe of Manasseh was divided into two half tribes that camped on either side of the Jordan. Picture the kidneys and the water that flows down from them. Even though his brother received the main blessing, Manasseh will be in the Holy City and will be represented in the 144,000. *"And of Joseph he said, Blessed of the LORD be his land, for the precious things of heaven, for the dew, and for the deep that coucheth beneath, and for the precious fruits brought forth by the sun, and for the precious things put forth by the moon, And for the chief things of the ancient mountains, and for the precious things of the everlasting hills, And for the precious things of the earth and fullness thereof, and for the good will of him that dwelt in the bush; let the blessing come upon the head of Joseph, and upon the top of the head of him that was separated from his brethren. His glory is like the firstling of his bullock, and his horns are like the horns of unicorns: with them he shall push the people together to the ends of the earth: and they are the ten thousands of Ephraim, and they are the thousands of Manasseh."* (Deuteronomy 33:13-17.)

Again we see the attributes of Joseph and how his sons were blessed because of him; just as we are blessed because of the merits of Christ. This tribe never completely drove out the Canaanites as was the case in most of Israel. (Joshua 17:12,13.) We tend to stop short of complete victory over sin in our lives and are satisfied to accept tribute keeping the sin around to harass us later.

Manasseh's tribe did rally to Asa *"in abundance for a reformation."* (II Chronicles 15:8,9.) They continued to contribute to the upkeep of the temple services in Jerusalem after they had joined the northern tribes in the division. They were among the remnant of Israel that humbled themselves and went down to Jerusalem to keep the Passover in the time of King Hezekiah.

Gideon was from the tribe of Manasseh. He was reluctant to lead the LORD's army, but when he was assured of the LORD's help, he went bravely into battle with only three hundred men. Read his story in Judges 6 and 7. All through their history, the men of this tribe were ready to take part in the reforms. They were leaders in breaking down the altars and the high places. (I Chronicles 31:1.) Like the tribe of Manasseh, the kidneys are always trying to get rid of the toxins in the body, and they work constantly to keep the blood pure.

Weakness found in the members of this tribe is, when they are loaded down with too much work, they tend to give up and quit. The kidneys do this same thing when they are constantly overloaded.

We need to keep this principal in mind, as we all share a load of work for our Master. When one is overloaded with burdens and strife reconstruct unhappiness can become a by-product. All of us need to never despair because Jesus is still on the throne. There is to be no despondency in God's service. Our faith is to endure the pressure brought to bear upon it. God is able and willing to bestow upon His servants all the strength they need. He will more than fulfill the highest expectations of those who put their trust in Him.

BENJAMIN

Genesis 49:27

"Benjamin shall raven as a wolf: in the morning he shall devour the prey, and at night he shall divide the spoil."

This was all Jacob said about his beloved son, Benjamin. After Joseph was sold, Jacob was so protective of Benjamin that, even after he was old enough to have fathered 10 sons, Jacob would still not let him go to Egypt the first time with his brothers. Did this make him a spoiled person, and self-willed, wanting his own way? Is this like the digestive system of the body when it is indulged, demanding its likes and dislikes regardless of the damage it does to the rest of the body?

This tribe represents the digestive system.

Benjamin was the youngest of Jacob's sons, and was named *"Benoni,"* which means *"Son of my sorrow,"* by his mother, Rachel, because she died shortly after his birth. His father changed his name to "Benjamin," which means *"Son of my right hand*." Read the story of the brothers in Egypt in Genesis 42:5.

The early history of this tribe fulfilled the likeness to a wolf. A wolf is easily frightened away if it is alone, but as a pack with their combined intelligence and teamwork, they are a most formidable foe. Wolves are persistent, plan well as a pack, cooperate together, and are willing to sacrifice for the good of the group. They were the chief enemy of the shepherds in Palestine, and Jesus warned to beware of wolves in sheep's clothing.

The book of Judges, chapters 19, 20 and 21 tell the story of the Benjamite's warlike nature and their senseless fight with the rest of the tribes to protect men that were wrong. Benjamin lost 25,100 men in their misguided effort and had only 600 men left.

Paul was of the tribe of Benjamin. He warred relentlessly against the church of Jesus trying to protect an organization that had so totally apostatized that they killed the Lord—the One they professed to worship. It wasn't until he was struck blind on the road to Damascus that he recognized what he was doing. Most of us are usually our own worst enemies. We bring trouble upon our heads by making wrong choices. Praise God, He does not give us up to our own devices, but in His mercy the tribe of Benjamin was rebuilt and completely changed their attitude. God showed His goodness and forgiveness with the tribe of Benjamin. He never gave up on them by rebuilding their number and purifying their attitude.

Then Moses' words could be seen in their actions: *"And of Benjamin he said, The beloved of the LORD shall dwell in safety by him; and the LORD shall cover him all the day long, and he shall dwell between his shoulders."* (Deuteronomy 33:12.)

Jonathan was of this tribe. He was a fierce warrior. He was such a loyal friend that David said his love was more faithful than the love of a woman. (II Samuel 1:26.) His father, Saul, was also of this tribe, and he was always trying to kill David. (I Samuel 18 and 19.) Thus, again, we see the struggle of good and evil within families and within self. Who will be in charge? If it is self, then we, like Saul, will set about to kill the LORD's anointed.

When the love of God softens and subdues the whole heart, then *the wolf shall lie down with the lamb and all shall dwell safely together*. This tribe remained with Judah when the tribes divided and the other ten became a separate kingdom. Like the relationship between Benjamin and Judah, the stomach and the brain are the two closest related organs. They have a great influence on one another and reap repercussions on one another. Again we can clearly see principals and object lessons applying to the Tribes of Israel and the body organs. *"For many walk, of whom I have told you often, and now tell you even weeping, that they are the enemies of the cross of Christ: Whose end is their destruction, whose God is their belly, and whose glory is in their shame, who mind earthly things."* (Philippians 3:18,19.)

DAN

Genesis 49:16 -18

"Dan shall judge his people, as one of the tribes of Israel. Dan shall be a serpent by the way, an adder in the path, that biteth the horse heels, so that his rider shall fall backward. I have waited for thy salvation, O LORD."

Dan was the fifth son of Jacob and Rachel's maid, Bilhah, was his mother. His name means "Judge." He was the lead tribe from the north side of the sanctuary. The Tribe of Dan was represented by the banner of the Serpent. Dan will not be represented in the Gates of the New Jerusalem. Instead the banner of the eagle of Levi's tribe will take the place of the banner of Dan.

This tribe in our study represents the muscular system. This system must be motivated by good judgment to know when to tighten a muscle and when to relax the opposing muscle for optimal movement of the body.

The first part of the blessing and prophecy from Jacob to his sons, Reuben and Dan, were of prosperity and benefit, but the end of these prophecies were fatal condemnations. Dan and his tribe would have been a tremendous blessing to the Lord and His people, but instead the Judge of Israel who was to be honored and respected, who was to help the righteous, became a serpent that was hated, despised and destroyed.

The horned adder might have some of the characteristics that were condemned of Dan. The horned adder is the color of sand, and he waits camouflaged in the sand, ready to bite an unsuspecting one as he passes by. Because he can not be easily seen or detected, like a backbiter, the attack comes only after the prey passes by. Backbiters usually have qualities of a self-appointed judge. They often exhibit keen insight and ability to weigh evidences. Unfortunately, they turn those insights into condemnation and a haughty and critical spirit. A backbiter's sole objective in life seems to be to right the wrongs that he so keenly perceives. He gives unsolicited advice, points out to others their defects, and promptly tells them where they have gone astray. When one Biblically assesses right and wrong, it needs to be done by regarding and reflecting principles and according to the Word of God. Sound constructive criticism builds, and backbiting, destructive criticism destroys. Whenever counsel is needed for constructive criticism, one needs to first be led to the Scripture and the Biblical principles of Matthew 18. By loving and counseling God's way, a brother may be saved, forever as an heir in Christ's Kingdom.

The tribe of Dan was given one of the smallest inheritances in Israel. Judges 18 tells of their deceitful dealing with Micah, and notes that they were all involved in the wrong kind of worship. They killed the people of Laish, who were living quietly and securely (verse 27), and burned their city. They set up a calf of gold, and prided themselves that they were not the slaves of convention and "old-time religion." Self is always behind this "holier than thine" line of thought.

When God's instruction is laid aside as irrelevant, and so called better modes of worship or lifestyle are promoted, we can be sure that man's exaltation is at the bottom of this perversion. It is dangerous to depart from the Word of the living God. Dan's name is not among the 144,000, nor will it appear on the gates of the Holy City.

> "And of Dan he said, Dan is a lion's whelp: he shall leap from Bashan."
> — Deuteronomy 33:22.

The muscular system has to exhibit good judgment from the brain to keep the body balanced. Every movement caused by a skeletal muscle, from lifting something heavy to scratching one's nose, involves at least two sets of muscles. When one muscle contracts, an opposite one is stretched. When judgment is lost in one's muscles, the movement is impaired. When a person has diseases of the muscles, it can totally immobilize one to the point that hardly anything can be done. Almost all of these diseases lead to death, because without movement of the muscles, one cannot breathe, swallow, or do any of the necessary body functions. It is the same when one loses spiritual judgment. It paralyzes the Christian experience and then spiritual death begins.

ASHER

Genesis49:20

"Out of Asher his bread [shall be] fat, and he shall yield royal dainties."

Asher was the eighth son of Jacob by Zilpah. Very little is said about him, except in the words of Jacob and Moses. He represents the endocrine system and very little was known about that until recent years. *"And of Asher he said, Let Asher be blessed with children; let him be acceptable to his brethren, and let him dip his foot in oil. Thy shoes shall be iron and brass; and as thy days, so shall thy strength be."* (Deuteronomy 33:24,25.)

Note that there are no criticisms or dire predictions made for Asher. He did take part in the sale of Joseph, showing that no matter how pleasing our personalities are, there is a constant danger of falling into sin if we are not watchful and connected to Christ.

The men of Asher are called *"choice and mighty men of valour"* in I Chronicles 7:40. They were known as *"expert in war."* Or as the margin says *"keeping their rank."* (I Chronicles 12:36.) The people who will make up this tribe seem perpetually blessed, and they try to keep everyone happy. The people of this tribe act as the diplomats in the church, and try to keep everything running smoothly. It might seem to others that they talk out of both sides of their mouth, but like the endocrine system, it functions in a similar fashion. It first puts out a hormone for an action, and then it produces another hormone to stop the first action from producing too much and causing an unbalanced system.

The people blessed with popularity and prosperity need much strength and humility from God to keep from depending upon self. *"As thy days are so shall thy strength be"* reminds us of our daily need for the grace of God. It is a daily job to gain strength to obtain victory over sin and self.

In the great battle of Megiddo, which is a type of the last battle of Armageddon, Asher stayed at home instead of joining the battle. (Judges 4:17.) When material possessions, reputation and safety were of prime concern, the battle was forgotten. The iron and brass shoes on the feet of Asher, which was an enduring principle, finally came to the forefront and the grace of God prevailed in the life of Asher. He was changed into someone who committed all to the cause of God, and Asher will be found among the tribes in the city.

NAPHTALI

Genesis 49:21

"Naphtali [is] a hind let loose: he giveth goodly words."

Naphtali was Jacob's sixth son and second child of Bilhah. The integumentary (in-teg-u-men-ta-ri) *system,* which includes skin, hair, nails, connective tissue, etc. is represented by this tribe.

The picture given by Jacob of this son is of a person who is like a deer. You cannot bind big heavy burdens upon a deer, because they travel quickly and cautiously. This animal is quick to sense danger and is watchful over her young. People who will be a part of this tribe are careful of other people and keep a sharp lookout for possible danger. Because they are not concerned with the mundane burdens as the tribe of Issachar, they tend to be impractical people. Because of this they take more time to study and counsel. The "*hind*" or "*hart*" was used anciently as a symbol of the object of love. *"My beloved is like a roe or a young hart."* (Song of Solomon 2:9.)

The skin represents the covering that Christ puts over us in the time of love, the robe of His righteousness. (Ezekiel 16:8 and Ruth 3:9.)

Zebulun and Naphtali were a people that jeoparded (risked) their lives unto the death in the high places of the field" in the battle of Meggido (Judges 5:18.) They will be found in the battle, covering their fellow Christians. They try to keep everyone together and are very faithful about visiting the neglected.

"And of Naphtali he said, O Naphtali, thou art satisfied with favour, and full of the blessing of the LORD: possess thou the west and the south."—Deuteronomy 33:23.

Covering the outside parameters in a war was not without danger. His territory was among the first to feel the inroads of the oppressor. The skin, when under attack from toxins, breaks out. The grace of God and the cleansing blood of Jesus is the only thing that keeps our covering from becoming *spotted* and *soiled*.

Chapter Thirteen

GOD'S DESIGN IN NUMBERS

Genesis 13:16

"And I will make thy seed as the dust of the earth: so that if a man can number the dust of the earth, [then] shall thy seed also be numbered."

In all of creation, a master design appears. Nothing is by chance, but all is perfectly designed and ordered. We are only going to investigate a tiny portion of this design, for this is a subject that will be studied by the redeemed throughout all eternity. It will never be exhausted. In our study of numbers, we will only be examining a small part of God's design in numbers. This will only be enough to whet your appetite on the subject, so that you will do your own study and research about what the numbers in the anatomy of the body represent spiritually.

In the Hebrew and Greek alphabets all numbers have a definite numerical value. There is a manifest design pervading the whole Bible by which the different writers writing at different times and places, and under different circumstances, use certain words a definite number of times. The actual number depends upon special significance of the word. Some words which the Holy Spirit places special emphasis occur a certain number of times.

These are: (1) a square number; (2) a cube; (3) a multiple of 7; (4) a multiple of 11.

There are **four** perfect numbers; **3** is the number of Divine perfection; **7** is the number of spiritual perfection; **10** is the number of ordinal perfection; **12** is the number of governmental perfection. The product of these four numbers 3x7x10x12 forms the great number of chronological perfection, **2520**.

As you study nature, you may observe how important order in number is in nature: in the gestation periods of different animals, in the order the leaves grow on plants, the kernels of corn on a cob, the number of bones and organs in the body, etc. Also note how numerical order in music is observed, in chemistry, in any branch of science that you study. We are only scratching the surface, and as you study, the Holy Spirit will keep adding to the wonder of that order in the study of numerology.

The following is a summary of the meanings of some of the numbers.

ONE - PRIMACY, UNITY

TWO - DIFFERENCE AND/OR SUPPORT

THREE - SOLID, COMPLETE, RESURRECTION, DIVINE PERFECTION

FOUR - CREATED WORKS, MATERIAL CREATION

FIVE - GRACE, REDEMPTION

SIX - SECULAR COMPLETENESS, MAN DESTITUTE OF GOD

SEVEN - COMPLETENESS, SPIRITUAL PERFECTION

EIGHT - SUPER-ABUNDANT, REGENERATION, SPECIAL NUMBER OF THE HOLY SPIRIT

NINE - JUDGMENT, FINALITY

TEN - COMPLETENESS OF ORDER, COMPLETE CYCLE

ELEVEN - DISORDER, DISINTEGRATION, IMPERFECTION

TWELVE - PERFECTION OF GOVERNMENT

THIRTEEN - REBELLION

FOURTEEN - DOUBLE MEASURE OF SPIRITUAL PERFECTION

FIFTEEN - ACTS WROUGHT BY THE ENERGY OF DIVINE GRACE

SEVENTEEN - PERFECTION OF SPIRITUAL ORDER

THIRTY - A HIGH DEGREE OF THE PERFECTION OF DIVINE ORDER, THE RIGHT MOMENT

FORTY - PROBATION, CHASTISEMENT

ONE:
PRIMACY, UNITY

Deuteronomy 6:4

"Hear O Israel, the LORD thy God is one LORD."

Unity, primacy, excludes all difference, for there is no second with which it can either harmonize or conflict. Exodus 20:3: *"Thou shalt have no other gods before Me"* = ONE GOD

One marks the beginning. Nothing is right that does not begin with God.

> *"Thus saith the LORD, the King of Israel, and His redeemer the LORD of hosts; I am the first, and I am the last and beside me there is no God."*—Isaiah 44:6

The **first** recorded words of Christ: "Wist ye not that I must be about my Father's business." (Luke 2:49.)

The **first** recorded words of His ministry: "It is written…" (Matthew 4:4.)

The **first** book (Genesis) contains Divine sovereignty and supremacy.

The **first** and great commandment which is written below:

> *"And Jesus answered him, The first of all the commandments is Hear O Israel: the Lord our God is one Lord; and thou shalt love the Lord thy God with all thy heart, and with all thy mind, and with all they strength; this is the first commandment."*—Mark 12:29,30

First occurrence of words are always important as they help us fix the meaning or point to some lesson concerning the word.

Hallelujah: First occurrence in OT Psalm 104:35; first occurrence in NT, Revelation 19:1-3—both in reference to judgment.

Prophet: First occurrence in Genesis 20:7, means one who witnesses for God as His spokesman.

Holy: First occurrence, Exodus 3:5, in reference to redemption. The creature cannot understand anything about holiness unless he is redeemed.

The Day of the LORD: First occurrence, Isaiah 2:12. Read the whole passage which has to do with the pride of man and the wrath of God.

The first question: OT, Genesis 3:9, "Where art thou?"—given to bring man to conviction, to show him that he was guilty, lost and ruined.

First question, NT: Matthew 2:2 "Where is He that is born?"

"Where is the Savior that I need?": This was given to show mankind how we were to be redeemed.

TWO:
DIFFERENCE

<u>Genesis 1:6</u>

"And God said, Let there be a firmament in the midst of the waters and let it divide the waters from the waters."

When a principal or concept discussed in the Bible is predicated by the number two, we can know that God wants us to understand that the principal or concept is a two sided issue. One side may be for good—one side may be evil.

The second <u>thing</u> recorded in the Biblical creation was light. Immediately there was difference and division, for God divided the light from the darkness.

The second <u>day</u> of creation had division for its great characteristic.

The second of any number of things always bears upon it the stamp of difference and generally of enmity. The first statement in the Bible says: *"In the beginning God created the heaven and the earth."* The second is: *"And the earth was without form and void."* The first speaks of perfection and order, the second of no form or order.

The second <u>book</u> of the Bible, Exodus, opens with the oppression of the enemy, and then introduces a Deliverer. The redemption song of Moses is found in the second book which mentions the Lord's triumph over His enemies. (Exodus 15).

In the NT, whenever there are two Epistles, one will find that the second has some special reference to the enemy:

<u>II Corinthians</u> contains a marked emphasis on the power of the enemy and the working of Satan. (II Corinthians 2:11; 11:14; 12:7.)

<u>II Thessalonians</u> contains a special account of the working of Satan in the revelation of the man of sin and the lawless one.

<u>II Timothy</u> gives the difference of the church in its ruin, and the church in its fullness.

<u>II Peter</u> tells of the righteous church and the apostasy to which we have fallen.

Things introduced to us in pairs in the Biblical passages so that one may understand principles by way of contrast or difference:

The two <u>foundations</u> to base one's faith upon: sand and rock. (Matthew 7:24-27.) Also, there were other twos for comparison and contrast:

The two <u>goats</u>, Leviticus 16.

The two <u>masters</u>, Matthew 6:24.

The two <u>praying men</u>, Luke 18:10.

The two <u>debtors</u>, Luke 8:41.

The two <u>sons</u> of Matthew 21:28 and Galatians 4:22.

A difference is also manifested in the people—notice below concerning those who cannot walk together:

"How can **<u>two</u>** walk together unless they be agreed?"—Amos 3:3

People

Cain and Abel are the example of two religions. Cain depended on his own works done his way; Abel depended on the merits and sacrifice of Christ and worshipped according to the instruction given by God.

Abraham and Lot both left Ur, they both left Egypt. But soon the difference between the two was manifested. Lot chose his own portion and Abraham's was chosen for him by God. Thus they were separated (Genesis 13)

Isaac and Ishmael both had the same father, but one was born after the spirit and one after the flesh.

Jacob and Esau both had the same parents, but one was righteous and one was profane.

Words which occur twice:

Enjoy (apolausis): I Timothy 6:17—God's enjoyment; Hebrews 11:25—Satan's enjoyment.

Choke (apopnigo): Matthew 13:7—enemy choking the seed; Luke 8:33—enemy choked in the sea.

Armor (panoply); Luke 11:22—armour of Satan; Ephesians 6:11-17—armor of God.

Son of perdition: John 17:12 and II Thessalonians 2:3.

Two in Testimony:

Two testimonies may be different but yet one may support, strengthen, and corroborate the other. Jesus said, *"The testimony of two men is true. I am one that bear witness of myself, and The Father that sent me beareth witness of ME."* (John 8:17,18.)

And it is written in the Law:

> **"...at the mouth of two witnesses, or at the mouth of three witnesses, shall the matter be established."**
> **— Deuteronomy 19:15**

The whole law itself hung on two commandments. (Matthew 22:40.)

The old covenant and the new covenant: Law and Grace, Faith and Works.

Examples of true witnesses are: God and His Only Begotten Son, Jesus Christ; Caleb and Joshua.

THREE:
SOLID, COMPLETE RESURRECTION

Mark 9:31

"For he taught his disciples, and said unto them, The Son of man is delivered into the hands of men, and they shall kill him; and after that he is killed, he shall rise the third day."

Three is the strongest and first geometrical figure. In construction, the triangle is the strongest form. Three lines are necessary to form a plane figure. Three dimensions of length, breadth, and height are necessary to form a solid figure. As two is the symbol of the square or plane content, so three is the symbol of the cube or solid contents. Three, therefore, stands for that which is solid, real, substantial, complete, and entire. All things that are specially complete are stamped with this number three.

The number three stands for the complete, Godly marriage. The marriage institution was one of two things that was brought forth from the Garden of Eden. In a complete Godly marriage contains the God, husband, and the wife, forming a perfect triangle.

God's character attributes are three: Omniscience, omnipresence, and omnipotence. There are three great divisions completing time—past, present, future. Thought, word, and deed complete the sum of human capability. Three degrees of comparison complete our knowledge of qualities such as, good, better, best or bad, worse, worst, etc.

Three propositions are necessary to complete the simplest form of argument—the major premise, the minor premise, and the conclusion.

Three kingdoms compose our ideas of matter—mineral, vegetable, and animal. Water, water vapor, and ice. Liquid, gas, and solid. Seed, plant, fruit are the cycles of plants.

When we turn to the Scriptures, this completion becomes Divine, and marks Divine completeness or perfection. The number three points us to what is ***real, essential, and perfect, - substantial, complete and divine.***

The word "fullness" occurs three times: Ephesians 4:19, ***"the fullness of God."*** Ephesians 4:13, ***"the fullness of Christ."*** Colossians 2:9, ***"the fullness of the Godhead."***

Three times is the blessing given in Numbers 6:23-26: "The LORD bless thee and keep thee; The LORD make His face to shine upon thee; The LORD lift up His countenance upon thee, and give thee peace."

Abraham was called the "friend" of God three times: II Chronicles 20:7; Isaiah. 41:8; James 2:23.

Three multitudes were miraculously fed: II Kings 4:42,43; Matthew 15:34-38; Mark 6:38-44.

Abraham brought his heavenly visitors three measures of meal: Genesis 18:6.

In the covenant God made with Abraham, three animals, each three years old, were used, plus two birds, which made five in all, marking it all as a perfect act of free grace. (Genesis 15).

The measure for the whole burnt offerings and for great special occasions was three tenths of flour.

There was darkness in the land of Egypt for three days (Exodus 10:22)

The Israelites were to go three days' journey from Egypt, thus denoting their *complete* separation from the world. Pharaoh wanted them to worship "in the land" as Satan would like for us to stay "in the world."

It was three days after the victory at the Red Sea that the first trial came. (Ex. 15:22.)

Three times a year did all the males have to appear before the Lord God for the three feasts (Exodus 223: 14-17).

The ten spies brought back three kinds of fruit, which testified to the perfect goodness of the land.

Gideon had 300 men to fight with. (Judges 7:6-8.)

The third book in the Bible is Leviticus, the book in which we learn about true worship. Here we have the LORD Himself, prescribing every detail of worship and leaving nothing to imagination or private devising. In the third commandment we see that the completeness of God is denoted in His name, which symbolizes His character. As Christians we take that name in vain when we do not reflect His character.

Three times the children of Israel said, *"All that the Lord hath spoken we will do,"* marking the completeness of the Covenant-making on the part of Israel and showing that man has never kept his part of the covenant without the power of the Holy Spirit.

Jordan was divided three times.

Completeness of **apostasy:** (Jude II), The way of Cain, The error of Balaam, The gainsaying of Korah.

Completeness of **temptation:** Christ in the wilderness was tempted three times for three sins that cover all sin, lust of the flesh, lust of the eyes, and the pride of life. (I John 2:16)

Completeness of Divine Judgment: *"MENE: God hath NUMBERED thy kingdom and finished it. TEKEL. Thou art WEIGHED in the balances and found wanting. PERES: Thy kingdom is DIVIDED and given to the Medes and Persians."*—Daniel 5:26-28.

Completeness of nature represents: Body, Mind, and Spirit.

Three is also the number of **Resurrection**. The third day the earth rose up out of the water, which was a symbol of that resurrected life that we have in Christ and in which alone we can do good works.

He raised three persons from the dead (the widow's son at Nain (Luke 7:11-17), Jairus' daughter (Matthew 9:18-26; Mark 5:21-43; Luke 8:40-56), and Lazarus (John 11:1-44)).

The inscriptions on the cross were in three languages (Greek, Latin, and Hebrew).

On the third day Jesus rose from the dead. It was in the third hour that He was crucified; it was for three hours that darkness shrouded the cross, from the sixth hour to the ninth hour. The darkness was so complete, hiding His Father's face that He cried, ***"My God, My God, why has Thou forsaken Me?"*** which shows *completely* that nothing of nature, nothing of life or intelligences of this world could give help in that hour of darkness. This shows us that we are also utterly incapable of delivering ourselves from the darkness of our natural condition. At the end of that dark period came the declaration ***"It is finished,"*** which tells us that now there is no such darkness for those who died with Christ and live in Him.

Revelation 1 is divinely marked by a series of three such as verse 1: given, sent, signified; verse 2: bare record of the word of God, testimony of Jesus, all that He saw; etc.

Three times the word *Christian* is found in the New Testament: Acts 11:26; Acts 26:28; I Peter 4:16.

Phrases that occur three times: *Walk worthy*: Ephesians 4:1, Colossians 1:10, I Peter 1:20. *Before the foundation of the world:* John 17:24, Ephesians 1:4, I Peter 1:20.

Jerusalem is 3300 feet higher than Jericho.

FOUR:
CREATED WORKS

Genesis 1:19

"And the evening and the morning were the fourth day."

God created the world on the fourth day of creation, and the following two days were spent populating it.

Four is the first number that is not a prime, the first that can be divided, the first square number, and therefore it marks a kind of completeness.

Four is the first number of the great elements—earth, air, fire, and water.

Four are the regions of the earth—north, south, east, and west.

Four are the divisions of the day—morning, noon, evening, and midnight.

Four are the seasons of the year—spring, summer autumn and winter.

There are four lunar phases.

In Genesis 2:10, 11, the one river of Paradise was parted and became into four heads (three now unnamed, plus one still known by original name, Euphrates).

The four beasts around the throne in Revelation always speak in connection with the Earth. They are the four heads of animal creation: the lion, or the wild beasts: the ox, of tame beasts; the eagle, of birds; the man, the head of all. Again we have three plus one: three animals plus one human. They are mentioned in connection with the Earth in Ezekiel 1.

The great prophetic world powers in Daniel are divided three plus one, the first three beasts are named, the fourth only described. (Daniel 7.) Four winds STROVE upon the great sea, the four beasts came up from the sea DIVERSE one from another; the story of man, strife and division. In Daniel's vision of the great image four different metals made up the body plus one combination of metal and clay.

The fourth book of the Bible is Numbers, which tells the story of the wilderness of Earth and of our pilgrimage through it.

The fourth commandment refers to the creation of the Earth and the sign of that creation.

In the blessings of the Lamb in Revelation 5, the blessings from the heavenly beings are seven, which is the number of spiritual perfection. The blessings from the earth creatures are four.

Four great records of Christ's life and death: Matthew, Mark, Luke (synoptic), and John, three plus one.

Four coverings of the Earthly Sanctuary: one vegetable (linen), plus three animal (goat's hair, ram's hair, and badger skins.)

The curtain of the Earthly Sanctuary contained: blue, purple, scarlet plus embroidered angels.

Four divisions of the priests and Levites: Aaron and his sons plus the sons of Gershon, Kohath and Merari.

Four of the unclean animals mentioned in Leviticus 11:3 that chewed the cud, but divided not the hoof plus one that divided the hoof, but chewed not the cud.

God's four judgments in the Earth: three inanimate, (sword, famine, pestilence), plus one animate, the noisome beast. (Ezekiel 14:21)

The four things found in Jeremiah 15:3 are three animate (dogs, fowls and beasts), plus one inanimate, the sword.

Four people are found in the furnace: Shadrach, Meshach, Abednego plus the Son of God.

Four people whose names were changed: Abram, Sarai, Jacob and Pashur.

In the parable of the sower (Matthew 3:3-9), there are four kinds of soil: three are unprepared, one is prepared.

The sphere of suffering is four-fold in II Corinthians 4: 8,9:

Troubled, but not distressed
Perplexed, but not in despair
Persecuted, but not forsaken
Cast down, but not destroyed

The rainbow is mentioned four times: Genesis 10; Ezekiel 1:28, Revelation 4:3; 10:1.

Four names of Satan in Revelation 12:0; the dragon (rebellious and apostate), the old serpent (seductive), the devil (accusing) and Satan (personal).

Four times Eve is mentioned in the Bible by name: Genesis 3:20 and 4:1; II Corinthians 11:3; I Timothy 2:13.

Four is the number of the world and represents man's weakness, helplessness, vanity and his utter inability to save himself.

FIVE:
REDEMPTION AND GRACE

Genesis 19:19

"Behold now, thy servant hath found grace in thy sight, and thou hast magnified thy mercy, which thou hast showed unto me in saving my life; and I cannot escape to the mountain, lest some evil take me, and I die:"

Five is four plus one, Divine strength added to and made perfect in the weakness of man. Omnipotence combined with the impotence of Earth; of Divine favor uninfluenced and invincible. It is therefore the number of GRACE, favor shown to the unworthy. Romans 3:24 says, "being justified freely" while John 15:25 translates that same word as "without a cause." So it may be said that God justified us for no reason other than the great love He has for us, not because we deserved it in any way.

In the covenant made with Abram, there were five sacrifices to stamp this transaction as a pure favor on God's part. When He changed Abram's name, He did it by inserting the fifth letter of the alphabet in the middle of it. At that same time Abraham was called upon to walk before God and be perfect. And God said **"I am the Almighty God** (*EL SHADDAI*, the all bountiful One)," showing Abraham and to us that He is able and willing to provide all our necessities and to give us the power we need to walk before Him and be perfect.

As the LORD AMLIGHTY He calls us out to be separate and be His sons and daughters. (II Corinthians 6:14-18, 7:1.)

The fifth *book* of the Bible is Deuteronomy and magnifies the grace of God and shows us that it was for His own Name's sake that He called, chose, and blessed us.

David chose smooth stones, four unused plus the One that was sufficient. David quoted from the book of Deuteronomy, the fifth book, (Deut 28:25,26) when He challenged Goliath in the name of the Lord. (I Samuel 17:45,46.) Jesus met the devil's temptations in the wilderness (Matthew 4) with scriptures from the book of Deuteronomy. (Deut 8:3; 6:16; 6:13.)

The children of Israel went out of Egypt five in a rank (Exodus 13:18 margin), in perfect weakness, helpless and defenseless, but they were invincible through the presence of Jehovah in their midst. He promises, "five of you shall chase a thousand." (Leviticus 26:8.)

"If God be for us, who can be against us?"—Romans 8:31.

"I had rather speak five words with understanding than ten thousand words in an unknown tongue." (I Corinthians 14:19.)

A few words spoken in the fear of God, in human weakness and depending on Divine strength and blessing, will be able to accomplish that which God has purpose; while words without end will be spoken in vain when spoken without God.

It was God's covenant of grace which He had made with Abraham, Isaac and Jacob that called forth His five-fold demand to Pharaoh which was stamped with 5 great facts and Pharaoh's objection to them: (Exodus 5.)

1. Jehovah and His Word: "Thus said Jehovah. God of Israel."—"Who is Jehovah that I should obey His voice?"
3. Jehovah's People: " Let My People go." "Who are they that shall go?" Moses said all the people and Pharaoh said only the men. Satan would like us to leave our families in Egypt.
4. Jehovah's Demand: Let them all go—No.
5. Jehovah's Feast: "That they may hold a feast unto Me—Leave something behind.
6. Jehovah's Separation: "In the wilderness"—"ye shall not go far away." Satan does not want us to get completely separated from the world.

These are God's demands for the people whom He has redeemed.

The tabernacle is marked by five; nearly every measurement is in multiples of five. Worship itself is all of grace. We must be called to worship. (Psalm 65:4.) We are told how to worship in the third book which has more divine utterances than any other book of the Bible. Count how many times the word Jehovah occurs. Worship springs from the will of God and is founded in grace so as we look at the tabernacle we will see how the number five and its multiples appear everywhere.

The outer court was 100 cubits long and 50 cubits wide. On either side were 20 pillars, and along each end were 10 pillars or 60 in all; that is 5 x 12, or grace in governmental display before the world. The pillars that held up the curtains were five cubits apart and five cubits high. The whole of the outer curtain was divided into squares of 25 cubits. Each pair of pillars thus supported an area of five square cubits of fine white linen, thus witnessing to the perfect grace by which alone God's people can witness for Him before the world. Their own righteousness is "filthy rags," (Isaiah 64:6), but Christ's righteousness is pure and white. This righteousness is based on atonement for the atonement for the altar of sacrifice was five by five. It was three cubits high which stamped it with the Divine perfection from which atonement flows.

The building was 10 cubits high, 10 cubits wide, and 30 cubits long.

The Holy of Holies was a perfect cube of 10 cubits; the Holy place was 20 cubits long. The tabernacle was formed of 48 boards measuring 4x12, significant of the nation as before God in the fullness of privilege on the Earth. There were five bars that held the 20 boards together.

The curtains, which covered the tent, were four in number. The first of linen, five on a side, 28 by 4: the second of goat hair of five plus six, 30 by 4; the third of ram's skin dyed red, no dimensions; and the fourth of badger's skins, no dimensions.

The entrance veils were three in number. The first was the gate of the court, 20 cubits wide and five high, hung on five pillars. The second was the door of the tabernacle, 10 cubits wide and 10 high, hung like the gate on five pillars. The third was the beautiful veil also 10 cubits x 10 cubits hung upon four pillars, which divided the Holy place from the Holy of Holies. One remarkable feature is that the dimensions of the veils were different but the area was the same. They were all 100 square cubits. There was only one GATE, one DOOR, and one VEIL; they each typified Christ as the only door of entrance for all the blessings connected with salvation. The GATE, the entrance into the courtyard, admitted the benefits of the atonement was wider and lower. "Enter in at the straight gate."—Matthew 7:13.

The DOOR, which admitted to the sanctuary worship, was both higher and narrower.

"I am the door: by me if any man enter in, he shall be saved, and shall go in and out, and find pasture." —John 10:9

The highest worship through the VEIL was only accessible to the high priest, the substitute for Christ. This veil was rent in two the moment that the true grace was manifested with the death of Jesus. We can now be in the presence of God in the person of Jesus Christ, our great High Priest.

"Having therefore, brethren, boldness to enter into the holiest by the blood of Jesus, By a new and living way, which he hath consecrated for us, through the veil, that is to say, his flesh."—Hebrews 10:19, 20.

The holy anointing oil had five parts; four spices (the amount of each all in multiples of five), plus one oil. (Exodus 30:23-25.)

At the consecration of the priest in Exodus 29:20 there were three acts each associated with the number five. The blood, and later the holy oil, was placed in three places:

1. On the tip of the right ear, which is a part of the five senses, consecrating his senses for God's use.
2. On the thumb of the right hand, which is, one of five digits, signifying that the priest was to do and act for God.
3. On the great toe of the right foot, which is one of five toes, signifying that the priest was to walk in God's way.

The holy incense had four ingredients plus one which salted it together. (Exodus 30:35 margin, mah-lach, to salt). The incense represents the merits of Christ and the smoke represents the prayers of saints. (Revelation 8:3,4.)

Five things that were missing from the second temple: The ark, the Shechinah glory, the fire from Heaven on the altar, the Urim and Thummim, (four tangible plus one, the Spirit of Prophecy that is an intangible thing).

A word that occurs five times is "parakletos," which is translated four times "Comforter" and one time "Advocate."

SIX:
SECULAR COMPLETENESS

Exodus 20:11

"For [in] six days the LORD made heaven and earth, the sea, and all that in them [is], and rested the seventh day."

Six is seven minus one, man's coming short of spiritual perfection. Man was created on the sixth day, so it has to do with man; it is the number of imperfection, the number of MAN as destitute of God, without God, without Christ.

Six days were appointed him for labor, with the seventh associated with God, as His rest day. Six is the number of labor, of man's labor as apart and distinct from God's rest. It marks the completion of creation as God's work here on earth, so the number is significant of secular completeness. Six is the number stamped on all that is connected with human labor. It is stamped upon the measures used in labor (inches 2 x 6 = 12 = 1 foot; 12 x 3 = 36 (6 x 6) = 1 yard) and the time used, 12 (2 x 6) hours of day, 12 hours of night.

Because of the curse, the number six denotes not only labor, but also of labor and sorrow and especially marks all the activity on this Earth that is not of God.

Cain's descendants are only given to the sixth generation.

The sixth commandment relates to the only sin for which forgiveness cannot be obtained from the party wronged, because the person is dead.

The number six is stamped upon the measurements of the Great Pyramid.

When 12 (the number of governmental perfection) is divided it indicates imperfection in rule and administration.

Solomon's throne had six steps. (I Kings 10:19.)

The shewbread was divided into two rows of six. (Leviticus 24:6.)

Six of the tribes were engraved on one stone and six on the other. (Exodus 28:9,10.)

Abraham's six intercessions for Sodom marked man's imperfection in prayer. (Genesis 18.)

In the 600th year of Noah's life the flood came.

The glory of the: LORD covered Mr. Sinai for six days and on the seventh day God called Moses into the cloud with Him. (Exodus 24:16.)

For six years the Israelites were to sow the land, and rest on the seventh. (Exodus 23:10,11.) For six days they were to work and rest on the seventh. (Ex. 28-11; 23:12.)

The princes of Israel brought six covered wagons and 12 oxen as an offering. (Numbers 7:3.)

A Hebrew slave was to serve six years and the seventh he was to be free. (Deuteronomy 15:12.)

There were to be six cities of refuge. (Numbers 35:13.)

The giant of Gath had six fingers on each hand and six toes on each foot, showing works and ways without grace. (II Samuel 21:20.)

Six years of Athaliah's reign. (II Kings 11:2,3). She slew all the seed of the royal household except Joash. He was hidden in the house of God during the six years. God promised that the king's son would reign, and during the six years the faithful held on to that promise and went about securing loyal adherents by repeating the promise. In the seventh year, Joash came to rule and all his enemies were destroyed.

This is a type of King Jesus who has been hidden for 6000 years and in the 7000-year period will deliver His faithful and destroys His enemies.

Man's enmity to the Person of Jesus Christ is branded with man's number.

Six-fold opposition to the work of God.

1. Grief—Nehemiah 2:10.
2. Laughter—Nehemiah 2:19.
3. Wrath, indignation, mocking and accusing.—Nehemiah 4:1-4.
4. Fighting and open opposition—Nehemiah 4:7-8.
5. Conference to stop work:
    a. Compromise, Nehemiah 6:1-2;
    b. Make plans together, v7;
    c. to weaken them from the work, v9.

   Nehemiah refused to parley with his enemies or meet them in conference.
6. False friends—Nehemiah 6:10-14. The greatest danger of all is Satan as an "angel of light" or from "wolves in sheep's clothing."

Six times Jesus was accused of having a devil:

1. Mark 3:22 and Matt. 12:24—Beelzebub and casting out devils.
2. John 7:20—Thou hast a devil.
3. John 8:48—Say we not well that … Thou hast a devil?
4. John 8:52—Now we know that Thou hast a devil.
5. John 10:20—He hath a devil and is mad.
6. Luke 11:15—He casteth out devils by Beelzebub.

The serpent has six names:

1. "nachash," a shining one—Genesis 3:1 and Job 26:13.
2. "akshoov," to bend back, lie in wait. Translated adder—Psalm 140:3.
3. "ephah," any poisonous serpent. Translated adder and viper—Job 20:16; Isaiah 30:6; 59:5.
4. "tsiph-ohnee," a small hissing serpent Isaiah 11:8, 59:5; Prov.32:32 Translated viper.
5. "tanneen," a great serpent or dragon (from root "to burn"; a venomous deadly serpent, from the heat and inflammation caused by its bite. Translated serpents in Numbers 21:8; Isaiah. 14:29; 30:6.

Six times our Lord was asked for a sign:

1. The Pharisees—Matt. 12:38; Mark 8:11.
2. The Sadducees—Matt. 16:1.
3. The disciples—Matt. 24:3; Mark 13:4.
4. The people—Luke 11:16.
5. The Jews—John 2:18.
6. The people—John 6:30.

Six earthquakes mentioned in the Bible that have happened:

1. Exodus 19:18.
2. I Kings 19:11.
3. Amos 1:1 (Zech. 14:5).
4. Matt. 27:51.

5. Matt. 28:2.
6. Acts 16:26.

Six people found Christ innocent:

1. Pilate—Luke 23:14
2. Herod—Luke 23:15
3. Judas—Matt. 27:3,4
4. Pilate's wife—Matt. 27:19
5. The dying thief—Luke 23:41
6. The centurion—Luke 23:47

Six classes of sins upon which were pronounced woes (Isaiah 5:8, 11,18-22):

1. Covetousness—selfishness
2. Drunkenness—dissipation
3. Scoffing—mocking
4. Dissembling—confusing truth
5. Conceit
6. Judging wrongly with mingled doctrine

Some of the enemies of God especially marked with 6:

1. Goliath: His height was six cubits, he had six pieces of armor (the Christian has seven), and his spearhead weighed 600 shekels of iron. THE PRIDE OF FLESHLY MIGHT.
2. Nebuchadnezzar: His image was 60 cubits high and six cubits broad and was worshipped when the music from six instruments was heard. PRIDE OF ABSOLUTE DOMINION.
3. Anti-Christ: His number is 666. PRIDE OF SATANIC GUIDANCE.

It is remarkable that the Romans did not use all the letters of their alphabet to represent numbers, as did the Hebrews and the Greeks. Six letters they did use were: D, C, L, X, V, and I. It is more remarkable and significant that the sum of the 6 numbers amounts to **"666."** In each of these pairs 500+100, 50+10, and 5 + 1, there is an addition of one to the five. It is the grace of God superseded by the corruption of man. The triple six marks the culmination of man's opposition to God. 666 was the secret symbol of the ancient pagan mysteries connected with the worship of the devil. It is today the secret connecting link between those ancient mysteries and their modern revival in Spiritualism. The efforts of the enemy are now directed towards uniting all into one great whole "Separation" is God's word for His people (Revelation 18:4; II Corinthians 6:17) and is the command of Christ, while "union and reunion" (*at the expense of truth)* is the command of the beast (Anti-Christ, Rev. 13).

1. D=500
2. C=100
3. L=50
4. X=10
5. V=5
6. I=1

The letter S in the Greek alphabet was the symbol of the figure six; but when it came to the six, another letter was introduced. The sixth letter "zeta"—but a different letter, a peculiar form of S, called "stigma," which means a mark, but especially a mark made by a brand as burnt upon slaves, cattle, or soldiers by their owners; or on devotees who thus branded themselves as belonging to their gods. The number 666 has another remarkable property. It is marked as the concentration and essence of the six by being the sum of all the numbers, which make up the square of six. The square of six is 36 and the sum of the numbers of 1 to 36 is 666. These numbers may be arranged in the form of a square with 6 figures each way, so that the sum of each six figures in any direction shall be another significant trinity of numbers, 111. The priests of the sun god used this chart.

SEVEN:
SPIRITUAL PERFECTION

Genesis 20:10

"But the seventh day [is] the Sabbath of the LORD thy God: [in it] thou shalt not do any work, thou, nor thy son, nor thy daughter, thy manservant, nor thy maidservant, nor thy cattle, nor thy stranger that [is] within thy gates."

The number seven in Hebrew is *shevah*. It is from the root *e* to be full or satisfied or have enough of. On the seventh day God rested from the work of Creation. It was full and complete and good and perfect. Nothing could be added to it or taken from it without marring it. Hence the word *(Shavath)* to desist, rest, and *Sabbath,* or day of rest. This root runs through various languages and means Sabbath or seventh day. It is seven that stamps with perfection and completeness that in connection with which it is used. Of time, it tells of the Sabbath and marks off the week of 7 days which, artificial as it may seem to be, is universal and memorial in its observance amongst all nations and in all times. It tells of the eternal Sabbath-keeping, which is always and for eternity for the people of God in all its everlasting perfection.

Another meaning of the root (Shavath) is to swear or make an oath. Its first occurrence is in Genesis 21:27-31, where a covenant was made using seven lambs. It was the security, satisfaction, and fullness of the obligation or completeness of the bond which caused the same word to be used for both the number seven and an oath. Beersheba, the well of the oath, is the standing witness of the spiritual perfection of seven.

In the creative works of God, seven completes the colors of the spectrum of the rainbow, and in music the notes of the scale. In each of these the eighth component is only a repetition of the first.

> "These six things doth the Lord hate: yea, seven are an abomination unto him: A proud look, a lying tongue, and hands that shed innocent blood, a heart that deviseth wicked imaginations, feet that be swift in running to mischief, a false witness that speaketh lies and he that soweth discord among brethren."—Proverbs 6:16-19

Psalm 12:6 tells us that the words of the LORD are purified 7 times.

The two genealogies of Jesus Christ demonstrate the perfection in the plan of God. The four gospels give different views of Christ:

In the Gospel of Matthew God says to us, "Behold Thy King" (Zech. 9:9).

In the Gospel of Mark He says, "Behold My Servant" (Isaiah 42:1).

In the Gospel of Luke He says, "Behold the Man" (Zech. 6:12).

In the Gospel of John He says, "Behold your God" (Isaiah 40:9).

Note the genealogies are found in Matthew and Luke. A servant need not produce his genealogy; neither can God have one. A King who must have one, and a Man who should have one. **Matthew** gives the **Royal** genealogy of Jesus as King; Luke gives the Human genealogy of Jesus as Man. That is why Matthew is a descending genealogy, while Luke's is an ascending one. Kings must be traced by their descent, with all power in the world being derived from God who sets up kingdoms and takes them down. This is why this one begins at Abraham, and come down to Joseph, the son of Heli. The genealogy of Luke starts from Joseph, and goes up to Adam and God. Man must be traced by his ancestors.

David's genealogy has both lists but here an important divergence takes place: In the Gospel of Matthew, we have Solomon listed as David's son. In Luke, Nathan is listed as his son. From this point therefore, there are two lines. **Matthew** gives the royal and legal line through Solomon; **Luke** gives the natural and lineal line through Nathan. The younger son, Solomon, was the chosen one according to the will of God. The latter is in the line according to natural descent as the result of the will of man and in the order of human birth. Both lines meet in Joseph's lineage, the son of Jacob by birth and the son of Heli by marriage with Mary, Heli's only daughter. Thus the two lines are united and exhausted in Jesus Christ; for by His death they both became extinct. He was the King of Israel by right and was declared to be the Son of God by resurrection from the dead.

In the genealogy of Luke 3, there are exactly 77 names, starting with God at the one end and Jesus at the other.

Although the genealogy as recorded in the Gospel of Matthew is constructed according to the recognized custom amongst the Jews, it is so arranged that it contains 42 generations (6x7). These 42 generations contain: 41 names that are named in Matt; and one, Jeconias, repeated twice.

Four missing rulers: Between Joram and Oziah, Matt. 1:8, Ahaziah, Joash, Amaziah, Jehoikim between Josias and Jechonias in Matt. 1:11.

In the Bible there are 21 names before Abraham (See Luke 3).

There are 66 names in this royal line through Solomon.

So while Jesus is the 77th name in the line, which comes through Nathan, the son of Mary, it is the 66th name in the line which comes through Solomon, the legal son of Joseph. 6 is the human number, 7 is the Divine, which epitomizes the fact that Jesus was both Son of God and Son of Man. When Elizabeth was six months pregnant with John the Baptist, Jesus' conception was announced to Mary (Luke 1:26).

"Seven times a day do I praise thee because of Thy righteous judgments."—Psalm 119:164.

Shem, Ham and Japheth were Noah's sons. The two sons, Shem and Japheth received their father's blessing and are mentioned together seven times; but six of those are in connection with Ham, whose posterity was cursed.

Mankind's seven-fold qualification for service as outlined in Judges 6:

1. Conviction as to his own humiliating condition, v. 15.
2. Willingness to inquire of God, v.13.
3. No confidence in the flesh, v.15.
4. Peace with God through grace and through His gift, vs. 18-23
5. Worship, v.24.
6. Obedience, v.25-27.
7. Power for great things, v.33-35; and chapter 7.

Genesis 12:2,3 tells of Abraham's seven-fold blessing.

Exodus 7:2-8 gives seven times the expression "I will," which tells what Jehovah would do for His people:

1. I will bring you out from under the burdens of the Egyptians.
2. I will rid you out of their bondage.
3. I will redeem you.
4. I will take you to Me for a people.
5. I will be to you a God.
6. I will bring you in unto the land.
7. I will give it to you for a heritage.

EIGHT:
SUPERABUNDANT REGENERATION RESURRECTION

Genesis 17:12

"And he that is eight days old shall be circumcised among you, every man child in your generations."

In Hebrew the number eight comes from the root word which means "to make fat," to super-abound as if a surplus above the "perfect seven." The seventh day was the day of completion and rest, so eight, as the eighth day, was over and above this perfect completion and was indeed the first of a new series, as well as being the eighth.

Jehovah's covenants with Abraham were eight in number; seven before Isaac was offered up and eighth when he had been received "as in a figure of speech" from the dead:

1. Gen. 12:1-3, sovereignty.
2. Gen. 12:7, the seed.
3. Gen. 13:14-17, Divine assurance.
4. Gen. 15: 13-21, limits of land.
5. Gen. 17:1-2, invincible grace.
6. Gen. 18:9-15, overcoming human failure.
7. Gen. 21:12, spiritual blessing
8. Gen. 22:15-18, resurrection blessing.

It will be noted that each covenant blessing is stamped with the character of its numerical significance.

Joseph's communications with his brethren were eight in number, seven times before Jacob's death and an eighth time after his death.

The feast of Tabernacles was the only feast which kept eight days, the eighth being distinguished from the seventh, Leviticus 23:34-39. This was the harvest feast.

In Isaiah 5:1,2 there are eight phrases describing the vineyard, but seven give the characteristics and one gives the result.

In Ezekiel's temple there were seven steps into the outer court (Ezekiel 30:22, 26), and eight steps leading from the outer court to the inner court (Ezekiel 40:31-34, 37). The seven led from labor to rest and the eighth from rest to worship.

Ephesians 4:4-6 includes seven unites, but the seventh is twofold: "God the Father" making eight in all.

The LORD Jesus was on a mountain eight times (excluding when the devil took Him to a mountain), seven times before the cross, and the eighth after He rose from the dead.

Colossians 3:12,13 contains seven graces, but in verse 14 is love, which completes and unites the others.

Abraham's sons were eight in number, seven born after the flesh and one by promise.

David was the eighth son of Jesse, but owing doubtless to the death of one son without an heir (and hence excluded from the genealogies), David would be called in I Chronicles 2:15 "the seventh."

There were eight commands to keep the Sabbath holy, seven by Jehovah and one by Moses.

Eight by itself is seven plus one, and is the number associated with resurrection and regeneration and the beginning of a new era or order. When the whole Earth was covered with the flood, it was Noah, the eighth person (II Peter

2:5) who was saved, and it was seven people who stepped out on to a new earth to begin a new order of things (I Peter 3:20).

Circumcision was to be performed on the eighth day because it was foreshadowing of the true circumcision of the heart.

The first-born was to be given to Jehovah on the eighth day.

The numerical total of the Greek letters in Jesus' name = 888. All of His names in the New Testament contain multiples of eight.

When you study the numerical values for numbers in Hebrew or Greek, you will find that in the original language, God and His followers are marked by multiples of eight, while the wicked and rebellious are marked by multiples of 13.

NINE:
JUDGMENT OR FINALITY

Luke 15:7

"I say unto you, that likewise joy shall be in heaven over one sinner that repenteth, more than over ninety and nine just persons, which need no repentance."

Nine is the square of three, and three is the number of Divine perfection as well as the number peculiar to products of God's Holy Spirit. The FRUIT of God's Spirit are nine graces: LOVE, JOY, PEACE, LONGSUFFERING, GENTLENESS, GOODNESS, FAITH, MEEKNESS and TEMPERANCE.

The GIFTS of the Spirit in I Corinthians 12:8-10 are also nine: WISDOM, KNOWLEDGE, HEALING, MIRACLES, PROPHECY, DISCERNMENT, TONGUES, and INTERPRETATION OF TONGUES.

Nine is the last of the single digits and thus marks the end. Nine is significant as the conclusion of a matter. It signifies completeness.

There are nine items the drought is called upon in the judgments of God in Haggai 1:11. Land, mountains, corn, wine, oil, crops, men, cattle, labor.

Nine people were "stoned" in the Bible.

TEN:
PERFECTION OF DIVINE ORDER

<u>Revelation 5:11</u>

"And I beheld, and I heard the voice of many angels round about the throne and the beasts and the elders: and the number of them was ten thousand times ten thousand, and thousands of thousands."

Ten starts a new series of numbers. It originates the system of calculation called "decimals." It implies that nothing is wanting; the number and order are perfect; the whole "cycle" is complete.

In anatomy the number 10 is everywhere: Ten ribs; ten digits on hands (works); ten toes (ways); ten nerves that go through the sacrum bone, etc.

Noah completed the antediluvian age in the 10th generation.

The ten commandments contain all that is necessary in law, encompassing all other laws and statutes.

The tithes (tenths) mark and recognize God's claim on the whole of everything.

The LORD's prayer has 10 clauses.

Ten virgins represent the whole of the people waiting for Christ's return.

There are ten parables in Matthew.

The unrighteous who shall not enter the kingdom are ten: *"Know ye not that the unrighteous shall not inherit the kingdom of God? Be not deceived: neither fornicators, nor idolaters, nor adulterers, nor effeminate, nor abusers of themselves with mankind, nor thieves, nor covetous, nor drunkards, nor revilers, nor extortioners, shall inherit the kingdom of God."*—I Corinthians 6:9,10.

The security of the saints is tenfold: *"For I am persuaded, that neither death, nor life, nor angels, nor principalities, nor powers, nor things present, nor things to come, nor height, nor depth, nor any other creature, shall be able to separate us from the love of God, which is in Christ Jesus our Lord."* —. Romans 8: 38,39.

There are 10 generations from Adam to Noah and ten from Noah to Abraham.

There are ten "I AM's" in John. (See John 6:35; 6:14; 6:51; 8:12; 8:18; 10:7,9; 11:25; 14:6; 15:1,5.) AMEN.

Chapter Fourteen

VISION—HEARING—SMELL TASTE—TOUCH

II Kings 4:10

"And Jabez called on the God of Israel, saying, Oh that thou wouldest bless me indeed, and enlarge my coast, and that thine hand might be with me, and that thou wouldest keep [me] from evil, that it may not grieve me! And God granted him that which he requested."

"Day by day, and hour by hour, there must be a vigorous process of self-denial and of sanctification going on within; and then the outward works will testify that Jesus is abiding in the heart by faith. Sanctification does not close the avenues of the soul to knowledge, but it comes to expand the mind, and to inspire it to search for truth, as for hidden treasure; and the knowledge of God's will advances the work of sanctification. There is a heaven, and O, how earnestly we should strive to reach it. I appeal to you students of our schools and colleges, to believe in Jesus as your Saviour. Believe that he is ready to help you by his grace, when you come to him in sincerity. You must fight the good fight of faith. You must be wrestlers for the crown of life. Strive, for the grasp of Satan is upon you; and if you do not wrench yourselves from him, you will be palsied and ruined. The foe is on the right hand, and on the left, before you, and behind you; and you must trample him under your feet. Strive, for there is a crown to be won. Strive, for if you win not the crown, you lose everything in this life and in the future life. Strive, but let it be in the strength of your risen Saviour."—E. G. White, Counsels on Education, p. 114

"Because the avenues to the soul have been closed by the tyrant Prejudice, many are ignorant of the principles of healthful living. Good service can be done by teaching the people how to prepare healthful food. This line of work is as essential as any that can be taken up. More cooking schools should be established, and some should labor from house to house, giving instruction in the art of cooking wholesome foods. Many, many will be rescued from physical, mental, and moral degeneracy through the influence of health reform. These principles will commend themselves to those who are seeking for light; and such will advance from this to receive the full truth for this time." "God wants His people to receive to impart. As impartial, unselfish witnesses, they are to give to others what the Lord has given them. And as you enter into this work, and by whatever means in your power seek to reach hearts, be sure to work in a way that will remove prejudice instead of creating it. *Make the life of Christ your constant study, and labor as He did, following His example."*—E. G. White, Counsels on Diet and Foods, p. 472.

KEEP FROM EVIL

As Jabez reached the crossroads of his life he made four requests of God that changed his life forever. God's people are at the crossroads today! We have the opportunity during this time to *"choose life not death."* (Deuteronomy 30:19.)

WHAT CHOICE WILL YOU MAKE?

In the Book of Revelation, Chapter 18, God's messenger tells us that Babylon is fallen. Verse four of this chapter tells us to *"come out of her my people."* God's people are to be a separate people, distinct in thoughts, actions, speech, habits, and character. They are to be separated from the world in every facet of their lives. Christ's prayed in John 17:15: *"I pray not that thou shouldest take them out of the world, but that thou shouldest keep them from the evil. They are not of the world, even as I am not of the world. Sanctify them through thy truth: thy word is truth."*

In order to be sanctified and preserved in righteousness, God's people *must* keep themselves pure and unspotted from the world. The things we <u>look</u> at *must* be pure, the things that we <u>hear</u> *must* be pure, and the things that we <u>touch</u> *must* be pure. The things that we <u>taste</u> *must* be that which will make the body pure and holy, and the air that we <u>breathe</u> *must* be pure, and untainted with the pollution of this world.

Adam and Eve had this perfect environment. God gave them the "Garden of Eden" in which to live. This environment was to keep them a pure and holy people.

It seemed a small matter to our first parents, when tempted, to transgress the command of God in one small act and eat of a tree that was beautiful to the eye and pleasant to the taste. To the transgressors this was but a small act, but it destroyed their allegiance to God and opened a flood of woe and guilt which has deluged the world. Who can know, in the moment of temptation, the terrible consequences that will result from one wrong, hasty step! Our only safety is to be shielded by the grace of God every moment, and not put out our own spiritual eyesight so that we will call evil, good, and good, evil. Without hesitation or argument, we must close and guard the ***avenues of the soul*** against evil.

God's people today, have *chosen to watch* things that pollute the mind. They have *chosen to taste and eat* food that pollutes the physical body. Many have *chosen to live in the cities* where the air is filled with the miasma of pollution. Our ears today are constantly bombarded with *sound that affects the mind, soul and body.* We are told in Leviticus 5:2,3: *"If a soul touch any unclean thing, whether [it be] a carcase of an unclean beast, or a carcase of unclean cattle, or the carcase of unclean creeping things, and [if] it be hidden from him; he also shall be unclean, and guilty. Or if he touch the uncleanness of man, whatsoever uncleanness [it be] that a man shall be defiled withal, and it be hid from him; when he knoweth [of it], then he shall be guilty."* Yet we <u>touch, watch, eat, breathe, and hear</u> many things that prevent us from gaining eternal life.

We are told: "It will cost us an effort to secure eternal life. It is only by long and persevering effort, sore discipline, and stern conflict that we shall be overcomers. But if we patiently and determinedly, in the name of the Conqueror who overcame in our behalf in the wilderness of temptation, overcome as He overcame, we shall have the eternal reward. Our efforts, our self-denial, our perseverance, must be proportionate to the infinite value of the object of which we are in pursuit."—E. G. White, "Testimonies to the Church" Vol. 3, page 324

Everyone is on life's journey, some to eternal damnation, some to eternal life. All of God's people are traveling to a city—the "New Jerusalem." There are many roads or *avenues* that we attempt to follow to get there. We are told: *"There are many ways of practicing the healing art; but there is only one way that Heaven approves. God's remedies are the simple agencies of nature that will not tax or debilitate the system through their powerful properties. Pure air and water, cleanliness, a proper diet, purity of life, and a firm trust in God, are remedies for the want of which thousands are dying; yet these remedies are going out of date because their skillful use requires work that the people do not appreciate. Fresh air, exercise, pure water, and clean, sweet premises, are within the reach of all, with but little expense; but drugs are expensive, both in the outlay of means, and the effect produced upon the system."*—E. G. White, Counsels on Diet and Foods, p. 301.

WILL YOU MAKE THE RIGHT CHOICE?

THE SENSES

Philippians 4:8

"Whatsoever things are true, whatsoever things [are] honest, whatsoever things [are] just, whatsoever things [are] pure, whatsoever things [are] lovely, whatsoever things [are] of good report; if [there be] any virtue, and if [there be] any praise, think on these things."

The structures of the body that perceive and receive signals from our environment are called *sense organs* or *receptors*. All sensory receptors are structures that are capable of converting environmental information (stimuli) into nerve impulses. Thus, all receptors are transducers, in that they convert one form of energy into another.

Sensory receptors are stimulated by specific stimuli. The eye is set in motion by light waves, the ear by sound waves. The eye does not respond to sound or the ear to light. God's Holy Word in the book of I Corinthians 12:6-12 tells us: *"And there are diversities of operations, but it is the same God which worketh all in all. But the manifestation of the Spirit is given to every man to profit withal. For to one is given by the Spirit the word of wisdom; to another the word of knowledge by the same Spirit; To another faith by the same Spirit; to another the gifts of healing by the same Spirit; To another the working of miracles; to another prophecy; to another discerning of spirits; to another [divers] kinds of tongues; to another the interpretation of tongues: But all these worketh that one and the selfsame Spirit, dividing to every man severally as he will. For as the body is one, and hath many members, and all the members of that one body, being many, are one body: so also [is] Christ."* This passage shows us that each individual has a job to do and is given the equipment necessary to do it. We are not all called to do the same job. Each individual or organ in the body has a particular task to accomplish.

Sound cannot occur unless there is someone to hear it. This is true because <u>sound</u> is something that is *received* (sensed) by the ear and *heard* (perceived) by the brain. If there is no ear to receive sound waves and no brain to translate those sound waves, there will be no perceived sound. The same is true of our other senses.

The different sensations are brought about when the signals received by the nerve fibers connect with specialized protons of the brain. We "see" an object not when its image enters our eyes but when that coded image stimulates the vision centers of the brain.

To help the reader grasp this idea, contemplate the following illustration. Visualize a relay race. The whistle is blown or the gun shot to signal the beginning of the race. This causes the first person to take off with a stick or baton in his hand. He is the ***generator potential***. Think of the stick that is carried as a stimulus or a message to be transmitted to the brain. One person can only run so far before he then hands the stick to the next person who then becomes the ***graded potential***. Because this person runs out of energy before the whole distance of the race is covered, he hands the message to the next runner (neurofibral node) who takes off with the message (nerve impulse) and hands it to the next runner (synapse of a nerve) who continues the race to the finish line (the proper portion of the brain to which the signal was directed).

Just as the runners all have two legs but these legs may be different lengths and sizes, so there are several types of sensory receptors but they all have certain features in common. Individual sensory receptors are each designed to receive a specific kind of stimulus yet they all interact with *afferent (toward the brain)* nerve cells. Sensory receptors can be nerve endings of a neuron or a specialized receptor cell.

Four kinds of sensory receptors are recognized on the basis of their location:

1. Exteroceptors: (L., "received from the outside") are receptors that respond to external environmental stimuli that affects the skin.
2. Teleceptors (Gr., "received from a distance") are located in the eyes, ears, and nose.

3. Interoceptors (L., "received from the inside") respond to changes in the body such as blood pressure, blood chemistry, etc.
4. Propriceptors (L., "received from one's own self") respond to deep body structures such as joints, muscles, vestibular apparatus of the ear, etc.

Stimuli derived from either the internal and external environments activate the sensory receptors. The responses of the receptors are converted into action potentials in first-order neurons that convey the impulses from the receptors to the central nervous system. The central nervous system then initiates the proper response to the message. Reflex activities are utilized, decisions are made and behavior is changed according to the response received from the central nervous system. The brain receives the nerve impulses, suppresses what is irrelevant, compares it with information already stored, and coordinates the final impulses to the effectors.

The sensory endings of receptors have two kinds of nerve endings:

1. Free nerve endings, which have no coverings.
2. Encapsulated endings, which are covered with various types of capsules.

"Those things, which ye have both learned, and received, and heard, and seen in me, do: and the God of peace shall be with you."—Philippians 4:9.

VISION

Ephesians 1:18

"The eyes of your understanding being enlightened; that ye may know what is the hope of His calling, and what the riches of the glory of His inheritance in the saints."

The spiritual lesson of the sight is understanding. The Bible equates seeing with understanding: Mark 8:17, 18; Mark 12:11; Luke 19:42; John 12:40.

When we don't agree with something or don't understand it, we say, "I can't see that." Jesus came to open the eyes of the blind, and that was not just physical eyesight. He came to give us understanding of God's love and His plan for us. As we study the sense of sight, it is clear how closely the actual mechanics of seeing parallel the spiritual understanding.

The wisest man that ever lived compares vision with law-keeping. *"Where there is no vision, the people perish: but he that keepeth the law, happy is he."* Proverbs 29:18

In describing the lost condition of Israel, Jeremiah says, *". . . the law is no more; her prophets also find no vision from the LORD."*—Lamentations 2:9

The first component necessary for vision is light. *"That was the true Light, which lighteth every man that cometh into the world."*

The study of the Bible will give strength to the intellect. Says the psalmist, *"The entrance of thy words giveth light; it giveth understanding unto the simple."*—Psalms 119:130. The question has often been asked me, "Should the Bible become the important book in our schools?" We are told: "It is a precious book, a wonderful book. It is a treasury containing jewels of precious value. It is a history that opens to us the past centuries. Without the Bible we should have been left to conjectures and fables in regard to the occurrences of past ages. Of all the books that have flooded the world, be they ever so valuable, the Bible is the Book of books, and is most deserving of the closest study and attention. It gives not only the history of the creation of this world, but a description of the world to come. It contains instruction concerning the wonders of the universe, and it reveals to our understanding the Author of the heavens and the earth. It unfolds a simple and complete system of theology and philosophy. Those who are close students of the word of God, and who obey its instructions, and love its plain truths, will improve in mind and manners. It is an endowment of God that should awaken in every heart the most sincere gratitude; for it is the revelation of God to man."—E. G. White, Counsels on Education, p. 105.

All light is not the True Light. *"Satan himself is transformed into an angel of light. Therefore it is no great thing if his ministers also be transformed as the ministers of righteousness; whose end shall be according to their works."* We must watch therefore, what light we are following.

How can we tell what is the true light? Isaiah defines light for us and tells us where there is no light. *"And when they shall say unto you, Seek unto them that have familiar spirits, and unto wizards that peep, and that mutter: should not a people seek unto their God? for the living to the dead? To the law and to the testimony: if they speak not according to this word, [it is] because [there is] no light in them"*—Isaiah 8:19,20.

People today are going to all kinds of sources for their light. "Thou art wearied in the multitude of thy counsels. Let now the astrologers, the stargazers, the monthly prognosticators, stand up, and save thee from [these things] that shall come upon thee. Behold, they shall be as stubble; the fire shall burn them; they shall not deliver themselves from the power of the flame: [there shall] not [be] a coal to warm at, [nor] fire to sit before it." Isaiah 47:13,14.

In Revelation 3, verses 15-17, John tells us the condition of the majority of God's people in the last days: *"I know thy works, that thou art neither cold nor hot: I would thou wert cold or hot. So then because thou art lukewarm, and neither cold nor hot, I will spue thee out of my mouth. Because thou sayest, I am rich, and increased with goods, and have need of nothing; and knowest not that thou art wretched, and miserable, and poor, and blind, and naked: I coun-*

sel thee to buy of me gold tried in the fire, that thou mayest be rich; and white raiment, that thou mayest be clothed, and [that] the shame of thy nakedness do not appear; and anoint thine eyes with eyesalve, that thou mayest see." Spiritual blindness comes as a result of refusal to follow light sent to us by God through His Spirit.

The Pharisees had the same problem in the New Testament *"And Jesus said, For judgment I am come into this world, that they which see not, might see; and that they which see might be made blind. And [some] of the Pharisees which were with him heard these words, and said unto him, Are we blind also? Jesus said unto them, If ye were blind, ye should have no sin: but now ye say, We see; therefore your sin remaineth."*—John 9:39-41.

The candlestick in the sanctuary gave light and the eye is the counterpart in the body. The eye has seven basic structures: Sclera and cornea, Choroid, Ciliary, Lens, Iris and pupil, Retina, Chambers.

The candlestick in the sanctuary gave light to the sanctuary just as the eye gives light to the brain where God wants to dwell.

"Since the Saviour shed His blood for the remission of sins, and ascended to heaven *'to appear in the presence of God for us'* (Hebrews 9:24), light has been streaming from the cross of Calvary and from the holy places of the sanctuary above. But the clearer light granted us should not cause us to despise that which in earlier times was received through the types pointing to the coming Saviour. The gospel of Christ sheds light upon the Jewish economy and gives significance to the ceremonial law. As new truths are revealed, and that which has been known from the beginning is brought into clearer light, the character and purposes of God are made manifest in His dealings with His chosen people. Every additional ray of light that we receive gives us a clearer understanding of the plan of redemption, which is the working out of the divine will in the salvation of man. We see new beauty and force in the inspired word, and we study its pages with a deeper and more absorbing interest."—E. G. White, Patriarchs and Prophets, p. 367.

The specialized receptors in our eyes constitute about 70 percent of the receptors of the entire body. The optic nerves contain about one-third of all the afferent nerve fibers carrying information to the central nervous system.

We have two eyes, but we perceive only one image. The number <u>two</u> stands for division and support. Each eye receives an image from a slightly different angle, and this creates the impression of distance, depth and three dimensionality. This is true of spiritual understanding. There are two sides of every question; faith and works, law and grace, mercy and justice, love and duty, etc. We must have the balance of both sides of a question to get true understanding.

In order to produce a single image, the six pairs (12) of extra ocular muscles attached to the eyes must move together with perfect coordination. Here again you can see that it takes 12 to make perfect government. Remember that the muscles represent judgment.

"The light of the body is the eye: if therefore thine eye be single, thy whole body shall be full of light. But if thine eye be evil, thy whole body shall be full of darkness. If therefore the light that is in thee be darkness, how great [is] that darkness!"—Mathew 6:22, 23. The word *single* here comes from a root word that means *braided together into a single thread* and from another word that means *without self-seeking.* In order to have true light we must humble ourselves and do the will of God.

Vision

Vision begins when "packets" of electromagnetic energy called photons are converted to neural signals that the brain can decode and analyze. (A photon is the smallest possible quantity of light.)

The <u>eyeball</u> is a sphere about one inch in diameter. It can be compared to a simple old-fashioned camera. Light passes through a lens; the external image is brought to a focus on the retina, which is somewhat like the film in the camera. Over a hundred million rods and cones convert light waves into electrochemical impulses, which are decoded by the brain.

"Jesus gave His life for the life of the world, and He places an infinite value upon man. He desires that man shall appreciate himself, and consider his future well-being. If the eye is kept single, the whole body will be full of light. If the spiritual vision is clear, unseen realities will be looked upon in their true value, and beholding the eternal world will give added enjoyment to this world."—E. G. White, Counsels on Stewardship, p. 136.

The Wall

The wall of the eyeball consists of 3 layers of tissue: the outer supporting layer, the vascular middle layer and the inner retinal layer.

The Supporting Layer

This layer consists mainly of a thick membrane of tough, fibrous connective tissue. It completely encloses the eyeball, except for the back portion, where small holes allow the fibers of the optic nerve to leave the eyeball on the way to the brain. The back segment, which makes up 5/6 of the tough outer layer is the opaque white sclera. This forms the white of the eye, gives the eyeball its shape and protects the delicate inner layers. The front part of this layer is transparent and is called the cornea. The cornea makes up about 1/6 of the circle and it bulges slightly. Light enters the eye through the cornea. It contains no blood vessels or lymphatic vessels, which is why it can be transplanted without rejection. There is no system to feed it, or to cleanse it. It is just a place where any kind of light can come through. It has 5 layers (5 is the number signifying grace), so God is telling us that He gives us the grace we need to discern the difference between true and false light.

Vascular Layer

This layer contains many blood vessels. The dark color of the middle layer is produced by pigments that help to light-proof the wall of the eye by absorbing stray light and reducing reflection. The back two thirds of this layer is a thin membrane called the *choroids*, which is a layer of blood vessels and connective tissue. It becomes thickened toward the front portion to form the *ciliary body,* which has fine hair like extensions. The smooth muscles in the ciliary body, contract to ease the tension on the *suspensory ligament* of the lens.

The front part of the choroids is a thin muscular layer called the *iris*, the colored part of the eyeball. The back part of the iris is always purple (the color of royalty; we belong to the King). The front part can be differently colored dependant upon our heredity.

In the center of the iris is an adjustable circular aperture, called the *pupil.* It appears black because most of the light that enters the eye is not reflected outward. The iris, acting as a diaphragm, is able to regulate the amount of light entering the eye because it contains smooth muscles that contract or dilate in an involuntary reflex in response to the amount of light available. This action is called adaptation.

Deuteronomy 32:9,10 tells us that God is concerned about His people *"For the LORD'S portion [is] his people; Jacob [is] the lot of his inheritance. He found him in a desert land, and in the waste howling wilderness; he led him about, he instructed him, he kept him as the apple of his eye."* The word for apple in the original Hebrew implies the pupil of the eye, the focus of the attention. All the attention of the universe has been focused on this world and the controversy here; all the efforts of heaven have been put forth to save the lost.

God's servant David asks the Lord in Psalm 17:8 *"Keep me as the apple of thy eye, hide me under the shadow of thy wings."* God also tells us on what to focus our attention *"Keep my commandments and live; and my law as the apple of thine eye. Bind them upon thy fingers, write them upon the table of thine heart."*—Proverbs 7: 2,3.

Intrinsic Muscles

Inside the eyes are three smooth intrinsic muscles: The *ciliary muscle* eases tension of the suspensory ligaments on the lens and allows the lens to change its shape for proper focusing. The *circular muscle* of the iris constricts the pupil, the *radial muscle* dilates it.

"The Bible is a casket containing jewels of inestimable value, which should be so presented as to be seen in their intrinsic luster. But the beauty and excellence of these diamonds of truth are not discerned by the natural eye. The lovely things of the material world are not seen until the sun, dispelling the darkness, floods them with its light. And so with the treasures of God's word; they are not appreciated until they are revealed by the Sun of Righteousness." —E. G. White, Counsels to Teachers, p. 422.

HEARING

Proverbs 20:12.

"The hearing ear, and the seeing eye, the LORD hath made even both of them."

The ear contains many structures that have foreign names. The list below will give you a basic understanding of a few of these names.

1. *Auris* means ear in Latin, so words that starts with *au*, such as auditory, have to do with the ear.
2. *Auricle* is the name given for the external ear. Cochlea (Kahk-lee-uh) means *snail* and concha means *shell*; some parts are named for their looks.
3. *Endo* means inside, so endolymph would mean the *fluid inside*; peri means around or about so perilymph would mean the *fluid on the outside* of the structure.
4. *Meatus* means passage or *opening* in Latin.
5. *Scala* means *stairway.*
6. *Vestibule* comes from the Latin for *entryway* so means the small space at the entry of something.
7. *Basilar* means at the *base or bottom* and Tectorial means *roof.*
8. *Tympani* means *drum.*

The above words are used in a variety of ways. For Example: Scala vestibule would mean *entry on the stairs* and scala tympani would mean *the drum on the stairs.*

The ear passage is carved out of the temporal bone. This bone is the principle of God's promises. (See Skeletal study.) Hearing is connected with promises in the Bible. Either promises of blessings or threatening. It all depends upon our attitude when the Lord speaks.

"And said, If thou wilt diligently hearken to the voice of the LORD thy God, and wilt do that which is right in His sight and wilt give ear to His commandments and keep all His statutes, I will put none of these diseases upon thee, which I have brought upon the Egyptians: for I am the LORD that healeth thee."—Exodus 15:26.

"He that turneth away his ear from hearing the law, even his prayer shall be abomination."—Proverbs 28:9.

"For ear trieth words, as the mouth tasteth meat."—Job 34:3.

"David declares, *'I love the Lord, because he hath heard my voice and my supplications. Because he hath inclined his ear unto me, therefore will I call upon him as long as I live.'* (Ps. 116:1, 2.) God's goodness in hearing and answering prayer places us under heavy obligation to express our thanksgiving for the favors bestowed upon us. We should praise God much more than we do. The blessings received in answer to prayer should be promptly acknowledged."—E. G. White, God's Amazing Grace, p. 324.

The ear is divided into three sections: ***the external ear, the middle ear, and the inner ear.***

THE EXTERNAL EAR

The external ear is called the *auricle* or *pinna*. The external ear is divided into 9 parts. Helix, Antihelix, Tragus, Antitragus, Lobule, Scaphoid fossa, Triangular fossa, Concha, and the Intertragic notch.

It is made of cartilage, which is the same material as bone but without the minerals that harden it. (The outer ear has no fixed principles; it collects all kinds of sound.) There are nine muscles, three arteries, three veins, and five nerves connected with the auricle.

The auditory canal conducts the sound to the ear. This passage is about 1½ inches long. The first third of it is composed of cartilage like the outer ear and the rest is carved out of the temporal bone. It is covered with skin, which contains oil glands, fine hairs and modified sweat glands. The glands secrete earwax. The hairs and wax make it difficult for tiny insects and other foreign matter to enter the canal. The wax also prevents the skin from drying out. (Without the lubrication and protection of the oil of the Holy Spirit, our spiritual hearing is in trouble.)

"And grieve not the Holy Spirit of God, whereby ye are sealed unto the day of redemption."—Ephesians 4:30.

In the newborn fetus there is no bone in this canal, which teaches the lesson that until a person is born again, there is no heavenly principle connected with his hearing.

The ear is attached to a ring of bone that has 3 layers and 3 nerves. When you next see an orchestra play, watch the person who plays the tympani drums. The sound waves hit the eardrums just like the drummer strikes the drum.

MIDDLE EAR

The middle ear is a small chamber between the tympanic membrane and the bone of the inner ear. It has an opening for the auditory (Eustachian, yoo-Stay-shun) tube connecting it with the passage of the nose. The Eustachian tube is made of partly bone and partly cartilage. The upper edge looks like a hook. This tube is necessary to bring air from the nostrils to press on the <u>inside</u> of the eardrum and equalizes the atmospheric pressure pressing on the eardrum from the <u>outside</u>. The pressure has to be equalized or it would be blown apart.

The Eustachian tube lies in the groove between the temporal (promises) and the sphenoid bone (which represents the final atonement or judgment.) The pressure must be balanced between these two principles. On *one* side we have the realization that we cannot be saved if we cling to one sin and on the other side we have the promise of God that we can do all things through Christ. This keeps us from being discouraged and blown apart spiritually.

There are 3 bones that make a bridge between the eardrum and the inner ear, the malleus *(hammer)*, the incus *(anvil)*, and the stapes *(stirrup)*. The 3 principles in hearing are <u>learn</u>, <u>fear</u>, and <u>obey</u>. The hammer has 5 parts, the anvil 3, and the stirrup 5. There is grace given in the learning and obeying and the only fear should be of the Lord.

"Gather the people together, men, and women, and children, and thy stranger that is within thy gates, that they may hear, and that they may <u>learn</u>, and <u>fear</u> the LORD your God, and observe to do all the words of this law:"—Deuteronomy 31:12.

"The fear of the LORD is the beginning of knowledge."—Proverbs 1:7.

We can choose to ignore God and follow the ways of the world. When we choose this lifestyle, it is man and the world that we learn of, that we are afraid of, and whom we obey. When we learn the ways of the world, we are afraid to upset people and want them to think well of us, so we live to suit them, dress to be like them, follow their standards, and eat what pleases them and us.

"Know we not, that to whom ye yield yourselves servants to obey, his servants ye are to whom ye obey; whether of sin unto death, or of obedience unto righteousness."—Romans 6:16.

Five ligaments connect the bones to the walls of the middle ear. There are two tiny muscles attached to the bones, one on the malleus and one of the stapes.

INNER EAR

The inner ear is called the labyrinth (Greek for maze) because of its structure of interconnecting chambers and passages. It consists of two main parts, one inside the other. The bony labyrinth is filled with a fluid called perilymph.

The inner membranous labyrinth contains all the sensory receptors for hearing and equilibrium and a fluid called endolymph.

HEARING

The cochlea is a bony tube wound 2¾ times in the form of a spiral. It is divided lengthwise into 3 ducts: the *scala vestibuli, the scala tympani,* and the *scala media* (or *cochlear duct).* The corti of the ear (named for the man who discovered it, whose name was Corti) goes the length of the tube.

According to *Gray's Anatomy,* the cochlea has a remarkable arrangement of cells like a keyboard of a piano. There are supporting cells and hair cells. The hair cells are arranged in rows along the length of the coil. There are about 20,000 outer hair cells and each hair cell has 40 to 60 sensory hairs. Isn't God generous! They are arranged in 3 rows. The inner hair cells (about 3500) are arranged in a single row. The hair cells have sensory hairs and a nasal body on one side. Each hair is enclosed within a hair cell which is embedded in the tectorial membrane.

Sound is conducted by the three mediums that make up matter: vapor (air), solid, and fluid. Sound comes through the air in the form of vibrations. These vibrations hit the ear drums making the three solid bones vibrate, which sends the mechanical waves through the fluid to hit the nerves which carry it on in electrical form to the brain to be interpreted there. The sounds that come into our ears carry information to the brain. They have a great effect on the decisions that we make and the information we record. Words are our main medium of communication and we gather words by sound or sight, or in cases of blindness, by touch.

"The humblest workers, in co-operation with Christ, may touch chords whose vibrations shall ring to the ends of the earth, and make melody throughout eternal ages."—E. G. White, The Ministry of Healing, p. 159.

"And the Word *was made flesh, and dwelt among us, (and we beheld his glory, the glory as of the only begotten of the Father), full of grace and truth."*—John 1:14.

To hear anything spiritually, it must first come through the Holy Spirit (vapor); then God's principles or laws are acknowledged (solid); then it sets the ideas (fluid) into motion to finally become a part of our memory.

Look up the following verses:

There are joyful sounds, Psalm 89:15; solemn sounds, Psalm 92:3; sounds of battle, Jeremiah 50:22; sounds of alarm, Joel 2:1; sound of the coming of the Lord, Matthew 24:31, I Thessalonians 4:16; sounds that are deceiving, Leviticus 26:36; sounds that are misleading, I Corinthians 13:1.

Numbers 10 gives instruction about the meaning of the different sounds of the trumpets.

The bells of the high priest's robe were given as an assurance to the people. As long as they were sounding, the people knew that God was accepting their repentance (Exodus 28:33-35).

When the people combined the right works, the building of the temple, the right sounds, the praise of the LORD, glory filled the temple (II Chronicles 5:13,14). Apply that principle to your body temple. Would you like to have your temple filled with the glory of the Lord?

The sacred temple of the body must be kept pure and uncontaminated, that God's Holy Spirit may dwell therein. We need to guard faithfully the Lord's property; for any abuse of our powers shortens the time that our lives could be used for the glory of God. Bear in mind that we must consecrate all—soul, body, and spirit—to God. All is his purchased possession, and must be used intelligently, to the end that we may preserve the talent of life. By properly using our powers to their fullest extent in the most useful employment, by keeping every organ in health, by so preserving every organ that mind, sinew, and muscle shall work harmoniously, we may do the most precious service for God.

"The active service of God is directly connected with the ordinary duties of life, even its humblest occupations. We are to serve God just where he puts us. He is to place us individually, and not we ourselves. Perhaps service in the home life is the place we are to occupy for a time, if not always. Then a preparation for that work should be obtained, that we may do our best in service for the Lord."—E. G. White, Youth's Instructor, April 14, 1898, par. 1.

THE PHYSIOLOGY OF HEARING

1. Sound waves are pressure waves that enter the external ear. After crossing the external auditory meatus, the waves reach the tympanic membrane.
2. Air molecules under pressure cause the tympanic membrane to vibrate.
3. Low-frequency sound waves produce slow vibrations, and high-frequency sound waves produce rapid vibrations. The vibrations move the malleus, on the other side of the membrane.
4. The handle of the malleus articulates with the incus, causing it to vibrate.
5. The vibrating incus moves the stapes as it oscillates into and out of the cochlea at the oval window.
6. The sound waves that reach the inner ear through the oval window set up pressure changes that vibrate the perilymph in the scala vestibuli.
7. Vibrations in the perilymph are transmitted across the vestibular membrane to the endolymph of the cochlear duct and also up the scala tympani.
8. The vibrations are transmitted to the basilar membrane, causing the membrane to ripple. The fundamental vibratory ripples result in the perception of pure tones. Overtones such as musical sounds, chords, and harmonics result from secondary vibrations superimposed on the fundamental vibrations of the spiral organ. The ripples in the long axis of the basilar membrane are concerned with the frequency and intensity of sound. The spiral organ is organized so that the high tones are encoded near the base, and the low tones are encoded near the apex of the cochlea. Loudness is associated with the amplitude of the vibrations (the amount of displacement of the basilar membrane).
9. Receptor hair cells of the spiral organ that are in contact with the overlying tectorial membrane are bent, causing them to generate graded receptor (generator potentials). These generator potentials excite the cochlear nerves to generate action potentials. When the hairs are displaced toward the basal body (axis of sensitivity), the hair cells are excited; when the hairs are displaced away from the basal body, the hair cells are inhibited.
10. The nerve impulses are conveyed along the cochlear branch of the Vestibulo-cochlear nerve. These fibers activate the auditory pathways in the temporal lobe of the cerebral cortex, where the appropriate sound is perceived.
11. Vibrations in the scala tympani are dissipated out of the cochlea through the round window into the middle ear.

THE SOUNDS WE HEAR

We can hear things that we cannot see, like the wind, the thunder, and the buzz of a bees wings. When we read the story of Elijah on Mount Horeb in I Kings 19, we are told that our ears must be opened by the Lord in order for Him to speak to us. It was not in mighty manifestations of Divine power, but by "a still small voice" that Elijah heard God's voice. God desired to impress Elijah that it is not always a work that makes the greatest demonstration that is most successful in accomplishing His purpose. While Elijah waited for the revelation of the Lord, a tempest rolled, the lightning flashed, and a devouring fire swept by—but God was not present. Then there came a still small voice, and the prophet covered his head before the presence of the Lord.

It is not always the most learned presentation of God's truth that convicts the soul. Not by eloquence or logic are men's hearts reached, but by the sweet influence of the Holy Spirit, which operates quietly yet surely in transforming and developing character. It is the still, small voice of the Spirit of God that has the power to change lives.

"The Lord GOD hath given me the tongue of the learned, that I should know how to speak a word in season to [him that is] weary: he wakeneth morning by morning, he wakeneth mine ear to hear as the learned. The Lord GOD hath opened mine ear, and I was not rebellious, neither turned away back."—Isaiah 50:4-5.

"He openeth also their ear to discipline, and commandeth that they return from iniquity. If they obey and serve [him], they shall spend their days in prosperity, and their years in pleasures. But if they obey not, they shall perish by the sword, and they shall die without knowledge."—Job 36:10-12.

We can choose what we want to hear. *"Hear ye and give ear; be not proud; for the LORD has spoken."*—Jeremiah 13:15.

The Bible gives us many instances of people who chose not to listen to Him.

"As for the word that thou hast spoken unto us in the name of the LORD, we will not hearken unto thee." —Jeremiah 44:16.

HEARING LOSS

Hearing ability declines when the basilar membrane loses some of its elasticity, either from hardening or from deposited matter, such as calcium deposits. We lose our ability to hear spiritually if we do not keep our hearts soft and dwell upon the love and sacrifice of God in our behalf. If we corrode our minds by listening to carnal things, eventually we will not be able to hear the promptings of the Holy Spirit. Isaiah tells us that part of the qualifications to be able to live with God is to shut our ears to the devil's noise. What are you listening to on the television set?

"He that walketh righteously, and speaketh uprightly; he that despiseth the gain of oppressions, that shaketh his hands from holding of bribes, that stoppeth his ears from hearing of blood, and shutteth his eyes from seeing evil."—Isaiah 33:15.

Current estimates are that some environmental noises are twice as intense now as they were in the 1960's, and increase daily as new technology is created. Approximately 10 million Americans use hearing aids, and many of these people have suffered hearing impairment through prolonged exposure to sounds that they did not think of as excessively loud.

Exposure to sounds of 110 decibels for 26 minutes can cause possible hearing loss. That is the noise of a jet takeoff 600 meters away, a live rock band, a loud audio player (some of these are higher than 110, a "boom car" being 130), etc. There are thousands of young people with hearing loss already.

Jesus cured deafness when he was here. He put his fingers (the symbol of His works or lifestyle) into a deaf man's ear (Mark 7:33). He will cure our spiritual deafness the same way. Let us not be dull of hearing as were the people who rejected His words (Acts 28:27).

"Let these saying sink down into your ears . . ."—Luke 9:44.

Noise is a stressor, constricting blood vessels, damaging ear tissue, reducing oxygen and nutrients to the ear, causing high blood pressure, and increasing heart rate. This will account for the rush that people get listening to loud music. It actually becomes addictive.

"For the time will come when they will not endure sound doctrine; but after their own lusts shall they heap to themselves teachers, having itching ears; And they shall turn away their ears from the truth, and shall be tuned into fables."—II Timothy 4:3,4.

BALANCE

The inner ear helps the body cope with changes in position and acceleration and deceleration. Signals coming from the eyes and joints of bones signal the head to move producing static balance. (Try standing on your toes with your eyes closed. Without your eyes to guide your body, you will begin to fall forwards.)

The vestibule is the central chamber of the labyrinth. Within the vestibule are two endolymph-filled sacs. The utricle (YOO-tri-cal; L. little bottle) and the saccule (SACK-yool: L. little sack). Each sac contains a sensory patch called a *macula*, which contains receptor hair cells embedded in a jellylike membrane. Loosely attached to the membrane and piled on top of it are hundreds of thousands of calcium carbonate crystals called *otoconia* (oto; Gr. Ear—conia, dust).

The bulge called the ampulla contains a patch of hair cells and supporting cells embedded in the crista ampullaris (crest of the ampulla). The hairs of the hair cells project into a gelatinous flap called the cupula (KYOO-pyu-luh; L. little cask or tub). The cupula acts like a swinging door, with the crista as the hinge. The free edge of the cupula brushes against the curved part of the ampulla. This causes a message to be transmitted along the vestibular nerve, to the vestibular nuclei in the brainstem and cerebellum.

Each of the three semicircular ducts is situated in a different plane, at right angles to each other and at least one duct is affected by every head movement. The sensory signals from these ducts are indicators of the positions and movements of the head. These inputs are critical in: (1) generating compensatory movements to maintain balance and an erect posture in response to gravity, (2) producing the coupled movements of the eyes that compensate for changes in the position of the constantly moving head, and (3) supplying information for the conscious awareness of position, acceleration, deceleration, and rotation. The saying, "He lost his head" means that somehow the person became unbalanced.

The vestibular tracts consist of pathways to the brainstem, spinal cord, cerebellum, and cerebral cortex. The 19,000 nerve fibers of each vestibular nerve have their cell bodies in the vestibular ganglion (group of cells) near the membranous labyrinth. The main fibers from this nerve pass into the upper medulla oblongata, which is the symbol for the altar of sacrifice. If we want to keep our spiritual balance we must stay near the cross, so that we may remember the great sacrifice that was given for us. Otherwise, we become unbalanced and fall either into legalism or cheap grace. Then we begin to charge God with being unbalanced.

"Yet ye say, The way of the Lord is not equal. Hear now, O house of Israel; Is not my way equal? Are not your ways unequal? When a righteous man turneth away from his wickedness that he hath committed iniquity, and dieth in them; for his iniquity that he hath done shall he die. Again, when the wicked man turneth away from the righteousness, and committeth iniquity, and dieth in them; for his iniquity that he hath done shall he die. Again, when the wicked man turneth away from his wickedness that he hath committed, and doeth that which is lawful and right, he shall save his soul alive. Because he considereth, and turneth away from all his transgressions that he hath committed, he shall surely live, he shall not die. Yet said the house of Israel, the way of the Lord is not equal. O house of Israel, are not my ways equal? Are not your ways unequal? Therefore I will judge you, O house of Israel, every one according to his ways, said the Lord God. Repent, and turn yourselves from all your transgressions, so iniquity shall not be your ruin. Cast away from you all your transgression, whereby ye have transgressed; and make you a new heart and a new spirit: for why will ye die, O house of Israel: For I have no pleasure in the death of him that dieth, saith the Lord God: wherefore turn yourselves, and live ye."—Ezekiel 18:25-32.

The Lord has to be allowed to cut away all the worldly noises that our ears love. *"To whom shall I speak, and give warnings, that they may hear? Behold, <u>their ear is uncircumcised</u>, and they cannot hearken: behold, the word of the LORD is unto them a reproach; they have no delight in it."*—Jeremiah 6:10.

"But this thing commanded I them, saying, Obey my voice, and I will be your God, and ye shall be my people: and walk ye in all the ways that I have commanded you, that it may be well unto you. But they hearkened not, nor inclined their ear, but walked in the imagination of their evil heart, and went backward and not forward."—Jeremiah 7:23,24.

The ear was a place to be marked. When a slave decided to stay with his master, here was a mark put in the ear. *"Then his master shall bring him unto the judges; he shall also bring him to the door, or unto the door post; and his master shall bore his ear through with an awl; and he shall serve him for ever."*—Exodus 21:6.

We can mark ourselves by having holes punched in our ears and hanging earrings there. Read Isaiah 3; The women who were decorating themselves with various jewelry including earrings were condemned by the Lord. Read Genesis 35:4; when the people wanted to put away their strange gods they took off their earrings. It is interesting to contrast this with Exodus 32:3, because the wording here is *brake off*, indicating a painful process. One group were trying to get closer to God; one group were about to make a strange god and needed to give up their earrings to do so. If we think in terms of "giving up" something it is hard to do. If we think in terms of "getting rid of something," it becomes a delight.

We reason sometimes that if the earrings are in the form of crosses, somehow we are marked for God, but remember that the pagans wore crosses thousands of years before Christ. They were a symbol of the sun god.

When a person was going to be consecrated to God they had blood put upon their ear, symbolizing the application of the blood and life of Christ to the hearing and balance.

"Then shalt thou kill the ram, and take of his blood, and put it upon the tip of the right ear of Aaron, and upon the tip of the right ear of his sons, and upon the thumb of their right hand, and upon the great toe of their right foot, and sprinkle the blood upon the altar round about."—Exodus 29:20.

In the cleansing of the leper, he had blood and oil put on his right ear (Leviticus 14: 14,17). This represented the sinner being cleansed from the leprosy of sin by having the life and death of Christ applied and the Holy Spirit as the oil to keep us ever aware of our need and to supply the power to keep us safe from the sounds of the enemy. Revelation tells us over and over *"He that has an ear, let him hear what the Spirit says . . ."*

SMELL

II Corinthians 2:14-17

"Now thanks be unto God, which always causeth us to triumph in Christ, and maketh manifest the savour of his knowledge by us in every place. For we are unto God a sweet savour of Christ, in them that are saved, and in them that perish. To the one we are the savor of death unto death; and to the other the savour of life unto life. And who is sufficient for these things? For we are not as many, which corrupt (margin: deal deceitfully with) the word of God: but as of sincerity, but as of God, in the sight of God speak we in Christ."

Everything living has an odor all of its own; each person smells different from every other person. That is how dogs can track one specific person just by smelling some of their clothing. We have a spiritual odor also; it is called the atmosphere of the soul. This atmosphere reflects our spiritual condition and attracts or repels people.

The lesson of smell is the lesson of the incense on the altar of the temple. Our sense of smell, of olfaction (Latin, meaning to smell and to make), is perhaps as much as 20,000 times more sensitive than our sense of taste. Adults can usually sense up to 10,000 different odors, and children can do even better. Unfortunately several poisonous gases, including carbon monoxide, are not detectable by our small receptors.

When you view the ear, you will note that the olfactory bulbs lie on top of the cribiform plate of the ethmoid bone.

The olfactory receptor cells (hairs) are located high in the roof of the nasal cavity. Each nostril contains a small patch of epithelium about the size of a thumbnail. It has 3 types of cells:

1. Receptor cells, which are olfactory neurons.
2. Sustenacular (supporting) cells.
3. Thin layer of small basal cells which undergo cell division and replace degenerating receptor and supporting cells.

Each receptor cell has a lifetime of about 30 days and has a short dendrite extending from its superficial end to the surface epithelium (lining). We have more than 25 million bipolar receptor cells (a hunting dog has about 220 million). They are among the most primitive neurons in the nervous system. They are chemo-receptor cells, a chemical detector. A transducer of a stimulus (changes it from a chemical to a nerve impulse) and the transmitter of the nerve impulse to the olfactory bulb.

Olfactory glands are located in between the neurons. They secrete a mucus-like fluid that covers the surface epithelium (lining). This fluid dissolves the odoriferous substances so that they can stimulate the receptor sites. In order for substances to be sensed, they must be volatile (able to turn into a vapor) and soluble (able to dissolve). Without these qualities the particles could not be carried into the nostrils by air currents, dissolve in the mucus-like coating on the olfactory epithelium and penetrate the lipid barrier surrounding the receptor cell.

There is no real understanding of how we perceive odors, but we can smell. Just as we do not understand the incarnation of Christ as demonstrated in the mystery of the incense. How can He be both human and divine? We may not understand it, but we can take advantage of it by becoming a partaker of the divine nature.

> *"Whereby are given to us exceeding great and precious promises: that by these ye might be partakers of the divine nature, having escaped the corruption that is in the world through lust."*—II Peter 1:4.

Olfaction is the only sense that does not project fibers into the thalamus before reaching the cerebral cortex of the temporal lobe.

We never smell God, He smells us. As you think of the record in the Bible of all the times God has appeared to man, there is never a smell connected with it. They see Him and hear Him. The Bible says the disciples touched Him (I John 1:1). We can taste and see that the Lord is good (Psalm 34:8). But there is not a smell mentioned. When Satan produces an apparition, there is sometimes a smell connected with that, such as roses.

The *ethmoid bone* has an important role to play in our sense of smell.

ETHMOID BONE—ALTAR OF INCENSE

The ethmoid bone represents the altar in the sanctuary. This is a exceedingly light and spongy bone, full of air spaces and canals and has a cubical form. It is located at the base of the cranium. This plate is located at the uppermost part of the ethmoid and projecting upward from it is a thick, smooth, triangular piece called the crista galli. Neurons of smell go through the ethmoid and the olfactory bulbs which lie on top of the cribiform plate. On each side of the crista galli, the cribiform plate is narrow and deeply grooved, to support the bulbs. If the cribiform plate is broken, you lose your sense of smell. The olfactory nerves all travel through the grooves and canals of the ethmoid.

In looking at a picture showing the nasal cavity, you will notice that the ethmoid bone sits right next to the sphenoid bone, just as the altar of incense in intimately connected with the Ark of the Covenant in the sanctuary. The altar of incense was the place of prayer, and this bone represents the principle of prayer.

"And the LORD said unto Moses, Take unto thee sweet spices, stacte, and onycha, and galbanum; these sweet spices with pure frankincense; of each shall there be a like weight. And thou shalt make it a perfume, a confection after the art of the apothecary, tempered **(margin: salted,)** *together pure and holy: And thou shalt beat some of it very small, and pure of it before the testimony in the tabernacle of the congregation. Where I will meet with thee: it shall be unto you most holy. And as for the perfume which thou shalt make, ye shall not make to yourselves according to the composition thereof: it shall be unto thee holy for he LORD. Whosoever shall make like unto that, to smell thereto, shall even be cut off from his people."*—Exodus 30:34-38.

The original word for tempered means to season with salt which makes the ingredients four plus one.

"And every oblation of thy meat offering shalt thou season with salt; neither shalt thou suffer the salt of the covenant of thy God to be lacking from thy meat offering: with all thine offerings thou shalt offer salt."—Leviticus 2:13.

The fifth ingredient is salt and 5 is the number of grace and redemption. There is no amount mentioned because grace always grows to meet the need. We are told what the salt represents in Colossians 4:6: *"Let your speech be always with grace, seasoned with salt, that ye may know how ye ought to answer every man."*

"And God is able to make all grace abound toward you: that ye always having all sufficiency in all things, may abound to every good work."—II Corinthians 9:8.

The other four ingredients represent the four facets of Christ's ministry that are represented by the four beasts around the throne and the four gospels: Christ as king in the Book of Matthew, the lion; Christ as Servant, as represented in the Book of Mark, the calf or ox; Christ as man in the Book of Luke, the man, Christ as divine in the Book of John, the eagle. These four symbols were also on the banners of the four lead tribes of Israel. (Numbers 3.)

The four ingredients were:

1. Stacte (from the root word, to speak by inspiration);
2. Onycha (from the root word to roar as a lion);
3. Galbannum (the choicest part);
4. Frankincense (from the root word, to be made white).

The first three are only mentioned once in the whole Bible, while frankincense is mentioned 17 times (This number represents the whole cycle plus spiritual completeness.) This spice was never to be used with any offering that brought iniquity to remembrance. (Numbers 5:15.) It was offered with all the meat offerings except the trespass offering (meat means food, not flesh).

The meat offerings were mostly offerings of consecration; oil (the Holy Spirit) and frankincense (Christ's divinity) had to be added to make it acceptable.

Even our prayers of praise and thanksgiving are not acceptable to the Father without the intercession of Jesus, our High Priest. Because the law is a transcript of His character and is holy, as God is holy, only God could be the sacrifice to atone for breaking that law.

Frankincense was also put on top of the shewbread. (Leviticus 24:7.) It is mentioned specifically in connection with Solomon's temple. (I Chronicles 9:29 and Nehemiah 13:5.) It was brought by the wise men to Christ as a recognition of His divinity. (Matthew 2:11.)

It was also part of the merchandise of Babylon, which could be found no more. (Revelation 18:13,14.) Babylon has tried to take the place of God by replacing His Mediator and changing His law and even saying. "You can be as God."

The incense was beaten; *"He was wounded for our transgressions, He was bruised for our iniquities; the chastisement of our peace was upon Him and with His stripes we are healed."*—Isaiah 53:5.

The incense was never to be duplicated for any other purpose. It was to be offered morning and evening with the prayers of the people.

When rebellious men offered incense, they were destroyed by fire from the LORD, but when offered by God's men, it stopped the plague. (Numbers 16.)

Revelation 8:3-5 depicts the Saints living at the end of the world. As there is no self righteousness in them, the Saints offer incense in the form of prayers pleading the merits of Christ as their only hope.

God is very displeased when we burn incense to other gods (Jeremiah 44). We can do this from the inner chambers of our souls while on the surface we appear to be worshipping God. Read about the 70 elders in Ezekiel 8 with smoke of the incense going up while they were an abomination unto God. This is trying to use the righteousness of Christ in the wrong way. When we use His sacrifice to excuse and cover up our sin, instead of confessing and forsaking it, we are burning incense to self.

Christ's life must be appropriated into our own life and we must have Christ in us, the hope of glory. To offer incense means that we are willing to submit our will to His, and to take Him as our example.

What smells go up to God from this earth? The first record of smelling is in Genesis 8:21. God smelled the burnt offerings and said they were a sweet savour, or as the margin says, a savour of rest. Why would the smell of burning flesh be pleasing to the Lord? Because He knew it meant that Noah and His family were taking advantage of the Sacrifice that was to be offered, that they could be saved eternally, not just from the flood.

When God says that the burning of the fat on the altar was a sweet savour, it was because the fat was a symbol of sin which had been separated from the sinner and disposed of. Every morning and evening, he smelled this odor along with the incense offered on the altar. These were all a symbol of the plan of salvation and His great Gift to us.

When the people forgot the meaning of the sacrifice and did not appreciate the fact that their sin was going to cost the life of Christ, the smell was not pleasing to God. The phrase "something smells rotten" or that "smells fishy" means that we don't think it is true.

"I hate, I despise your feast days, and I will not smell in your solemn assemblies."—Amos 5:21.

"I will make your cities waste and bring your sanctuaries unto desolation, and I will not smell the savour of your sweet odours."—Leviticus 26:31.

"When ye come to appear before me, who hath required this at your hand, to tread my courts? Bring no more vain oblations, incense is an abomination unto me, the new moons and Sabbaths, the calling of assemblies, I cannot away with; it is iniquity (margin: *grief*), *even the solemn meeting. Your new moons and your appointed feasts my soul hateth: they are a trouble unto me; I am weary to bear them. And when ye spread forth your hands, I will hide mine eyes from you: yea when ye make many prayers, I will not hear: your hands are full of blood. Wash you, make you clean; put away the evil of your doings from before mine eyes; cease to do evil; Learn to do well; seek judgment, relieve the oppressed, judge the fatherless, plead for the widow. Come now, and let us reason together, saith the LORD: though your sins be as scarlet, they shall be as white as snow; though they be red like crimson, they shall be as wool. If ye be willing and obedient, ye shall eat the good of the land: But if ye refuse and rebel, ye shall be devoured with the sword: for the mouth of the LORD hath spoken it."*—Isaiah 12-20.

Let us ever be aware of the smell that is ascending from our spiritual life up to our heavenly Father. Let it be a sweet smell that makes Him happy.

TASTE

Job 6:30

"Is there iniquity in my tongue? cannot my taste discern perverse things?"

The receptors for taste and smell are both chemoreceptors, and the two sensations are clearly interrelated. A person whose nasal passages are blocked cannot taste food as effectively. To a certain degree sight is also used in tasting, which explains why we expect a certain taste from an object.

The surface of the tongue is covered with many small papillae (puh-PILL-ee; nipple or pimple). This is what gives the tongue its bumpy appearance. The papillae are most numerous on the back surface of the tongue and are also found on the roof of the mouth, throat, and the back surface of the epiglottis.

The three main types are:

1. *Fungiform* (L., mushroom-like), which each contain one to eight taste buds, and are near the tip of the tongue;
2. Ten to twelve *circumvallate* (L., wall around), which contains 90 to 250 taste buds apiece;
3. *Filiform* (L., threadlike), which are pointed structures near the anterior two-thirds of the tongue, and do not necessarily contain taste buds.

Located within the crevices of most papillae are approximately 10,000 receptor organs or the sense of taste called taste buds. Each taste bud contains about 25 receptor cells. The supporting cells act as reserve cells, replenishing the receptor cells when they die. Mature receptor cells have a life of only about 10 days.

Extending from the free end of each receptor cell are short taste hairs (microvilli) that project through the tiny outer opening of the taste bud. Before a substance can be tasted, it must be in solution. Saliva containing ions or dissolved molecules of the substance to be tasted enters the taste pore and interacts with receptor sites on the taste hairs.

Each cell has many different types of receptor sites, even though all taste cells are structurally just the same. There are four basic sensations, which can vary when combined: *salt, sweet, sour,* and *bitter.* Salt and sweet are perceived mostly on the tongue, but bitter and sour are perceived on the palate.

The spiritual lesson of the taste is the lesson of experience. Any time hair or hair cells are involved, experience is involved. *"O taste and see that the LORD is good: blessed is the man that trusteth in Him."*—Psalm 34:8. *"But we see Jesus, who was make a little lower than the angels for the suffering of death, crowned with glory and honour; that he by the grace of God should taste death for every man."*—Hebrews 2:9.

The first taste mentioned in the Bible is the taste of manna. (Exodus 16:31.) God intended manna to be a sweet experience for His people, learning to trust in Him to supply all their needs. Read Exodus 16. *"And had rained down manna upon them to eat, and had given them of the corn of heaven. Man did eat angels' food: he sent them meat (food) to the full."*—Psalm 78:24,25. Would you like to have had the opportunity to eat food directly from heaven? Maybe not!!! The people murmured and complained about this angel food. *"And the mixed multitude that was among them fell a lusting; and the children of Israel also wept again, and said, Who shall give us flesh to eat? We remember the fish, which we did eat in Egypt freely; the cucumbers, and the melons, and the leeks, and the onions, and the garlic; But now our soul is dried away: there is nothing at all, beside this manna, before our eyes."*—Numbers 11:4-6. God gave them what they asked for and many died as a result. Read all of Numbers 22 for the rest of the story.

The bitter taste represents disappointments. The children of Israel ate lamb and unleavened bread and bitter herbs to celebrate the Passover. The lamb represented Christ and His sacrifice to free us from the bondage of sin; the unleavened bread represented putting away all sin; and the bitter herbs pointed back to the bitterness of the bondage in

Egypt. So when we feed upon Christ, it should be with contrition of heart, because of our sins. *"And they shall eat the flesh in that night, roast with fire, and unleavened bread; and with bitter herbs shall they eat it."*—Exodus 12:8.

Sweet represents our pleasant experiences. *"Pleasant words are as an honeycomb, sweet to the soul, and health to the bones."*—Proverbs 16:24. *"How sweet are they words unto my taste! Yea, sweeter than honey to my mouth."*—Psalm 119:103. *"And I took the little book out of the angel's hand, and ate it up; and it was in my mouth sweet as honey: and as soon as I had eaten it, my belly was bitter."*—Revelation 10:10. Read Ezekiel 3; he had the same experience as John. God's word was sweet to him, but when the people would not listen and repent, that was bitter.

Salt represents grace, as you learned from the study on Smell. If a person adds too much salt to their food, it becomes unpalatable. The spiritual counterpart is if we try to cover up sin with God's grace, it does not work. Sin must be given up before it can be covered. There are many that keep trying to add grace to their sin to be right with God. Grace will not cover one *unforsaken* sin.

Sour represents the experience in sin. *"In those days they shall say no more, The fathers have eaten a sour grape, and the children's teeth are set on edge. But every one shall die for his own iniquity: every man that eateth the sour grape, his teeth shall set on edge."*—Jeremiah 31:29,30. *"Their drink is sour: they have committed whoredom continually: her rulers with shame do love, Give ye."*—Hosea 4:18e.

Do we ever get our taste mixed up? *"Woe unto them that call evil good and good evil; that put darkness for light, and light or darkness; that put bitter for sweet and sweet for bitter."*—Isaiah 5:20. In Genesis 3, the story is told of how Satan convinced Eve that bitter would be sweet. He said it would be sweet to disobey God, but the experience turned bitter for the whole world. *"Bread of deceit is sweet to a man; but afterwards his mouth shall be filled with gravel."*—Proverbs 20:17. Read Job 20:12-14. *"If so be ye have tasted that the Lord is gracious."*—I Peter 2:9.

TOUCH

<u>I Chronicles 16:22</u>

"[Saying], Touch not mine anointed, and do my prophets no harm."

Tiny receptors in the skin detect stimuli that the brain interprets as light touch, itch, touch-pressure, vibration, heat, cold, tickle and pain.

Light touch is when the skin is touched but not deformed. Receptors for light touch are mostly in the outer layer of skin, especially the tips of the fingers and toes, the tip of the tongue, and the lips.

There are root hair plexuses around hair follicles. When hairs are bent, they act as levers, and the slight movement stimulates the free nerve endings surrounding the follicles. (L. little sac, the tube that encloses the hair root and bulb). These act as detectors of touch and movement. That is why a tiny insect can be felt crawling along a hairy arm even if its feet never touch the skin.

The sensory area of the brain specialized for touch and temperature is located in the general sensory region of the parietal lobe. Light touch utilizes at least three neural pathways from the spinal cord to the cerebral cortex.

In contrast to such sensitive areas of the skin as the fingertips, the torso (especially the back) and back of the neck are relatively insensitive to light touch. A test measures the minimal distance that two stimuli must be separated to be felt as two distinct stimuli. When there are many receptors, two distinct points may be felt when they are separated by only two or three millimeters, while some areas are far apart and the points may have to be separated by as much as 70mm before they can be felt as two points. This means that there is virtually no spot on the fingertips that is insensitive to light touch.

Touch-pressure results in a deformation of the skin. The difference between light touch and deep pressure on your skin can be shown by touching something such as a pencil and then squeezing it tightly. The skin will be indented by the deep pressure. Sensations of this manner last longer and are felt over a wider area, than sensations of light touch. Receptors for deep pressure are lamented (*L. thin plates*) corpuscles. They measure changes in pressure rather than pressure itself. They are found in all areas of the body that are regularly subjected to pressure.

The first instance in the Bible of God touching man is when He made Adam and Eve with His very own hands. Imagine the love He put into His creation. God's touch can be a blessing or a curse depending upon the receiver. *"Then the LORD put forth his hand, and touched my mouth. And the LORD said unto me, Behold, I have put my words in thy mouth."*—Jeremiah 1:9.

Sometimes we do not know the difference between the touch of God and the touch of Satan. When Satan touched Job, he thought it was God. *"So went Satan forth from the presence of the LORD, and smote Job with sore boils from the sole of his foot unto his crown."*—Job 2:7. *"Have pity upon me, have pity upon me, O ye my friends; for the hand of God hath touched me. Why do ye persecute me as God, and are not satisfied with my flesh?"*—Job 19:21,22. There is wickedness in our world, but all suffering is not the result of a perverted course of life. Job is brought distinctly before us as a man whom the Lord allowed Satan to afflict, to prove Job's loyalty to God before the whole universe. Job did not have the whole picture, but even in the darkness, he trusted God with all of his heart, and God rewarded his faithful servant.

The Bible talks about angels touching men. A light touch is described in Acts 12:7, when the angel came to awaken Peter to release him from prison. A destroying touch is spoken of in Acts 12:23, when the angel smote Herod. The same word, smote, is used in both cases, but what a difference in the results. The same angel who had come from the royal courts to rescue Peter had been the messenger of wrath and judgment to Herod. The angel smote Peter to arouse him from slumber; it was a different stroke that he smote the wicked king, laying low his pride and bringing upon him the punishment of the Almighty. Herod died in great agony of mind and body, under the retributive judgment of God.

Five times in answer to prayers was Daniel touched by a heavenly being. Each time it was a touch that comforted and strengthened him

Man touched God when He was on the earth, some of these touches were for blessings, and others were for cursing.

"And when the men of that place had knowledge of him, they sent out into all that country round about, and brought unto him all that were diseased; and besought him that they might only touch the hem of his garment: and as many as touched were made perfectly whole."—Matthew 14:35,36.

"And a women having an issue of blood twelve years, which had spent all her living upon physicians, neither could be healed of any, Came behind Him, and touched the border of His garment: and immediately her issue of blood stopped."—Luke 8:43,44.

"And the whole multitude sought to touch him: for here went virtue out of him, and healed them all."—Luke 6:19.

The wicked touched Jesus when he was here:

"Then did they spit in his face, and buffeted him; and others smote him with the palms of their hands,"—Matthew 26:67.

"And one shall say unto him, What are these wounds in thine hands? Then he shall answer, Those with which I was wounded in the house of my friends."—Zechariah 13:6.

"And when they had plaited a crown of thorns, they put it upon his head, and a reed in his right hand: and they bowed the knee before him, and mocked him, saying, Hail, King of the Jews! And they spit upon him and took the reed, and smote him on the head."—Matthew 27:29,30.

The responses to heat and cold are caused by naked nerve endings. Think about how your face can stand much more cold than your fingers. Teeth are usually more sensitive to cold than to heat because of the difference in receptors for cold and heat.

Free nerve endings are the most widely distributed receptors in the body, and are involved with pain. The next most numerous are those for touch, cold, and heat. Most pain receptors are probably chemoreceptors that release chemicals after a local trauma activates them.

Pain is a warning signal that alerts the body that something is wrong. Pain receptors are specialized free nerve endings that are present in most parts of the body. The intestines and brain tissue have no pain receptors.

There are different types of pain: fast-conducted, sharp, prickling pain; slow-conducted, burning pain; deep, aching pain in joints, tendons and viscera.

Some tissues are more sensitive to pain than others are. If you stick a needle into the skin, it produces great pain, but if you probe into a muscle with a needle, the pain is not as great. A puncture in an artery is very painful; a puncture in a vein is almost painless. A kidney stone that distends the ureter (the tube from the kidneys into the bladder) produces excruciating pain. On the other hand, the intestines are not sensitive to pain if they are cut or burned, but are sensitive if they are distended or contracted (cramps) because they affect surrounding areas.

Pain impulses are conveyed by two or more pathway systems, including one which relays pain impulses from the spinal cord to the thalamus and one which relays pain impulses to the reticular formation and thalamus. The perception of pain occurs in the thalamus, but the judgment of the type of pain and its intensity occurs in the parietal lobe of the cerebral cortex. When we remember that the thalamus is the center control of the brain, just as God is the center control of the universe, and that the spinal cord and the reticular formation represent the angels communicating with heaven, carrying the reports of what is happening here and getting instruction on what to do for man, you get the whole picture of the pain information relay.

The Bible shows us God is in His high and Holy place, not in a state of inactivity, not in silence and solitude, but surrounded by ten thousand times ten thousand and thousands of thousands of holy intelligences, all waiting to do His will. Through channels which we cannot discern He is in active communication with every part of His dominion. God is bending from His throne to hear the cry of the oppressed. Not a sigh is breathed, not a pain felt, not a grief pierces the soul, but the throb vibrates to the Father's heart. To every sincere prayer He answers "Here am I." He uplifts the distressed and downtrodden. In all our afflictions, He is afflicted. In every temptation and every trial the angel of His presence is near to deliver. The angels of glory find their joy in giving, giving love and tireless watch-care to souls that are fallen and unholy.

"In all their affliction He was afflicted, and the angel of His presence saved them: in His love and in His pity He redeemed them; and He bare them, and carried them all the days of old."—Isaiah 63:9.

Referred pain is felt often at a place far removed from where the problem really is. For instance the pain of a heart attack may be felt in the left shoulder, arm and armpit.

Phantom pain is felt in a limb or organ that is not there. The message is coming from the relay nerves around where it was.

Proprioception (L. one's self + receptor) comes into play when we sleep on one arm too long and are caused to move by the pain that results. Messages coming from joints and ligaments that are put under painful stress make us aware of our bodily parts without actually seeing them.

We can identify objects through touch; this ability is called *stereognosis* (STEHR-ee-og-NO-sis: Gr. *steros,* solid, three dimensional; *gnosis,* knowledge). This ability depends on the sensations of touch and pressure, as well as on sensory areas in the parietal lobe of the cerebral cortex.

There are 3 ascending tracts that carry messages for general sensory receptors. The afferent (toward the brain) nerves that convey highly localized and discriminative sensations are larger, have more myelin, and conduct faster than those that convey less-defined sensations.

Chapter Fifteen

VICTORY OVER SIN

I Corinthians 15:7

"But thanks [be] to God, which giveth us the victory through our Lord Jesus Christ."

"All should guard the senses, lest Satan gain victory over them; for these are the avenues of the soul.

"You will have to become a faithful sentinel over your eyes, ears, and all your senses if you would control your mind and prevent vain and corrupt thoughts from staining your soul. The power of grace alone can accomplish this most desirable work.

" Satan and his angels are busy creating a paralyzed condition of the senses so that cautions, warnings, and reproofs shall not be heard; or, if heard, that they shall not take effect upon the heart and reform the life.

"My brethren, God calls upon you as His followers to walk in the light. You need to be alarmed. Sin is among us, and it is not seen to be exceedingly sinful. The senses of many are benumbed by the indulgence of appetite and by familiarity with sin. We need to advance nearer heaven."—E. G. White, Adventist Home, p. 401.

SATAN'S STRATEGY IS TO CONFUSE THE SENSES

"Satan's work is to lead men to ignore God, to so engross and absorb the mind that God will not be in their thoughts. The education they have received has been of a character to confuse the mind and eclipse the true light. Satan does not wish the people to have a knowledge of God; and if he can set in operation games and theatrical performances that will so confuse the senses of the young that human beings will perish in darkness while light shines all about them, he is well pleased.

"Satan Cannot Enter the Mind Without Our Consent.—We should present before the people the fact that God has provided that we shall not be tempted above what we are able to bear, but that with every temptation He will make a way of escape. If we live wholly for God, we shall not allow the mind to indulge in selfish imaginings.

"If there is any way by which Satan can gain access to the mind, he will sow his tares and cause them to grow until they will yield an abundant harvest. In no case can Satan obtain dominion over the thoughts, words, and actions, unless we voluntarily open the door and invite him to enter. He will then come in and, by catching away the good seed sown in the heart, make of none effect the truth."—E. G. White, Adventist Home, p. 102.

CLOSE EVERY AVENUE TO THE TEMPTER

"All who name the name of Christ need to watch and pray and guard the avenues of the soul, for Satan is at work to corrupt and destroy if the least advantage is given him.

"It is not safe for us to linger to contemplate the advantages to be reaped through yielding to Satan's suggestions. Sin means dishonor and disaster to every soul that indulges in it; but it is blinding and deceiving in its nature, and it will entice us with flattering presentations.

"If we venture on Satan's ground, we have no assurance of protection from his power. So far as in us lies, we should close every avenue by which the tempter may find access to us.

"Who can know, in the moment of temptation, the terrible consequences which will result from one wrong, hasty step! Our only safety is to be shielded by the grace of God every moment, and not put out our own spiritual eyesight so that we will call evil, good, and good, evil. Without hesitation or argument we must close and guard the avenues of the soul against evil. Every Christian must stand on guard continually, watching every avenue of the soul where Satan might find access. He must pray for divine help and at the same time resolutely resist every inclination to sin. By courage, by faith, by persevering toil, he can conquer. But let him remember that to gain the victory Christ must abide in him and he in Christ."—E. G. White, Adventist Home, p. 402.

AVOID READING, SEEING, OR HEARING EVIL

"The apostle [Peter] sought to teach the believers how important it is to keep the mind from wandering to forbidden themes or from spending its energies on trifling subjects. Those who would not fall a prey to Satan's devices must guard well the avenues of the soul; they must avoid reading, seeing, or hearing that which will suggest impure thoughts. The mind must not be left to dwell at random upon every subject that the enemy of souls may suggest. The heart must be faithfully sentineled, or evils without will awaken evils within, and the soul will wander in darkness.

"Everything that can be done should be done to place ourselves and our children where we shall not see the iniquity that is practiced in the world. We should carefully guard the sight of our eyes and the hearing of our ears so that these awful things shall not enter our minds. When the daily newspaper comes into the house, I feel as if I want to hide it, that the ridiculous, sensational things in it may not be seen. It seems as if the enemy is at the foundation of the publishing of many things that appear in newspapers. Every sinful thing that can be found is uncovered and laid bare before the world."—E. G. White, Adventist Home, p. 403.

"Those who would have that wisdom which is from God must become fools in the sinful knowledge of this age, in order to be wise. They should shut their eyes, that they may see and learn no evil. They should close their ears, lest they hear that which is evil and obtain that knowledge which would stain their purity of thoughts and acts. And they should guard their tongues, lest they utter corrupt communications and guile be found in their mouths."—E. G. White, Adventist Home, p. 403.

Study Guide

FOREWORD

1. For what are we to strive?
2. "Many will be rescued from physical, mental, and moral degeneracy through the influence of __________.

KEEP FROM EVIL

1. In order to be sanctified and preserved in righteousness, what are God's people to do?
2. What three things will enable us to be overcomers?

THE SENSES

1. On what six things are we to think upon?
2. "The structures of the body that perceive and receive signals from our environment are called __________.
3. True or False? "Sound cannot occur unless there is someone to hear it." Please Explain.
4. "Individual sensory receptors are each designed to receive a specific kind of stimulus yet they all interact with __________________ nerve cells."
5. Name the four different types of sensory receptors.
6. What conveys the impulses from the receptors to the central nervous system.
7. In response to these impulses, what four things does the central nervous system do?

VISION

1. What is the spiritual lesson of sight? Please explain in your own words.
2. What is the first component necessary for vision?
3. In reference to Revelation 3:17, give your definition of the "eyesalve" that we are to anoint ourselves with.
4. What are the seven basic structures of the eye?
5. True or False? The receptors in our eyes constitute about 50 % of the receptors of the entire body.
6. What gives us our impression of distance?
7. When does vision begin?
8. Give the names of the three layers of tissue found in the wall of the eyeball and divine each.
9. True or False? The pupil of the eye is always colored black. Explain your answer.
10. Define "adaptation."
11. According to Proverbs 7:2,3—what are we to do with thc commandments?
12. What are the three intrinsic muscles inside the eye? Define them.

HEARING

1. The ear passage is carved out of the _____________ bone.
2. What happens when someone refuses to "hear" the law?
3. Name the three sections of the ear and tell how each functions.
4. What is the cochlea and give its function.
5. Sound is conducted by three mediums. Name them and tell how they work.
6. In your own words describe the physiology of hearing.
7. What has the power to change lives?
8. What causes our hearing ability to decline?
9. True or False? "Exposure to sounds of 110 decibels for 26 minutes always causes hearing loss."
10. "Noise is a _______________." Explain your answer.
11. Define the "vestibule" of the eye.

12. What is the "cupula"?
13. Name the four the "vestibular tracts."
14. As a member of God's family, is it wrong to wear earrings? Why?

SMELL

1. True or False? Everything living has an odor.
2. Our sense of smell, is about __________ times more sensitive than taste.
3. Name the types of cells found in the epithelium of our nostrils.
4. What is the function of the olfactory glands?
5. How do we burn incense to other gods in our souls?

TASTE

1. The receptors for taste and smell are both _____________. Define your answer.
2. What are the three types of papillae?
3. True or False? Mature receptor cells have a life of about 20 days.
4. Before a substance can be tasted it must be ________________.
5. What are the four basic sensations? Give a definition of what they represent.
6. What is the spiritual lesson of taste?

TOUCH

1. What are the eight stimuli of the skin that the brain interprets?
2. True or False? There is virtually no spot on the fingertips that is insensitive to light touch.
3. On some areas of the body two distinct points of light touch may be felt as close as ______________ .
4. "Receptors for deep pressure are ____________________."
5. Thought question. How can we know the difference between the touch of God and the touch of Satan?
6. The responses to heat and cold are caused by ________________ .
7. Define "pain."
8. What is deferred pain?
9. Define "stereognosis."
10. The afferent nerves the convey highly localized and discriminative sensations are __________, _________________, and ______________ than those that convey less-defined sensations.

Chapter Sixteen

WATER

Isaiah 58:10, 11

"And [if] thou draw out thy soul to the hungry, and satisfy the afflicted soul; then shall thy light rise in obscurity, and thy darkness [be] as the noon day: And the LORD shall guide thee continually, and satisfy thy soul in drought, and make fat thy bones: and thou shalt be like a watered garden, and like a spring of water, whose waters fail not."

The fluids of the body make up 60% of the total body weight. There are many different fluid systems in the body: blood, lymph, cerebrospinal fluid, tears, urine, and sweat. All these fluids are composed of water with different substances added.

The transformation of food into good blood is a wonderful process, and all human beings should be intelligent upon this subject. In order that the digestive fluids may be called into action, the saliva must become mixed with the food. Then the teeth must do their work carefully and thoroughly. Each organ of the body gathers its nutrition to keep its different parts in action. The brain must be supplied with its share; and the bone with its portion. The great Master Builder is at work every moment. He supplies every muscle and tissue, from the brain to the end of the fingers and toes, with life and strength.

Day by day the human structure performs its work under the great Master Architect who superintends every function of the body seeking to make it into a glorious temple for Himself.

Anytime the body becomes diseased or is injured, blood is attracted to that point. When the blood is attracted to the weakest point, there is a wearing of the channel through which the blood flows. It is necessary that the blood flows evenly throughout the entire body so that all parts are nourished. Unless we, as God's agents, submit to have the peace of God rule in our heart, there will be a rush of blood to the brain which will disqualify us for labor. The Lord will not, cannot, help His servants unless we will co-operate with God; unless we will stop worrying and trust in the Lord. —E G White, Manuscript Releases vol 3, p. 308, para. 3.

Water can represent both kinds of truth; the truth of God or the truth of Satan. As usual Satan has the counterfeit. We decide what truth is by the facts we think we have and the spirit by which we interpret those facts. If we have the Spirit of truth, the Holy Spirit, we can discern what is the real truth. If we have the spirit of counterfeit truth, the Devil, we will interpret facts with his interpretation. The parable of the talents in Matthew 25 contains a prime example of this; two of the servants looked at the loan of talents as a blessing, and one looked at it as a problem. They all had the same master and the same facts, but had not the same spirit and so came to different conclusions. Read this story and see how important it is to have the right spirit.

The first mention of water in the Bible is in Genesis 1: *"And the earth was without form and void; and darkness was upon the face of the deep. And the spirit of God moved upon the face of the waters. And God said, Let there be light: and there was light."*—Genesis 1:1-3.

As a result of God's Spirit moving upon the waters, the light appeared and the waters were divided from the waters.

In John 4, Jesus offered the living water; He invites us: *"And the Spirit and the bride say, Come. And let him that heareth say, Come. And let him that is athirst come. And whosoever will, let him take the water of life freely."*—Revelation 22:17.

We must be washed by clean water, and the Bible tells us clearly what that is and what the results will be.

"That he might sanctify and cleanse it with the washing of water by the word. That he might present it to himself a glorious church, not having spot, or wrinkle, or any such thing: but that it should be without blemish."—Ephesians 5:26.

"Let us draw near with a true heart, in full assurance of faith, having our hearts sprinkled from an evil conscience, and our bodies washed with pure water."— Hebrews 10:22.

"Then will I sprinkle clean water upon you, and ye shall be clean: from all your filthiness, and from all your idols, will I cleanse you. A new heart also will I give you, and a new spirit will I put within you: and I will take away the stony heart out of your flesh, and I will give you an heart of flesh. And I will put my spirit within you, and cause you to walk in my statutes, and ye shall keep my judgments, and do them."—Ezekiel 36:25-27.

Management of the available reserves of water in the body becomes the responsibility of a very complex system. This complex multilevel water rationing and distribution process remains in operation until the body receives unmistakable signals that it has gained access to adequate water supply. Since every function of the body is monitored and gauged to the flow of water, "water management" is the only way of making sure that adequate amounts of water and its transported nutrients reach the more vital organs first and that they will have sufficient supplies of this water to deal with any new "stresses" placed upon them.

Thinking that tea, coffee, alcohol, and manufactured beverages, are desirable substitutes for the purely natural water needs of the daily "stressed" body is a catastrophic mistake. It is true that these beverages contain water, but what else they contain are dehydrating agents. They rid the body of the water it has plus dehydrates more water from the reserves of the body! Today, modern life-style encourages people to be dependent on all sorts of beverages that are commercially manufactured. Children are not educated to drink water, they become dependent on sodas and juices. This is a self-imposed restriction on the water needs of the body.

Hydrolysis means water loosening in Greek. Hydrolysis takes place when a molecule of water interacts with a substance to break up that substancc. (Think about how water loosens the dried food and breaks it up when you soak the dirty dishes.) The original substance is then rearranged, together with the water, into different molecules.

The bonds of sucrose (table sugar) are broken (hydrolyzed) by water to form two simple sugars, glucose and fructose. Glucose and fructose have the same number and kinds of atoms, but in different arrangements. Different chemical substances are like different faces; they may have the same components, but by being arranged differently, they make different substances. Faces have two eyes, one nose, and one mouth, but are all arranged differently to make different-looking people.

Hydrolysis is important to the body. By means of hydrolysis, large molecules of proteins, nucleic acids, and fats are broken into simpler, smaller, more usable molecules. Generally speaking, almost all digestive and degradative processes in the body occur by hydrolysis.

Apply this spiritually, to the way that truth of God works in our characters. It changes our ideas and thoughts and makes us "a new creature."

"And from thence they went to Beer: that is the well whereof the LORD spake unto Moses, Gather the people together, and I will give them water. Then Israel sang this song, Spring up, O well; sing ye unto it: The princes digged the well, the nobles of the people digged it, by the direction of the lawgiver, with their slaves."—Numbers 221:16-18.

Waters are also defined as peoples, multitudes, and nations and tongues (Rev. 17:15), which covers all the ideas in the world.

Isaiah 17:13, nations rush like many waters.

Jeremiah 51:54, Babylon's waves do roar like great waters, a noise of their voice is uttered.

Babylon's water (idea): *"O thou that dwellest upon many waters, abundant in treasures, thine end is come and the measure of thy covetousness."* —Jeremiah 51:13.

"A drought is upon her waters; and they shall be dried up: for it is the land of graven images and they shall be dried up: for it is the land of graven images and they are mad upon their idols."—Jeremiah 50:38.

To most of us "dry mouth" is the only accepted sign of dehydration of the body. This signal is the last outward sign of extreme dehydration. The damage occurs at a level of persistent dehydration that does not necessarily demonstrate a "dry mouth" signal.

Naturally, chronic dehydration of the body means persistent water shortage that has become established for some time. Like any other deficiency disorder, such as vitamin C deficiency in scurvy, vitamin B deficiency in beriberi, or iron deficiency in anemia, the most efficient method of treatment of the associated disorders is by supplementation of the missing ingredient. Accordingly, if we begin to recognize the health complications of chronic dehydration and their prevention, the cure is obvious—DRINK MORE WATER!

Revelation 16:12 tells of the drying up of the great river Euphrates that was the lifeline of Babylon, and points forward to the time when everyone in the world will see that the great rivers of philosophy and man's vain imaginations will be dried up. It will finally be clear that only the truth of God was really truth.

"But the wicked [are] like the troubled sea, when it cannot rest, whose waters cast up mire and dirt. [There is] no peace, saith my God, to the wicked."—Isaiah 57:20,21.

Read Ezekiel 47 about the waters issued out of the temple from under the threshold of the house eastward; notice the different depths of the water. Our knowledge of God is progressive. First we just wade and then, when our faith and trust grow, we jump in and swim. Verse 9 states that anywhere the river went here would be healing and life. Verse 12 shows that the water issued out of the sanctuary.

"Thy way, O God, is in the sanctuary: who is so great a God as our God?"—Psalm 77:13.

All real truth and all help (Psalm 20:2) come out of the sanctuary because that is where God dwells. We should study the earthly sanctuary because the plan of salvation and the character of God are explained in its services. Read Psalm 73. David was perplexed about why the philosophy of the wicked worked and made them so prosperous until he went into the sanctuary of God and saw their end result. Then he saw how foolish it was to envy them. Jeremiah says of God's people: *"They shall come with weeping, and with supplications will I lead them: I will cause them to walk by the rivers of waters in a straight way, wherein they shall not stumble."*—Jeremiah 31:9.

"Be astonished, O ye heavens, at this, and be horribly afraid, be ye very desolate, saith the LORD, for my people have committed two evils; they have forsaken me the fountain of living waters, and hewed them out cisterns, broken cisterns that can hold no water. Is Israel a servant? Is he a home-born slave? why is he become a spoil? The young lions roared upon him, and yelled, and they made his land waste: his cities are burned without inhabitant. Also the children of Noph and Tahapanes have broken the crown (or feed on thy crown). Hast thou not procured this unto thyself, in that thou hast forsaken the LORD thy God, when he led thee by the way? And now what at hast thou to do in the way of Egypt, to drink the waters of Sihor? Or what has thou to do in the way of Assyria, to drink the waters of the river: Thine own wickedness shall correct thee, and thy backslidings shall reprove thee: know therefore and see that it is an evil thing and bitter, that thou hast forsaken the LORD thy God, and that my fear is not in thee, saith the Lord God of hosts."—Jeremiah 2:12-19.

"How canst thou say, I am not polluted, I have not gone after Baalim? (the sun god) see thy way in the valley, know what thou hast done..." v.23. *"I (king of Assyria) have digged and drunk strange waters, and with the sole of my feet have I dried up all the rivers of besieged places (fenced places)."*—II Kings 18:24. This whole passage, verses 20-28, sounds like the Lord was talking to Satan, who was the motivating force of the Assyrian king. Satan has offered strange truth and tried to dry up the rivers of truth from the Lord in the fenced places, those that were walled in by His law. It has always been the purpose of the enemy to muddy up the truth about the law of God, to confuse people with the idea that they can be saved by keeping the law or that they can be saved by ignoring it. Either lie is a fatal delusion.

Isaiah 8:7 refers again to the river of Assyria flooding over Judah because they refused the waters of Shiloah. Genesis 49:10 tells us that Shiloah is Christ.

"Forasmuch as this people refuseth the waters of Shiloah that go softly, and rejoice in Rezin and Remaliah's son; Now therefore, behold the Lord bringeth up upon them the waters of the river, strong and many, even the king of Assyria, and all his glory: and he shall come up over all his channels, and go over all his banks: And he shall pass through Judah; he shall overflow and go over, he shall reach even to the neck; and the stretching out of his wings shall fill the breadth of thy land, O Immanuel. Associate yourselves, O ye people, and ye shall be broken in pieces; and give ear, all ye of far countries: gird yourselves, and ye shall be broken in pieces. Take counsel together, and it shall come to nought; speak the word, and it shall not stand: for God is with us. For the LORD spake thus to me with a strong hand, and instructed me that I should not walk in the way of this people, saying, Say ye not, A confederacy to all them to whom this people shall say, A confederacy; neither fear ye their face, nor be afraid."—Isaiah 9:6-12.

When falsehoods overwhelm the Earth, God allows falsehood to come as a test. If His professed people love not the truth, they will be drowned by this false water. We must love to obey God and hang on to the truth no matter what the majority do.

"And the serpent cast out of his mouth water as a flood after the woman, that he might cause her to be carried away of the flood. And the earth helped the woman, and the earth opened her mouth, and swallowed up the flood which the dragon cast out of his mouth."—Revelation 12:15, 16.

"For this shall everyone that is godly pray unto thee in a time when thou mayest be found: surely in the floods of great waters they shall not come nigh unto him. Thou art my hiding place; thou shalt preserve me from trouble; thou shalt compass me about with sons of deliverance. Selah."—Psalm 32:6,7.

"Because thou hast kept the word of my patience. I also will keep thee from the hour of temptation, which shall come upon all the world, to try them that dwell upon the earth."—Revelation 3:10. God will destroy the serpent in the sea, Isaiah 27:1.

Read II Kings 5:1-19. When Naaman was asked to wash in Jordan, which was muddy, he said that there were better rivers in his country of Syria. Sometimes the philosophy and doctrine of the world seem clearer to the natural heart than does the way of God.

The Jordan was muddy. What makes the truth of the Bible unclear is that man is always tromping around in it bringing in his own ideas until it gets muddy. If we search diligently for the clean springs of water we will be able to find them. It is the only truth there is and the only way to be cleansed of our sin (leprosy). Jesus was baptized in the Jordan. This baptism shows how *we* can be cleansed by being totally submersed in the truth.

"Seemeth it a small thing unto you to have eaten up the good pasture, but ye must tread down with your feet the residue of your pastures? And to have drunk of the deep waters, but ye must foul the residue with your feet? And as for my flock, they eat that which ye have trodden with your feet: and they drink that which ye have fouled with your feet." Ezekiel 34:18,19.

Chapter Seventeen

BLOOD

Leviticus 17:11

"For the life of the flesh [is] in the blood: and I have given it to you upon the altar to make an atonement for your souls: for it [is] the blood [that] maketh an atonement for the soul."

The average person has about 5 liters of blood in their bodies. Blood is composed of plasma and formed elements (blood cells) and uses about 5% of the water in the body.

The main constituent in blood is plasma, which is 90% water, 8% proteins, and 2% other formed elements, and it provides the solvent for dissolving and transporting nutrients.

Since we have already covered the subject of water, we will now discuss the proteins in our blood. Proteins are divided into three different classes:

1. Albumins: These are 60% synthesized in the liver, they promote water retention in the blood, which in turn maintains normal blood volume and pressure. If the amount of albumin in the plasma decreases, fluid leaves the bloodstream and accumulates in the surrounding tissue, causing *edema*. Albumin also acts as a carrier of molecules by binding to molecules of other substances, such as hormones, that are transported by proteins.
2. Fibrogen: Proteins are about 4% fibrogen. Fibrogen is produced by the liver, and it is essential for blood clotting.
3. Globulins: Approximately 36% of the proteins consists of globulins, which are divided into three classes.

   a) Alpha and beta globulins transport lipids and fat soluble vitamins in the blood. One form of these are called *low-density lipoproteins*, (LDL) and they transport cholesterol from its site of syntheses in the liver to various body cells. Others called *high-density lipoproteins*, (HDL) removes the cholesterol and triglycerides from the arteries, preventing their deposition there.
   b) The other part of the globulins is called "gamma" globulins and are the immunoglobulins. These are antibodies that help prevent diseases such as measles, mumps, etc. There are five (the number of grace) classes of antibodies. They combine with foreign bodies (the antigen), such as bacterium, virus, protein, or cancer cells that it "recognizes" as foreign and disposes of these through the elimination system. Approximately 2% of these are electrolytes, amino acids, glucose and other nutrients.

Electrolytes are inorganic molecules that separate into ions when they are dissolved in water, either positively charged (cations) or negatively charged (anions). The major cation of plasma is sodium, which has an important effect on fluid movements and helps determine the total amount of fluid outside the cells. Potassium, calcium, phosphate, io dide, and magnesium are some other electrolytes.

Nutrients included in the plasma are glucose, amino acids, and lipids.

Several metabolic wastes are transported by the plasma, especially lactic acid, along with some nitrogenous waste products from protein metabolism.

Oxygen, nitrogen, and carbon dioxide are the principal gases dissolved in the plasma. Carbon dioxide and oxygen are transported in the plasma by red blood cells. Nitrogen is transported in the dissolved state.

FORMED COMPONENTS

The formed components are made in the bone marrow, and consist of red and white blood cells.

Red blood cells are called *erythrocytes,* and they make up half the volume of the human blood. The human body has about 25 trillion of these red blood cells. If five or six red blood cells were placed in a row, they would reach across the period at the end of this sentence. Each cell is shaped like a disc, slightly concave on top and bottom, like a dough-nut without the hole poked completely through. It cannot manufacture, but it must rely on what it already has stored, of proteins, enzymes, and RNA. Its job is to transport oxygen and carbon dioxide.

Each red blood cell is made up of *hemoglobin*, an oxygen-carrying globular protein. Each hemoglobin is 5% heme, an iron-containing pigment, and 95% globin, a protein. The iron atom is the binding set for oxygen and plays a key role in binding oxygen and releasing it to the cells at the right time. The function of hemoglobin depends on its ability to combine with oxygen where the oxygen concentration is high (in the lungs) and release oxygen where the concentration is low (in body tissues). Hemoglobin also carries waste carbon dioxide from the tissues to the lungs, where the carbon dioxide is exhaled. By removing carbon dioxide, hemoglobin also helps maintain a stable acid-base balance in the blood.

Before birth, erythrocytes are produced in the liver and spleen; after birth they are produced in the bone marrow. (Remember, bones represent laws and principles.) They are derived from large cells called *committed stem cells* at the rate of about 10 billion cells an hour. The rate is influenced by the spleen, which stores blood and destroys old red blood cells. These cells die in about 80–120 days. The kidneys are the main controller, as they produce a hormone that responds to a decrease in oxygen in the blood which calls for the erythrocytes to be manufactured more quickly. This is only a brief overview; the processes are very intricate and involve many more systems. Again we see how the body is many members, all cooperating for the good of the whole.

White blood cells are called *leukocytes.* Their function is to serve as scavengers and destroy micro-organisms at infection sites. They help remove foreign molecules, and remove debris that results from dead or injured tissue cells. They are able to move about independently and pass through blood vessel walls into the tissues. Unlike the red blood cell, they do have nuclei and are able to reproduce. Their chemical processes are much more complex than those of the red blood cell. They are made in the bone marrow and in lymphoid tissue. There are two basic classifications: *granu-locytes* (named this because they have large numbers of granules in their cytoplasm) and *agranulocytes* (without gran-ules).

There are three kinds of granulocytes:

1. *Neutrophils,* which make up about 60% of granulocytes, are phagocytes that engulf and destroy microor-ganisms and other foreign materials. The neutrophil may also be destroyed while doing its job. Dead mi-croorganisms and neutrophils make up the thick whitish fluid we call pus.
2. *Eosinophils* are phagocytes, which are active during allergy attacks, cancer, parasitic infections, and some autoimmune diseases. They contain plasminogen, which helps dissolve blood clots.
3. *Basophils* contain heparin (an anticoagulant), histamine (which dilates general body blood vessels and constricts blood vessels in the lungs) and it slows *reacting substance A* of allergies, which produces some of the allergic symptoms such as bronchial constriction. Basophils play an important role in providing im-munity against parasites.

There are two types of agranulocytes:

1. *Monocytes* are mobile phagocytes. They circulate in the bloodstream for about 30 to 70 hours, leave through a capillary wall and enter the tissue, where they enlarge to 5 to 10 times their size, and become phagocytic macrophages. They form a key portion of the *reticuloendothelial* system (*reticular* means *meshed or in the form of a network* and *thelium* means *a cellular layer)* which lines the liver, lungs, lymph nodes, thymus gland, and bone marrow. They clean up microorganisms and cellular debris; they also produce a *colony-stimulating factor* which stimulates the bone marrow to produce more monocytes and neutrophils and process specific antigens.
2. *Lymphocytes* move sluggishly and do not travel the same routes through the bloodstream as other leukocytes. They come from bone marrow and then invade lymphoid tissues where they establish colonies. These colonies then produce additional lymphocytes without involving the bone marrow. Most of these are found in the body's tissues, especially in lymph nodes, the spleen, the thymus gland, tonsils, adenoids, and the lymphoid tissue of the gastro-intestinal tract. They can leave the blood more easily than other cells and they differ from other leukocytes in being able to re-enter the circulatory system. Some lymphocytes live for years recirculating between blood and lymphoid organs. They are not phagocytes, which means they do not clean up the debris like other white blood cells.

Platelets *(thrombocytes)* are so named because they are flat like a plate. They start the intricate process of blood clotting. They are more numerous than leukocytes, and are produced by the bone marrow at the rate of 200 billion per day. They have no nuclei and cannot divide, but have a complex metabolism and internal structure. After entering the blood stream, they pick up and store chemical substances that can be released later to help seal vessel breaks. Platelets stick to each other and to the collagen in connective tissue, but not to red or white blood cells. This makes them able to clot blood without interfering with circulation. WE ARE WONDERFULLY AND FEARFULLY MADE!

The first mention of blood in the Bible was after the first person was killed. God told Cain that Abel's blood *"crieth unto me from the ground."* (Genesis 4:10.) When the blood is all spilled out, death occurs. There is no more oxygen or nutrients carried to the cells, and everything in the body dies.

In the Bible, blood was sacred, because in it was the life of the being and it was required for the remission of sin. After the flood, God gave man permission to eat flesh, but He forbade the eating of blood and the shedding of man's blood. (Genesis 9:3-6.) Six times in the Old Testament and three times in the New Testament, it is written that it is forbidden to eat blood. (See Leviticus 17:10-14; 3:17; 7:26; 19:26; Deut. 12:16; 12:23; Acts 15:20; 29; & 21:25.)

It was because of the innocence of the sacrifice that the blood could atone for man's sin. Because of the sinless life of Christ, He can be our Substitute. He takes the death that we deserve and gives us the life that He deserves. WHAT A BARGAIN FOR US! We can wash our stained garments and make them white in His blood. (See Rev. 7:14). There is no way that you can wash a garment in blood and have it come out white. It is physically impossible. There is only ONE BLOOD that can make something white and that is the blood of Christ. The reason that the Israelites could not be made clean in the blood of bulls and goats is that they forgot that the animal blood was the symbol of the shedding of the blood of the Lamb of God, which takes away all sin. (Hebrews 9 & 10.) *"Then Jesus said unto them, Verily, verily, I say unto you, Except ye eat the flesh of the Son of man, and drink his blood, ye have no life in you. Whoso eateth my flesh, and drinketh my blood, hath eternal life; and I will raise him up at the last day. For my flesh is meat indeed, and my blood is drink indeed. He that eateth my flesh, and drinketh my blood, dwelleth in me, and I in him. As the living Father hath sent me, and I live by the Father: so he that eateth me, even he shall live by me. This is that bread which came down from heaven: not as your fathers did eat manna, and are dead: he that eateth of this bread shall live for ever."*—John 6:53.

Jesus used the symbols of the blood and the bread when He instituted the Lord's supper. To the death of Christ we owe even this Earthly life. The bread we eat is the purchase of His broken body. The water we drink is bought by His spilled blood. No one eats his daily food, but that he is nourished by the body and the blood of Christ.

Christ's words were much more true of our spiritual nature. It is by receiving the life for us poured out on Calvary's cross, that we can live the life of holiness. And this life we receive by His word, and by doing those things which He has commanded. Thus we become one with Him. As we partake of the symbols of His broken body and blood, faith contemplates our Lord's great sacrifice and the soul assimilates the spiritual life of Christ. That soul will receive spiritual strength from this service which forms a living connection by which the believer is bound up with Christ and thus bound up with the Father's love.

Chapter Eighteen

LYMPH

Ecclesiastes 1:7

"All the rivers run into the sea; yet the sea [is] not full; unto the place from whence the rivers come, thither they return again."

The above text exactly describes the journey of the lymph as it empties into the blood stream.

Approximately one-sixth of the entire body is composed of the space between cells. This space is referred to as the interstitium, and the fluid contained within the space is referred to as interstitial fluid. This fluid flows into the lymphatic vessels and becomes lymph. Every solid structure in the body is bathed in this interstitial fluid which bathes and nourishes the body tissue. This is the internal sea in which the cell floats. When it is inside the lymphatic capillaries, it is called lymph (Latin for clear water). Lymph is similar to blood except that it does not contain red blood cells as does most of the proteins found in blood. We will describe its functions in the study on the lymphatic system.

Lymphatic vessels, which drain waste products from the tissues, usually run parallel to arteries and veins. The lymphatic vessels transport the lymph to lymph nodes which filter the lymph. The cells responsible for filtering the lymph are called macrophages. These large cells engulf and destroy foreign particles including bacteria and cellular debris.

The lymph nodes also contain B lymphocytes, the white blood cells that are capable of initiating antibody production in response to the presence of viruses, bacteria, yeast, and other organisms.

"Thy way is in the sea, and thy path in the great waters, and thy footsteps are not known."—Psalm 77:18.

"O that thou hadst hearkened to my commandments! Then had thy peace been as a river, and thy righteousness as the waves of the sea."—Isaiah 48:18.

Chapter Nineteen

CEREBROSPINAL—INTRACELLULAR—WASTE

<u>Psalms 46:4</u>

"There is a river, the streams whereof shall make glad the city of God, the holy place of the tabernacles of The Most High."

This river that makes us glad is represented by the fluid in the brain and spinal canal. We will study this more extensively in the study on the nervous system. The river of life begins in the throne room of our brains and runs down the spinal column between the two trees of life in the cerebellum.

"And he showed me a pure river of life, clear as crystal, proceeding out of the throne of God and of the Lamb." —Revelation 22:1

The cerebrospinal fluid is an ultrafiltrate of blood; this fluid contains no red blood cells or the largest protein molecules. It has only traces of protein, oxygen, carbon dioxide, and a few white blood cells. Sodium, potassium, calcium, magnesium and chloride ions, and glucose are also found in the cerebrospinal fluid in solution. This fluid cushions the brain and actually causes it to float inside the cranium.

The intracellular fluid is inside the cells and contains approximately 40 per cent of the water in the body. It is basically the same as the fluid outside, except in the cell there is more potassium and less sodium than in the fluid outside of the cell.

Waste fluid is also present in the body in the form of sweat, urine, and moisture from the lungs. We will study these fluids in the systems to which they apply.

Chapter Twenty

BONES OF THE HEAD AND NECK

CRANIAL BONES

Ephesians 6:17

"And take the helmet of salvation, and the sword of the Spirit, which is the word of God."

Cranium means helmet. In Ephesians 6:17, we are told to take the helmet of salvation. As we study the cranial bones, the plan of salvation is clear. The cranial bones represent: FRONTAL—The seal of God; TEMPORAL—The promises of God; OCCIPITAL—Self denial; PARIETAL—The righteousness of Christ; ETHMOID—The altar of incense (prayer); SPHENOID—The Ark of the Covenant (final blotting out of sin and the atonement with God).

OCCIPITAL

The occipital bone represents our foundation stone of self-denial which was fully manifested in the cornerstone that the builders rejected, Jesus Christ. The Christian must build on this foundation if he would build a strong symmetrical character and be well balanced in his religious experience. It is in this way that man will be prepared to meet the demands of truth and righteousness as they are represented in the Bible.

The life of Christ, the LORD of Glory, is our example. He came from heaven where all was riches and splendor. He came to this sinful world and He laid aside His royal crown, His royal robe, and clothed His divinity with humanity. WHY? He came to this world to manifest the Father's character of love to mankind. He did not rank Himself with the wealthy or the lordly of the Earth. The mission of Christ was to reach the very poor of Earth. He, Himself, worked from His earliest years as the Son of a carpenter. *Self-denial.* Did He not know its meaning? The riches and glory of heaven were His own, but for our sakes *He* became poor that *we* through His poverty might become rich. The very foundation of His mission was self-denial and self sacrifice. The world was His. He made it; yet in a world of His own creation—the Son of man had not where to lay His head. He said, *"The foxes have holes, and the birds of the air have nests; but the Son of man hath not where to lay His head."*—Matthew 8:20.

The hole in the middle of the occipital bone is where the brainstem (the representative of the altar of sacrifice), meets the spinal cord. The joint between the first cervical vertebrae, the atlas and the occipital, is what makes us able to nod yes. We should say "yes" to Christ when He says, *"If any man will come after me, let him deny himself and take up his cross daily, and follow Me. For whosoever will save his life shall lose it: but whosoever will lose his live for My sake, the same shall save it."*—Luke 9:23,24.

There are 12 pairs of muscles attached to this bone (12 x 2 = 24) representing perfect government times support.

PARIETAL

The parietal bones represent righteousness. They are connected with three bones—the temporal (promises), the occipital (foundation), and the frontal (seal of God's character). The righteousness of Christ allows these three to be tied together, with the sphenoid (atonement throne of God) holding them all together.

Righteousness is holiness, likeness to God, and *"God is love."* (I John 4:16.) It is conformity to the law of God, for *"all Thy commandments are righteousness."* (Psalm 119:172), and *"love is the fulfilling of the law."* (Romans 13:10.)

Righteousness is love, and love is the light and life of God. The righteousness of God is embodied in Christ. We receive righteousness by receiving Him. When the sinner believes that Christ is his personal Saviour, then according to His unfailing promises, God pardons his sin and justifies him freely. The repentant soul realizes that his justification comes because Christ, as his substitute and surety, has died for him and is his atonement and righteousness.

There is one muscle attached to this bone.

FRONTAL

The frontal bone is two bones at birth and becomes fused with maturity. In Exodus 26:9, it is the part of the eleven white curtains that is doubled in the forefront. This bone forms the forehead. This principle is the "seal of God's character," a sign that we belong to Him. *"And I saw another angel ascending from the east, having the seal of the living God: and he cried with a loud voice to the four angels, to whom it was given to hurt the earth and the sea, Saying, Hurt not the earth, neither the sea, nor the trees, till we have sealed the servants of our God in their foreheads."*—Revelation 7:2-3.

This same seal is called a mark in Ezekiel 9. In Revelation 2:17, it is called a white stone with a new name written. In Revelation 3:12 we are told *"... I will write upon him the name of my God, and the name of the city of my God: and I will write upon him my new name."* These are the children of God who are represented as being in the Philadelphian condition who have the "open door" which no man can shut. (Revelation 3:8)

The frontal bone articulates (joins) with 12 bones. The principles represented by the 12 bones that touch the frontal bone are principles needed to receive the seal of God. (<u>Two</u> parietals, <u>two</u> nasal, <u>two</u> maxillae, <u>two</u> lacrimal, <u>two</u> zygomatic, the spenoid, and the ethmoid).

TEMPORAL

The temporal bones, together with parts of the sphenoid bone, form part of the sides and base of the cranium. Each temporal bone has four parts which is the creature number. This represents the promises of God made to His creatures.

"Thy lips are like a thread of scarlet, and thy speech is comely: Thy temples are like a piece of a pomegranate within thy locks" "As a piece of a pomegranate are thy temples within thy locks."—Song of Solomon 4:3; 6:7. The pomegranate is red and is full of many seeds, which are encased in red juice. In Psalm 126:6, we are told: *"He that goeth forth and weepeth, bearing precious seed, shall doubtless come again with rejoicing bringing his sheaves with him."* The promises of God are made available by the blood of Christ. This fruit was on the bottom of the High Priest's blue robe, a sign that promises are intimately tied with obedience.

Promises can be positive blessings or they can be curses. Whatever God promises, He will do according to the fulfillment of the provision of the promise. (Read of the blessings and curses in Deuteronomy 28.)

The temporal bone joins with the sphenoid, mandible, zygomatic, occipital and parietal. As you see the significance of these bones, you will see how the promises are necessary to all parts.

Encased within the temporal bone are three tiny hearing bones (discussed in the ear lesson) and the sensory receptors for hearing and balance. Without the promises of God, it is impossible to find balance or to hear correctly in this mixed-up world.

The mastoid process (protrusion) of the temporal bone is located behind the ear and provides the point of attachment for a neck muscle. The neck represents attitude and our attitude depends on our faith in God's promises. The mastoid processes contain air spaces (sinuses) which connect with the middle ear where the balance mechanisms are located. Sometimes infections within the inner ear can spread and cause mastoiditis. We should keep all our empty spaces full of the Holy Spirit, not debris.

In Judges 4, is a type of the last big battle of Armageddon. Sisera, the captain of the enemy, had a tent peg driven through his temples. Read about this battle and see how all the enemy's plans and promises of destruction of God's people came to nought when God fought for His people.

SPHENOID (ARK OF THE COVENANT)

As you study this bone, you can see the wings on each side of the ark. (Each wing has three surfaces) Exodus 25:9-22. The sphenoid articulates with all the other seven bones of the cranium and five of the face, (2 zygomatic, 2 palatine, and the vomer), making 12 bones in all which serves as an anchor to keep them all in place. There are eleven pairs of muscles attached to this bone.

The sphenoid bone is 2 bones at birth, and it is not until some time later, that it becomes one bone. The cella tursica of the sphenoid bone is where the optic nerve crosses and the pituitary gland sits. This is a type of the ten commandments in the Ark of the Covenant. The Ark of the covenant was housed in the Most Holy Place. Into this chamber only the High Priest was allowed to go and only once a year on the great Day of Atonement. (Leviticus 16) The glory, the very presence of God, was manifested here. All of this typifies the experience that we must go through in the last days just prior to Christ's return. Early Writings p. 55 tells how God's people will leave the Holy experience and enter into the Most Holy experience just before Christ comes. If we do not do this, we will be found worshiping Satan without knowing it.

This bone represents the principle of atonement with God. Before the people can be judged righteous, their lives must be investigated to see if they still want to belong to God. As the books of record are opened in the judgment; the lives of all who have believed on Jesus come in review before God.

Beginning with those who first lived upon the Earth, our Advocate presents the cases of each successive generation and closes with the living. Every name is mentioned, and every case closely investigated. Names are accepted, and names are rejected. When any have sins remaining upon the books of record, unrepented of and unforgiven, their names will be blotted out of the Book of Life, and the record of their good deeds will be erased from the book of God's Remembrance. The LORD declared to Moses: *"Whosoever hath sinned against Me, him will I blot out of My book."*—Exodus 32:33; Ezekiel 18:24.

All who have truly repented of sin and by faith claimed the blood of Christ as their atoning sacrifice, have had pardon entered against their names in the books of heaven. As they become partakers of the righteousness of Christ, their characters are found to be in harmony with the Law of God and their sins will be blotted out. They, themselves, will be accounted worthy of eternal life. (Isa. 43:25; Rev. 3:5; Matt. 10: 32,33; Acts 3:19.)

Gray's Anatomy tells us that the cella tursica of the sphenoid bone helps to form the septum of the nose. There are two large irregular cavities in this area called the sphenoidal sinuses. These are filled with air. They do not exist in children, but they increase in size as age advances. When we have sinus infections, these air spaces are filled with debris and it is hard to breathe. These infections usually cease when we stop eating sugar and dairy foods. When the spaces for the Holy Spirit are filled with the debris of the world, we find it hard to breathe spiritually.

ETHMOID (ALTAR OF INCENSE)

The ethmoid bone represents the altar of incense. This is an exceedingly light and spongy bone full of air spaces and canals and has a cubical form. It is located at the base of the cranium. It contains the "cribiform plate." This plate is located at the uppermost part of the ethmoid. Projecting upward from it is a thick, smooth, triangular piece called the crista galli which can be seen in a picture of this bone. The crista galli is usually inclined to one side and many have thought that it looks like the praying figure of Christ in the garden of Gethsemane. It is difficult to see exactly how this looks and how the plates stand up from a side view. Neurons of smell go through the ethmoid and the olfactory bulbs lie on top of the cribiform plate. On each side of the crista galli, the cribiform plate is narrow and deeply grooved to support the bulbs. If the cribiform plate is broken, one loses the sense of smell. The olfactory nerves all travel through the grooves and canals of the ethmoid.

The ethmoid bone sits right next to the sphenoid bone, just as the altar of incense is intimately connected with the Ark of the Covenant in the sanctuary. The altar of incense was the place of prayer, and this bone represents the principle of prayer.

The ethmoid bone articulates with 15 bones (grace multiplied by divine perfection): the sphenoid, the frontal, and all of the face except the mandible and the zygomatic. No muscles are attached to this bone.

FACIAL BONES

II Corinthians 3:18

"But we all, with open face beholding as in a glass the glory of the LORD, are changed into the same image from glory to glory, even as by the Spirit of the LORD."

The face represents the spirit or character, which shows in the face. When Moses came down from spending 40 days with God on the mountain, his face was shining so brightly he had to cover it because the people were afraid. (Exodus 34:29.) We are told this will be the condition of God's people just before Christ returns: "Servants of God, endowed with power from on high with their faces lighted up, and shining with holy consecration, went forth to proclaim the message from heaven."—E G White, Early Writings, p. 278 para. 1.

The bones of the face represent the principles necessary to develop our character and to change us into the image of God. MAXILLARY—Presence of Holy Spirit; ZYGOMATIC—Truth; MANDIBLE—Punishment or chastisement; NASAL—Conviction; PALATINE—Love of God and love to man; INFERIOR NASAL CONCHA—Faith; VOLMER - Works; LACRIMAL—Repentance.

MANDIBLE

The mandible is the largest and strongest facial bone and is the only movable bone in the skull. It is the jawbone. The jawbone in the Bible is used as a form of punishment. (Judges 15:15-19.) God told Satan, as the dragon in Ezekiel 29:1-4, that He would put hooks in his jaws, which meant that He would thwart Satan's plans to kill God's people. Satan is a roaring lion out to devour us. Again in Ezekiel 38:4, God is talking to Satan who is represented as Gog (Revelation 20:9) and promises hooks in his jaws.

The mandible has three parts; the body and two *rami* (branches). At the end of the rami there are two processes separated by a deep depression called the mandibular notch. The lower teeth are located in the upper part of this bone. The mandible forms the chin which no other animal has. "Keep your chin up" and "taking it on the chin" are two idioms that tell a story of courage. God is not trying to save the soul of anything except man. He does not chastise the animals.

The same jawbone that killed the Philistines, served as a source of water and comfort for Samson. *"My brethren, count it all joy when ye fall into divers temptations; Knowing [this], that the trying of your faith worketh patience. But let patience have [her] perfect work, that ye may be perfect and entire, wanting nothing."*—James 1:2-4.

God made man perfectly holy and happy; and the fair Earth, as it came from the Creator's hand, bore no blight of decay or shadow of the curse. It is transgression of God's law—the law of love—that has brought woe and death. Yet even amid the suffering that results from sin, God's love is revealed. It is written that God cursed the ground for man" sake. (Genesis 3:17.) The thorn and the thistle—the difficulties and the trials that make his life one of toil and care—were appointed for mans benefit as a part of the training needful in God's plan. This was appointed for His uplifting of mankind from the ruin and degradation that sin has wrought.

The only bone that articulates (unites by means of a joint) with the mandible is the temporal bone which represents *the promises of God.* Sometimes it is difficult to accept trials but we need to remember, *"There hath no temptation taken you but such as is common to man: but God [is] faithful, who will not suffer you to be tempted above that ye are able; but will with the temptation also make a way to escape, that ye may be able to bear [it]."*—I Corinthians 10:13.

The Father's presence encircled Christ, and nothing befell Him but that which infinite love permitted for the blessing of the world. Here was His source of comfort, and it is also *our* source of comfort. He who is imbued with the Spirit of Christ abides in Christ. Whatever comes to him comes from the Saviour who surrounds him with His presence. Nothing can touch him except by the LORD's permission. When we are going through a hard time, the only thing we can do is to hang on to the promises of God. *"Ye have not yet resisted unto blood, striving against sin. And ye have forgotten the exhortation which speaketh unto you as unto children, My son, despise not thou the chastening of the Lord, nor faint when thou art rebuked of him: For whom the Lord loveth he chasteneth, and scourgeth every son whom he receiveth. If ye endure chastening, God dealeth with you as with sons; for what son is he whom the father chasteneth not? But if ye be without chastisement, whereof all are partakers, then are ye bastards, and not sons. Furthermore we have had fathers of our flesh which corrected [us], and we gave [them] reverence: shall we not much rather be in subjection unto the Father of spirits, and live? For they verily for a few days chastened [us] after their own pleasure; but he for [our] profit, that [we] might be partakers of his holiness. Now no chastening for the present seemeth to be joyous, but grievous: nevertheless afterward it yieldeth the peaceable fruit of righteousness unto them which are exercised thereby."*—Hebrews 12:4-11.

In old age, the mandible shrinks and is greatly diminished in size. Hopefully by that time, we have received enough chastening to have learned how to trust and obey our Father.

At the junction of the temporal and mandible, we find the tempromandibular joint. This joint is what allows this bone to move forward and backward, up and down, and side to side. It is the only moveable bone in the skull. The rest are fixed, just like principles, and unmovable. God's punishment can be adjusted according to need.

If there is a problem at this joint, it is known as TMJ dysfunction. It can cause a lot of pain and trouble talking and chewing. This represents our problem trying to reconcile chastening and promises.

MAXILLAE

The upper jaw is formed by these two bones. Each has a hollow body containing a large sinus and four processes. This represents the help of the Comforter in forming our character.

The maxillae forms part of the *"orbit"* which protects the eye (understanding).

Each side of the maxilla articulates with eight bones: FRONTAL—the seal of God; ETHMOID—prayer; VOLMER—works; LACRIMAL—repentance; NASAL—conviction; PALATINE—love; ZYGOMATIC—truth; and the INFERIOR NASAL CONCHAE—faith. This bone touches in some way every bone in the face except the mandible (punishment). This shows how vital the Holy Spirit is to our character development. The upper teeth of the maxillae and the lower teeth in the mandible come together to chew our food. Blessings and adversities are both needed to nourish us to that we may grow.

The promised blessing of the Comforter, if claimed by faith, would bring all other blessings in its train, and is to be given liberally to the people of God. Through the cunning devices of the enemy, the minds of God's people seem to be incapable of comprehending and appropriating the promises of God. They seem to think that only the scantiest showers of grace are to fall upon the thirsty soul. The people of God have accustomed themselves to think that they must rely upon their own efforts, and that little help is to be received from Heaven. The result is that they have little light to communicate to other souls who are dying in error and darkness. The church has long been contented with little of the blessing of God. They have not felt the need of reaching up to the exalted privileges purchased for them at infinite cost. Their spiritual strength has been feeble, their experience dwarfed, and their character crippled. They are not able to present the great and glorious truths of God's Holy Word in a way that would convict and convert souls through the agency of the Holy Spirit. The power of God awaits their demand and reception. Every defect in the character may be overcome by the help of the Holy Spirit.

The lengthening of the face, just before adolescence, is caused by the growth of the maxillae. As we mature in our Christian experience, we should let the Spirit use us more and more.

NASAL

These two bones are small, oblong bones that form the bridge of the nose. They are in between the maxillae and represent <u>conviction</u>. They vary in size and form in different individuals and are covered by muscles (judgment). Two sets of muscles are attached to the nasal bones. These are muscles that cover the head from front to back and the cartilage of the nose. There are two kinds of conviction; a conviction of sin and a conviction of principle.

When the heart yields to the influence of the Spirit of God, the conscience will be quickened and the sinner will discern something of the depth and sacredness of God's holy law which is the foundation of His government in Heaven and on Earth. The "light," which lighteth every man that cometh into the world, illumines the secret chambers of the soul, and the hidden things of darkness are made manifest. (John 1:9.) Conviction takes hold upon the mind and heart. The sinner has a sense of the righteousness of Jehovah and feels the terror of appearing in his guilt and uncleanness before the Searcher of hearts. He sees the love of God, the beauty of holiness, and the joy of purity, and he longs to be cleansed and restored to communion with Heaven. In order to see his guilt, the sinner must test his character by God's great standard of righteousness. *"By the law is the knowledge of sin"*—Romans 3:20. It is a mirror which shows the perfection of a righteous character and enables him to discern the defects in his life.

The other type of conviction is defined in the dictionary as: "The state of being convinced, or a fixed or strong belief." Our convictions need daily to be reinforced by humble prayer and reading of the Word. While we each have an individuality, we should hold our convictions firmly. We must hold them as God's truth and in the strength which God imparts. If we do not do this, they will be wring from our grasp as a person wrings his nose.

LACRIMAL

The lacrimal bones are thin, and are the smallest of the face bones. They are located behind the maxilla. They help form the orbit of the eye and contain the lacrimal sac that collects excess tears from the surface of the eye. They represent the principle of repentance.

The prayer of David, after his fall, illustrates the nature of true sorrow for sin. His repentance was sincere and deep. There was no effort to palliate his guilt; no desire to escape the judgment threatened. David saw that enormity of his transgression; he saw the defilement of his soul, and he loathed his sin. It was not for pardon only that he prayed, but for purity of heart. He longed for the joy of holiness—to be restored to harmony and communion with God. (Psalms 32:1,2; 51:1-14.) A repentance such as this, is beyond the reach of our own power to accomplish. This kind of repentance is obtained only from Christ who ascended up on high and has given gifts unto men.

Many think that they cannot come to Christ unless they first repent, and that repentance prepares for the forgiveness of their sins. It is true that repentance does precede the forgiveness of sins for it is only the broken and contrite heart that will feel the need of a Saviour. But must the sinner wait till he has repented before he can come to Jesus? Is repentance to be made an obstacle between the sinner and the Saviour? The Bible does not teach that the sinner must repent before he can heed the invitation of Christ, *"Come unto me all ye that labor and are heavy-laden, and I will give you rest."*—Matthew 11:28. It is the virtue that goes forth from Christ that leads to genuine repentance. Acts 5:31 states that He gives repentance. We can no more repent without the Spirit of Christ to awaken the conscience than we can be pardoned without Christ. We must behold the Lamb of God upon the cross of Calvary, and as the mystery of redemption begins to unfold to our minds, the goodness of God leads us to repentance.

Ask Him to give you repentance and to reveal Christ to you in His infinite love, in His perfect purity. When the light from Christ shines into our souls, we shall see how impure we are. We shall discern the selfishness of motives, and the enmity against God that has defiled every act of life. Then we shall know that our own righteousness is indeed as filthy rags, and that the blood of Christ alone can cleanse us from the defilement of sin and renew our hearts in His own likeness. The soul thus touched will hate its selfishness, abhor its self-love, and will seek through Christ's righteousness for the purity of heart that is in harmony with the law of God.

INFERIOR NASAL CONCHAE

These bones are in the side walls of the nose and are curled like a scroll so that they can provide more surface for the membrane that keeps the inside of the nose moist. This is our main air passage. These bones are thin, and spongy. These bones represent faith. They articulate with prayer, the Holy Spirit, repentance, and the love of God.

There are no muscles attached to these bones. To a person with no faith, there seems to be no judgment in faith. Faith is believing without real evidence. Faith is having our eyes upon the unseen, not the seen. *"Now faith is the substance of thing hoped for and the evidence of things not seen." "But without faith it is impossible to please Him; for he that cometh to God must believe that He is, and that He is a rewarder of them that diligently seek Him."*—Hebrews 11:1,6.

True faith and true prayer—how strong they are! They are as two arms by which the human suppliant lays hold upon the power of Infinite Love. Faith is trusting in God and believing that He loves us, and that He knows what is for our best good. And then instead of our own way, it leads us to choose His way. In place of our ignorance, it accepts His wisdom; in place of our weakness, His strength; in place of our sinfulness; His righteousness. Our lives, and ourselves, are already His. Faith acknowledges His ownership and accepts its blessing. Truth, uprightness, and purity, are pointed out as secrets of life's success. It is faith that puts us in possession of eternal life. Every good impulse or aspiration is the gift of God. Faith received from God is the only thing that produces true growth and efficiency.

"This is the victory that overcometh the world, even our faith."—I John 5:4. It is faith that enables us to look beyond the present, with its burdens and cares, to the great hereafter where all that now perplexes us shall be made plain. Faith sees Jesus standing as our Mediator at the right hand of God. Faith beholds the mansions that Christ has gone to prepare for those who love Him.

VOMER

The vomer is shaped like a plowshare and forms part of the nasal septum. It represents our service for Jesus. The call to us to place all on the altar of service. This comes to each one of us. We are not all bidden to sell everything we have or to leave family and friends but God asks us to give His service the first place in our lives. He asks us to allow no day to pass without doing something to advance His work in the Earth. He does not expect from all the same kind of service. One may be called to ministry in a foreign land, another may be asked to give his means for the support of the gospel work. Others may be called to be teachers at the International Institute of Original Medicine. God accepts the offering of each. It is the consecration of the life and all its interests that is necessary. Those who make this consecration will hear and obey the call of Heaven.

To everyone who becomes a partaker of His grace, the LORD appoints a work for the welfare of others. Individually we are to stand in our lot, saying "here am I, send me." Whether a man be a minister of the Word or a physician, whether he be a merchant or a farmer, professional man or mechanic, the responsibility rests upon him. It is his work to reveal to others the Gospel of their salvation. Every enterprise in which he engages should be a means to this end.

The *volmer* divides the nose, the air (spirit) passage; if one side is stopped up, we can't breathe or talk correctly. There are always two sides to every question and in our service we should always be aware that if both are not used, we get lopsided. Mercy / Justice; Faith / Works; Love / Duty. We need the Spirit in both nostrils.

PALATINE

These two bones lie behind the upper jaw and form the roof of the mouth (back part of the hard palate), parts of the floor and walls of the nasal cavity, and the floor of the orbit for the eye. It is wedged in between the sphenoid (place of atonement) and the maxillae (the presence of the Spirit). It is L shaped, has a vertical plate and a horizontal plate, which represents our two directions of love. It represents love to God and love to man. It articulates with the place of atonement, the presence of the Spirit, prayer, faith, and the service bones.

LOVE is a four letter word and there are two other words that also contain four letters that can best describe what love means. Our love to God can be best demonstrated by OBEY. And our love to man by GIVE.

OBEY—*"If ye love me keep my commandments"*—John 14:15. *"And hereby we do know that we love Him, if we keep His commandments, He that saith, I know Him and keepeth not His commandments, is a liar, and the truth is not in him. But whosoever keepeth His word, in Him verily is the love of God perfected: hereby know we that we are in Him."*—I John 2:3-5.

GIVE—*"He that loveth his brother abideth in the light, and there is none occasion of stumbling in him."*—I John 2:10. Our Saviour's joy was in the uplifting and redemption of fallen men. For this He counted not His life dear unto Himself, but endured the cross despising the shame. So the angels are ever engaged in working for the happiness of others. This is their joy. The Spirit of Christ's self-sacrificing love is the spirit that pervades heaven and is the very essence of it's bliss. This is the spirit that Christ's followers will posses; the work that they will do. When the love of Christ is enshrined in the heart, like sweet fragrance it cannot be hidden. Its holy influence will be felt by all with whom we come in contact.

ZOGOMATIC

These are the cheek bones. They act as a tie by connecting the maxilla with the frontal bone above and the temporal bone behind. Truth is the principle represented. It forms part of the orbit of the eye, the place where light enters. Truth is very seldom popular and people are always ready to receive a comfortable lie. *"Even him, whose coming is after the working of Satan with all power and signs and lying wonders. And with all deceivableness of unrighteousness in them that perish; because they received not the love of the truth, that they might be saved. And for this cause God shall send them strong delusion, that they should believe a lie; That they all might be damned who believed not the truth, but had pleasure in unrighteousness."*—II Thessalonians 2:9-12.

Between righteousness and sin, love and hatred, truth and falsehood, there is an irrepressible conflict. When one presents the Truth, he is drawing away the subjects of Satan's kingdom, and the prince of evil is aroused to resist it. Persecution and reproach await all who are imbued with the Spirit of Christ. The character of the persecution changes with the times, but the principle—the spirit that underlies it—is the same. As men seek to come into harmony with God, they will find that the offense of the cross has not ceased. Principalities and powers and wicked spirits in high places are arrayed against all who yield obedience to the law of Heaven.

HYOID

The hyoid bone represents the principle of separation. It is set in the neck, apart from the rest of the bones and is U shaped. It does not touch any other bone. It has five parts (the number of grace and redemption). It supports the tongue and provides attachment sites for the muscles used in speaking and swallowing. It is attached to the temporal (promise) bone by muscles and ligaments.

"Be ye not unequally yoked together with unbelievers: for what fellowship hath righteousness with unrighteousness? and what communion hath light with darkness? And what concord hath Christ with Belial? or what part hath he that believeth with an infidel? And what agreement hath the temple of God with idols? for ye are the temple of the living God; as God hath said, I will dwell in them, and walk in [them]; and I will be their God, and they shall be my people. Wherefore come out from among them, and be ye separate, saith the Lord, and touch not the unclean [thing]; and I will receive you, And will be a Father unto you, and ye shall be my sons and daughters, saith the Lord Almighty Having therefore these promises, dearly beloved, let us cleanse ourselves from all filthiness of the flesh and spirit, perfecting holiness in the fear of God"—II Corinthians 6:14–7:1.

The only correct standard of character is the holy law of God. It is impossible for those who make that Law the rule of life to unite in confidence and cordial bothered with those to turn the truth of God into a lie and regard the authority of God as a thing of naught. Between the disobedient and the one who is faithfully serving God, there is a great fixed gulf. Upon the most momentous subjects, their thoughts and sympathies are not in harmony. One class is ripening as wheat for the garner of God, the other is ripening as tares for the fires of destruction. How can there be unity of purpose or action between them? *"Know ye not that the friendship of the world is enmity with God? Whosoever therefore will be a friend of the world is the enemy of God."*—James 4:4.

If we are Christians, having the Spirit of Him who died to save men from their sins, we shall love the souls of our fellow men too well to countenance their sinful pleasures by our presence or our influence. We cannot sanction their course of action by associating with them or partaking in their feasts and their councils. Such a course, so far from benefiting them, would only cause them to doubt the reality of our religion.

THE NECK

Hebrews 2:9-18

"But we see Jesus, who was made a little lower than the angels for the suffering of death, crowned with glory and honour; that he by the grace of God should taste death for every man. For it became him, for whom [are] all things, and by whom [are] all things, in bringing many sons unto glory, to make the captain of their salvation perfect through sufferings. For both he that sanctifieth and they who are sanctified [are] all of one: for which cause he is not ashamed to call them brethren, Saying, I will declare thy name unto my brethren, in the midst of the church will I sing praise unto thee. And again, I will put my trust in him. And again, Behold I and the children which God hath given me. Forasmuch then as the children are partakers of flesh and blood, he also himself likewise took part of the same; that through death he might destroy him that had the power of death, that is, the devil; And deliver them who through fear of death were all their lifetime subject to bondage. For verily he took not on [him the nature of] angels; but he took on [him] the seed of Abraham. Wherefore in all things it behooved him to be made like unto [his] brethren, that he might be a merciful and faithful high priest in things [pertaining] to God, to make reconciliation for the sins of the people. For in that he himself hath suffered being tempted, he is able to succor them that are tempted."

Because there are so many important lessons in the neck, we will present this special section. The neck is the connecting link between the head and the body. Our connecting link between Heaven and Earth is Christ. He restored our relationship with our heavenly Father.

Notice that Christ is not ashamed to call us brethren who are sanctified. This is what makes Christ our brother and His Father our Father. Sanctification means "separation, a setting apart" in the original Greek. Sanctuary means "a place set apart." It is God's will that our body temple be set apart for a holy use. That is the lesson He was trying to teach in the sanctuary in the wilderness.

Christ will declare the name of His Father to the brethren in the midst of the church. Church comes from the Greek word, *ekklesia,* which means "that which is called out." God has a church. It is not the great cathedral, neither is it the national establishment, neither is it the various denominations. It is the people who love God and keep His commandments. *"Where two or three are gathered together in my name, there am I in the midst of them"*—Matt. 18:20. "Where Christ is even among the humble few, this is Christ's church, for the presence of the High and Holy One who inhabiteth eternity can alone constitute a church."—E G White, The Upward Look, p. 315 para. 5.

Man cannot make an artificial neck. It cannot be replaced or faked or supplemented. The neck is not an independent unit. It has to be attached to the head and the body or there is no life. It is interesting that many of God's followers have been killed by beheading, and we are promised that it will happen again at the end of time. If Satan cannot separate us from our spiritual Head, he will try to separate us from our physical head by chopping our neck. *"And I saw thrones, and they sat upon them, and judgment was given unto them: and [I saw] the souls of them that were beheaded for the witness of Jesus, and for the word of God, and which had not worshipped the beast, neither his image, neither had received [his] mark upon their foreheads, or in their hands; and they lived and reigned with Christ a thousand years."*—Revelation 20:4.

The neck can represent separation and/or unity. The false separation principle is the beheading, in separating the head from the body, and the Father from His children. True separation is to come apart from the world which will eventually unite all of God's people. We are told in Revelation 3 that unless we separate ourselves from the Laodicean condition we will be spewed out of God's mouth. *"Wherefore come out from among them, and be ye separate."*—II Corinthians 6:17. *"And I heard another voice from heaven, saying, Come out of her, my people, that ye be not partakers of her sins, and that ye receive not of her plagues. For her sins have reached unto heaven, and God hath remembered her iniquities."*—Revelation 18:4,5.

Christ is the only way to properly get your head on straight. No life-giving principles can flow to the body if they have to pass through other people. We must have a direct link to our Heavenly Father and to the throne room above. People who depend on some denomination here on Earth to obtain salvation will never receive it. When the true crisis of life comes, they will have no support. Just as Asa, who was having a problem with his feet (which represents our ways or lifestyle), went to a worldly physician to be healed, and the result was that he died. (II Chronicles 16:12 & Jeremiah 17:5.) "There are many ways of practicing the healing art; but there is only one way that Heaven approves. God's remedies are the simple agencies of nature, that will not tax or debilitate the system through their powerful properties. Pure air and water, cleanliness, a proper diet, purity of life, and a firm trust in God, are remedies for the want of which thousands are dying; yet these remedies are going out of date because their skillful use requires work that the people do not appreciate. Fresh air, exercise, pure water, and clean, sweet premises, are within the reach of all, with but little expense; but drugs are expensive, both in the outlay of means, and the effect produced upon the system."—E G White, Counsels on Diet and Foods, p. 301 para. 2.

Worldly physicians who teach worldly philosophy is man-centered and teaches love of self. Psychology is a man-made. Worldly support is opposed to Christ and His principles of self-sacrifice. "Those who have the truth for these last days will bear a message adapted to the poor. One would think that the Gospel was inspired in order to reach this class. Christ came to the Earth to walk and work among the poor. To the poor He preached the Gospel. His work is the Gospel worked out on medical missionary lines—in justice, mercy, and the love of God which is the sure fruit borne because the tree is good. And today in the person of His believing, working children, who move under the guidance of the Holy Spirit, Christ visits the poor and the needy, relieving want and alleviating suffering."—E G White, Letter 83, 1902; Medical Ministry, p. 243 para.2

The hyoid bone in the neck represents this separation principle. It is set all by itself in the neck, apart from the rest of the bones. Our attitude shows in our neck. A stiff or hard neck denotes rebellion. *"But they obeyed not, neither inclined their ear, but made their neck stiff, that they might not hear, nor receive instruction."* Jeremiah 17:23. *"Ye stiff-necked and uncircumcised in heart and ears, ye do always resist the Holy Ghost: as your fathers did, so do ye."*—Acts 7:51.

The neck is also the area in which wisdom is revealed. The neck contains the seven cervical vertebrae which represents the principle of wisdom. *"Wisdom hath builded her house, she hath hewn out her seven pillars."*—Proverbs 9:1. James 3:17 tells us what the seven principles of true wisdom are: "But the wisdom that is from above is first pure, then peaceable, gentle, and easy to be entreated, full of mercy and good fruits, without partiality, and without hypocrisy." The reason that man does not understand spiritual things is that he does not have that connecting link between the Head who is wisdom, and the body. *"Happy is the man that findeth wisdom, and the man that getteth understanding."*—Proverbs 3:13. *"The fear of the LORD is the beginning of wisdom; a good understanding have all they that do His commandments: His praise endureth forever."*—Psalm 111:10.

There are 9 groups of muscles in the neck, with a total of 33 muscles. There are four muscles in the neck that move the head. These muscles attach to the sternum, clavicle, and skull bones. The nerve that controls the neck is the cranial nerve. It not only controls the movement but also controls the voice production. This nerve is a mixed nerve and originates in two different places. The spinal root of the cranial nerve comes from the first five cervical spinal levels. It joins with the bulbar root from the brainstem and together they control the movement of the head and shoulders. This represents the connection that Christ has in controlling the movements taking place in Heaven. This nerve joins the vagus nerve which helps to control head, neck, chest and abdomen, showing that Christ has a vital control of things happening on Earth.

Sound is produced in the neck. We should have wisdom in our speech and let Christ control our words. *"For in many things we offend all. If any man offend not in word, the same is a perfect man, and able also to bridle the whole body."*—James 3:2.

Sound is produced by a complex coordination of muscles. We shall study more about this in the Respiratory System. "The mouth of the just bringeth forth wisdom: but the froward tongue shall be cut out. The lips of the righteous know what is acceptable: but the mouth of the wicked speaketh frowardness." Proverbs 10:31,32

Coughing takes place in the neck and is the effort to clear foreign objects out of our neck and lungs, the Spirit chamber. Foreign particles cause much trouble if they enter there. Christ is now cleansing the Most Holy Place in Heaven and in our mind from all the pollution. When He goes in, the dirt comes out.

Swallowing represents believing, and also is involved with the neck. That which has been mulled over (chewed) has to be swallowed. As you breathe and swallow, keep in mind that if it were not for Jesus' life and death as our substitute, this would not be possible. *"Having therefore, brethren, boldness to enter into the holiest by the blood of Jesus, By a new and living way, which he hath consecrated for us, through the veil, that is to say, his flesh; And [having] an high priest over the house of God; Let us draw near with a true heart in full assurance of faith, having our hearts sprinkled from an evil conscience, and our bodies washed with pure water."* —Hebrews 10: 19-22.

In the process of digestion, food and liquid pass through the lips, which are the gates of our mouth. Solids pass through the soft palate into the oral pharynx. The epiglottis shuts off the air passage, and the food enters the esophagus which leads into the stomach. The epiglottis is a veil that protects.

The jugular vein and the carotid arteries are located in the neck. When these arteries become clogged, the brain does not receive enough blood and clear thinking becomes a problem. When we cease to apply the blood of Jesus, He leaves us to the deceptions that we choose. The sacrificial animals were killed by cutting their throat and they bled to death from the drained blood vessels in the neck.

The thyroid gland is also located in the neck. This important gland has many jobs. It has two lobes connected with a bridge (isthmus), which represents the two main pillars of God's government, Mercy and Justice. Our bridge is Jesus Christ. Through Jesus, God's mercy was manifested to men, but mercy does not set aside justice. The law reveals the attributes of God's character, and not a jot or tittle of it could be changed to meet man in his fallen condition. God did not change His law, but He sacrificed Himself, in Christ, for man's redemption. "*God was in Christ, reconciling the world unto Himself.*"—II Corinthians 5:19.

The law requires righteousness. It also demands a perfect character. The sinner cannot give what he does not possess. But Christ, coming to the Earth as man, lived a holy life, and developed a perfect character. Christ fulfilled the requirements of the law and imparted this to man through His blood. His sacrifice stands for the life of men. Thus they have remission of their sins that are past through the forbearance of God. More than this, Christ imbues men with the attributes of God. He builds up the human character after the similitude of the divine character, a goodly fabric of spiritual strength and beauty. Thus the very righteousness of the law is fulfilled in the believer in Christ.

The parathyroids, which represent the four living creatures around the throne, are located on the back of the thyroid. Read Ezekiel 1-10 and Revelation 4-6, and you will see the judgment connected with these four creatures. *"Mercy and truth are met together; righteousness and peace have kissed each other."*—Psalm 85:10.

Nerves of the Neck

There are eight pairs of cervical spinal nerves. This shows the superabundance of Christ's provision for us. Besides these nerves, there is a network of nerves called the cervical plexus. This consists of six superficial branches and six deep branches, 12 on each side of the neck, 24 in all, thc same number as the 24 elders around the throne. These are Christ's helpers from the Earth.

Information about daily cycles of light and dark is detected by the eyes and conveyed via the optic nerve to the hypothalamus. From there, sympathetic nerves convey the signal to the superior cervical nerve ganglia in the upper part of the neck and then to the pineal gland which represents God's timing.

The phrenic nerves descend into the thorax to innervate the diaphragm. Problems with this cause hiccups, and paralysis of the diaphragm, etc.

When the word "neck" is researched in the Bible, it is interesting to see all the things that are placed on the neck, both by man and by God. The yoke on the neck is a symbol of service or slavery depending with whom you are yoked. Jesus said His yoke is easy, but the yoke of transgression is hard.

Keep in mind all of the work going on in the neck and by the grace of God. May we keep our necks soft and pliable and open to the Saviour's influence.

Study Questions

Chapter Seventeen

1. Name the different fluid systems in the body.
2. The transformation of ______ into ____________ is a wonderful process, and all should be intelligent upon this subject.
3. What happens to our bodies when blood does not flow evenly throughout the system?
4. What does the parable of the talents in Matthew 25 teach us?
5. What does correct water management consist of?
6. True or False? Juices are a good substitute for water drinking. Why?
7. What is hydrolysis and why is it important?
8. Dry mouth is the _________ sign of ________________.
9. Why do we need to study the Biblical Sanctuary?

Chapter Eighteen

1. Name three proteins found in the blood and describe their function.
2. What are electrolytes and what happens when they become unbalanced?
3. What are erythrocytes and what do they do?
4. What is the function of white blood cells?
5. What causes our blood to clot when we are cut?
6. Explain the following phrase. "No one eats his daily food, but that he is nourished by the body and the blood of Christ."
7. What is interstitial fluid and describe what would happen if it became contaminated.
8. Describe cerebrospinal fluid, intracellular fluid and waste fluid.
9. The occipital bone represents self denial. Give your viewpoints in a short essay on how we can develop a life of self denial and what will be the results.
10. The frontal bone is where we will be sealed by God to receive eternal life. How will this sealing take place?
11. Without the promises of God it is ______________ to find ____________ or to hear correctly in this mixed up world.
12. The sphenoid bone represents atonement with God. How would you explain this to a non-believer?
13. What function does the cribiform plate perform? This is a research question.
14. What is the only moveable bone in the skull? Describe it function.
15. The upper jaw is formed by what two bones?
16. There are two kinds of conviction; conviction of _______ and conviction of ________________.
17. True or False. The lacrimal bones are thick and the largest of the face bones.
18. Must the sinner wait till he has repented before he can come to Jesus? Expand your answer into a short paragraph.
19. Describe the function of the inferior nasal conchae.
20. To everyone who becomes a partaker of God's grace is appointed a _________ for the welfare of ______________.
21. In your own words describe how we develop a loving character.
22. What are the zogomatic bones and what is represented by them?
23. Whosoever therefore will be a friend of the world is the ____________ of God. Explain how this is so.
24. Our bodies need to be set apart for ___________ use.
25. What does ekklesia mean?
26. What does God's church consist of? Elaborate. Does God have a special people here on this earth that represent Him?

27. What is the only way of practicing the healing art that God approves of?
28. The nerve that controls the neck is the _____________ nerve.
29. The thyroid gland has 2 lobes. What do they represent?
30. What is the cervical plexus and what does it represent?

Chapter Twenty-One

THE DEFENSE SYSTEMS

Psalms 51:2

"Wash me thoroughly from mine iniquity, and cleanse me from my sin."

We shall study the lymphatic system and the immune system together because they work together. The lymphatic system is the transportation, training, and maintenance system and the immune system is the front line soldier. The soldiers of the immune system travel in both transportation systems: the lymphatics and the blood vessels.

In review of the tribes of Zebulun and Gad, one can clearly see the difference in their personalities, which is the basic difference between these two systems. Zebulun, representing the lymphatic system, is quiet and unobtrusive, trying to keep everything cleaned up. Most people are not aware of the importance of the lymphatic system. Gad, representing the immune system, is defensive and ready to fight. Lymphatic quietly cleans up; immune goes to war.

The spiritual lesson of the lymphatic system is the cleansing from sin. *"As it is written, there is none righteous, no, not one."*—Romans 3:10. *"If we confess our sins, He is faithful and just to forgive us our sins and to cleanse us from all unrighteousness."*—I John 1:9. *"Wherewithal shall a young man cleanse his way? By taking heed thereto according to Thy word."*—Psalm 119:9.

The spiritual lesson of the immune system is victory over sin. *"But thanks [be] to God, which giveth us the victory through our Lord Jesus Christ."*—I Corinthians 15:57.

The plan of salvation must, of necessity, include not only forgiveness of sin, but complete restoration. Salvation from sin is more than forgiveness of sin. Forgiveness infers that sin has incurred and is conditioned upon a breaking away from it. Sanctification is a separation from sin, and indicates deliverance from its power and victory over it. The first is a means to neutralize the effect of sin; the second is a restoration of power for complete victory.

Sin, like some diseases, leaves man in a deplorable condition by being weak, despondent, and disheartened. He has little control over his mind. His will is compromised, and with the best of intentions, he is unable to do what he knows to be right. He feels that there is no hope. He knows that he has himself to blame, and remorse fills his soul. To his bodily ailments is added the torture of conscience. He knows that he has sinned and is to blame. His desire is to rid his body of both the physical and spiritual ailments. "God has given us His holy precepts, because He loves mankind. To shield us from the results of transgression, He reveals the principles of righteousness. The law is an expression of the thought of God; when received in Christ, it becomes our thought. It lifts us above the power of natural desires and tendencies, above temptations that lead to sin. God desires us to be happy, and He gave us the precepts of the law that in obeying them we might have joy."—E G White, Desire of Ages, p. 308.

The sinner receives the good news of the Gospel and he realizes that even though his sins are as scarlet, they shall be as white as snow. All is forgiven. He is "saved." What a wonderful deliverance it is! His mind is at rest. No longer does his conscience torment him. He has been forgiven. His heart wells up with praise to God for His mercy and goodness to him.

As a disabled ship is towed to port and becomes "safe" from the storm, so man is "saved" but not sound. Repairs need to be made on the ship before it is pronounced seaworthy, and the man needs reconstruction before he is fully restored. This process of restoration is called sanctification, and includes in its finished product body, soul, and spirit. The restoration of the physical body is not complete until the coming of Christ when "this mortal puts on immortality." (I Corinthians 15:53). When the work is finished, the man is "holy" and completely sanctified and restored to the image of God. As we study these systems, keep this spiritual warfare in mind.

In our study of the lymphatic and immune systems, we shall also have to keep in mind that other systems of the body play a very important role in the protection of the body. The skeletal system manufactures most of the soldiers used by the immune system to fight disease. The skin is involved and protects all of our body organs from contamination. All the body systems work together to help keep us safe.

Chapter Twenty-Two

THE LYMPHATIC SYSTEM

Proverbs 22:6

"Train up a child in the way he should go: and when he is old, he will not depart from it."

The lymphatic system works quietly behind the immune system to train and prepare the body to fight disease. It also transports the soldiers and maintains a back up for the bodies defense against disease. In the military we would call this system the quartermaster. It contains the lymph vessels, lymph nodes, the lymph fluid, leukocytes, lymphatic organs, and specialized lymphoid tissue.

The lymphatic system:

1. Returns to the blood excess fluid and proteins from the spaces around cells,
2. Plays a major role in the transport of fats from the tissue surrounding the small intestine to the blood,
3. Filters and destroys microorganisms and other foreign substances, and
4. Aids in providing long-term protection for the body.

Lymph

In the study of the fluids of the body, we learned that every solid structure in the body is bathed in fluid. This fluid comes from the water, small proteins, and other materials that are forced out of the blood capillaries into the spaces between cells. This interstitial fluid bathes and nourishes surrounding body tissues. This fluid is constantly being repaired and purified by the lymphatic system.

Lymph Vessels

The lymph vessels are similar to blood vessels and follow the same paths. The big difference between them is that blood vessels are a closed system and go to and from the heart. Lymph vessels are one way and have an open end so that they can pick up the fluid and transport it toward the blood vessels.

Tiny capillaries called *lymphatic capillaries*, which are most abundant near the innermost and outermost surfaces of the body, gather the lymph fluid for transport. This happens in the dermis of the skin, and in the mucosal and submucosal layers of the respiratory and digestive systems. These tiny capillaries are also numerous beneath the mucous membrane that lines the body cavities and covers the surface of organs. Very few lymphatic capillaries are found in muscles, bones, or connective tissue. There are none in the central nervous system or the cornea of the eye.

Specialized lymphatic capillaries called *lacteals* extend into the intestinal villi. Lacteals absorb fat from the small intestine for transportation into the blood and for distribution throughout the body.

The lymphatic capillaries join with other capillaries to become larger collecting vessels called lymphatics. These lymphatics are the larger vessels that resemble veins, but their walls are thinner. They contain more valves, and they pass through specialized masses of tissue called lymph nodes. Lymphatics are arranged into two sets: a deep set within the body, and a superficial set by the skin. These lymphatics join with one another and form two large ducts. They are the right lymphatic duct and the thoracic duct. These ducts empty their contents into the subclavian veins above the heart.

Just as the rivers carry water into the seas and through evaporation this water is transported back to the land, so the fluid of the body is left around the cells and the lymph vessels pick up the excess fluid and return it back to the blood vessels. *"All the rivers run into the sea; yet the sea is not full; unto the place from whence the rivers come, thither they return again. All things are full of labor; man cannot utter it:"*—Ecclesiastes 1:7,8.

Primarily three forces accomplish the actual movement of lymph:

1. The action of circular and some longitudinal smooth muscles in the lymphatic vessels.
2. The squeezing action of skeletal muscles during normal body movement. (Lymph flow may increase 5- to 15-fold during vigorous exercise).
3. By running parallel with the venous system in the thorax, where a subatmospheric pressure exists. This pressure gradient creates a "pull factor" called the auxiliary respiratory pump that aids lymph flow.

Lymph Nodes

Scattered along the lymphatic vessels like beads on a string are small bean-shaped masses of tissue called lymph nodes. These nodes were known as lymph glands until it was discovered that they did not secrete anything. The nodes filter out harmful substances from the lymph trapping them in a mesh of reticular fibers where they are destroyed by lymphocytes and macrophages. Lymph nodes are also the initiating sites for the specific defenses of the immune response. These nodes are found in seven locations:

1. Axillary,
2. Supra trochlear (by elbow),
3. Cervical,
4. Intestinal,
5. Inguinal,
6. Iliac,
7. Lumbar.

The largest concentrations are at the neck, armpit, thorax, abdomen, and groin. Seven show the perfection of this system as it was created.

Tonsils

Tonsils are masses of lymphatic nodules enclosed in a capsule of connective tissue. No lymph enters the tonsils. There are three sets of tonsils: Two pharyngeal, two palatine and two lingual. These are gate keepers placed at the entrance of the body where foreign things try to come in. *"Thy watchmen shall lift up the voice; with the voice together shall they sing: for they shall see eye to eye, when the LORD shall bring again Zion."*—Isaiah 52:8. They contribute soldiers (lymphocytes) in this strategic area to kill the enemy before it can invade the body. It is through the mouth that most germs enter the body. That is why these three gate keepers are so important. *"He that keepeth his mouth keepeth his life: but he that openeth wide his lips shall have destruction."*—Proverbs 13:3. *"Thus saith the Lord GOD; No stranger, uncircumcised in heart, nor uncircumcised in flesh, shall enter into my sanctuary, of any stranger that [is] among the children of Israel."*—Ezekiel 44:9.

The lesson of the tonsils is the lesson of watching and keeping. Three keepers of the door are mentioned in II Kings 25:18 and Jeremiah 52:24: *"And he commanded them, saying, This [is] the thing that ye shall do; A third part of you that enter in on the sabbath shall even be keepers of the watch of the king's house; And a third part [shall be] at the gate of Sur; and a third part at the gate behind the guard: so shall ye keep the watch of the house, that it be not broken down."*—II Kings 11:5,6. To keep the gates and to keep watch is a very important position. The word "keep" in Hebrew is "shamar," and means "to hedge about (as with thorns)," i.e. guard; to protect; attend to; beware; be circumspect; to take heed (to self); keep; mark; look narrowly; observe; preserve; regard; reserve; save; sure; watch. Shamar is used over 100 times in the Bible in regard to keeping the Law of God. *"Wait on the LORD and keep His way, and He shall exalt thee to inherit the land: when the wicked are cut off, thou shalt see it."*—Psalm 37:34.

Thymus

The thymus is the "college" that trains the lymphocytes to become T cells and gets them ready to do their special jobs. No lymph fluid enters the thymus. It is most active in childhood and early adolescence. "To effect a permanent change for the better in society, the education of the masses must begin in early life. The habits formed in childhood and youth, the tastes acquired, the self-control gained, the principles inculcated from the cradle, are almost certain to determine the future of the man or woman. The crime and corruption occasioned by intemperance and lax morals might be prevented by the proper training of the youth." {MYP 233.2}

Out of the millions and millions of T cells sent to the thymus, most die and only a small percentage of them, (which are still tens of millions) graduate and leave to work in the body. *"Many are called, but few are chosen."*—Matthew 22:14. Only when strife for supremacy is banished, when gratitude fills the heart, and love makes fragrant the life—it is only then that Christ is abiding in the soul, and we are recognized as laborers together with God.

Spleen

The spleen has five functions and especially represents Christ's grace and redemption in this system. It is dark purple on the outside, which is the color of self-sacrifice. It has white pulp, the color of Christ's robe of righteousness and it has red pulp, the color of sacrifice.

The five functions of the spleen are:

1. It produces red blood cells before birth. Remember that after birth, the blood cells are produced in the bones (principles). Before we are born again into the family of God, God's grace is pleading with us to accept the offer of salvation. By His grace we have the blood of Christ available. Even though we do not understand His principles, He loves us. *"For when we were yet without strength, in due time Christ died for the ungodly. For scarcely for a righteous man will one die: yet peradventure for a good man some would even dare to die. But God commendeth his love toward us, in that, while we were yet sinners, Christ died for us."*—Romans 5:6-8.
2. It stores newly formed red blood cells and platelets to be released in times of injury, emergency, or during bursts of physical activity. *"To an inheritance incorruptible, and undefiled, and that fadeth not away, reserved in heaven for you, Who are kept by the power of God through faith unto salvation ready to be revealed in the last time."*—I Peter 1:4,5.
3. It releases blood, especially in an emergency. It can release 200mL of blood into the general circulation in one minute. *"Seeing then that we have a great high priest, that is passed into the heavens, Jesus the Son of God, let us hold fast [our] profession. For we have not an high priest which cannot be touched with the feeling of our infirmities; but was in all points tempted like as [we are, yet] without sin. Let us therefore come boldly unto the throne of grace, that we may obtain mercy, and find grace to help in time of need."*—Hebrews 4:14-16.
4. Antigens in the blood of the spleen activate lymphocytes that develop into cells that produce antibodies. *"Thou therefore, my son, be strong in the grace that is in Christ Jesus."*—II Timothy 2:1. *"But if thou shalt*

indeed obey His voice, and do all that I speak; then I will be an enemy unto thine enemies, and an adversary unto thine adversaries."—Exodus 23:22.

5. Macrophages, which are abundant in the spleen, help remove damaged or dead red blood cells and platelets, microorganisms, and other debris from the blood as it circulates through the spleen. It recirculates the iron from the hemoglobin of dead red blood cells back to the bone marrow to produce new red blood cells. *"Always bearing about in our body the dying of the Lord Jesus, that the life also of Jesus might be manifest in our body."*—II Corinthians 4:10.

Peyer's Patches—Nodules

Peyer's patches, which are called nodules, are clusters of unencapsulated lymphoid tissue found in the tonsils, small intestine, and appendix. They are also found along the bronchi of the respiratory tract. They generate plasma cells that secrete antibodies in large quantities in response to antigens in the intestine. Such plasma cells do not remain together, but are distributed along the length of the intestine in the following way: Inactive B cells migrate from bone marrow to aggregated lymph nodules. Then the B cells that are activated by exposure to antigens, leave the lymph nodules, migrate to lymph nodes, and then later enter the blood stream through the thoracic duct. The B cells are then carried by the bloodstream throughout the body and eventually come to reside as plasma cells beneath the mucosal surface of the intestine. Here they recognize an enormous variety of specific antigens and produce corresponding antibodies primarily against bacteria and some viruses.

Chapter Twenty-Three

THE IMMUNE SYSTEM

I Chronicles 7:4

"And with them, by their generations, after the house of their fathers, [were] bands of soldiers for war, six and thirty thousand [men]: for they had many wives and sons. And their brethren among all the families of Issachar [were] valiant men of might, reckoned in all by their genealogies fourscore and seven thousand."

The lesson of the immune system is the lesson of victory over sin. The nonspecific defenses do not involve the production of antibodies. The specific defenses involve the formation of antibodies, which help destroy foreign substances.

Nonspecific Defenses

The first line of defense is restricting entrance. The skin, hair follicles, and sweat glands contain many microorganisms that are barred from entering the body by the cells of the skin. The openings to the outside, such as the nasal passages, mouth, lungs, and digestive system are lined with mucous membranes which put out sticky mucus. This mucus traps many harmful microorganisms before they can get too far into the body. Sweat washes microorganisms from the pores and skin surface. Tears wash away foreign substances from the eyes. Nasal hairs filter the air before it enters the upper respiratory tract, and cilia of the tract sweep bacterial and other particles trapped in mucus toward the digestive tract for final elimination. When microorganisms do enter the stomach, the extreme acidity of the stomach kills most of them.

God's first line of defense is to try and shelter us from the wiles of the Devil. He asks us to come apart and be separate so that we will not be attracted to the world. (I Corinthians 6:17.) He sets up barriers all around us with His laws and principles to try and hedge us in and to keep the enemy out. Read Isaiah 5:1-3.

Inflammation occurs in response to tissues being damaged by punctures, abrasions, burns, bites, foreign objects, infections, or toxins. This causes the tissues to become irritated and red and warm to the touch caused by an increase of blood flow to the area. Swelling also sometimes occurs caused by an increase of interstitial fluid, which creates pain as a result of the pressure on receptors and nerves and chemical by-products that irritate nerve endings.

When infection occurs, large numbers of neutrophils are fighting and release pyrogens (chemical substances that raise body temperature by help from the hypothalamus). This helps fight infection by increasing the blood flow and leukocyte activity in the area while inhibiting the growth of some infectious organisms.

Phagocytosis (FAG-o-sie-TOE-sis; Gr. *phagein,* to eat + *cyto,* cell) involves three leukocytes (white blood cells), monocytes, macrophages, and neutrophils. These soldiers consume foreign substances or dead tissue. Drainage by lymphatics of tissue fluid, cell debris, proteins, and dead microorganisms from the damage site brings the foreign proteins into contact with the lymphocytes in the lymph nodes. This contact *initiates* the specific immune defenses.

Specific Defenses

An antigen is a substance against which an antibody is produced. It is a large protein or polysaccharide molecule that is ordinarily foreign to body tissues.

Antibodies are proteins produced by B cells in response to an antigen. They are specialized to react with antigens, triggering a complex process called immunity.

Immunity protects the body by destroying the invaders. About two trillion lymphocytes in the body produce about 100 million trillion antibodies during a lifetime. What is even more impressive than this is the fact that not all antibodies are alike. When we are in trouble, the LORD has a thousand ways to provide help for us that we can not even imagine.

The specific defenses, which constitute the immune response, discriminate between foreign substances (antigens) by forming specific proteins called antibodies and/or specific cells that react with the foreign substances and help destroy them.

Warriors

There are seven types of warriors in the immune reaction:

1. B cells: These cells change into an antibody-secreting plasma cell when stimulated by an antigen.
2. Plasma cells: These cells secrete antibodies that mark foreign substances for destruction.
3. Helper T cells: These cells activate the B cells after they encounter specific antigens by releasing a B-cell growth factor. This is necessary for appropriate responses of natural killer T cells and suppressor T cells to antigens. The responses also activate *lymphokines* that activate the microphages.
4. Killer T cells: The killer T cells attack any foreign microorganism.
5. Suppressor T cells: Suppressor cells suppresses the autoimmune responses, excessive antibody production; regulates activities of natural killer T cells, and inhibits development of B cells into plasma cells.
6. Memory T cells: These cells remember and identify antigens during later infections.
7. Macrophage: These macrophages ingest microorganisms and present antigens to the T cells to initiate specific immune responses.

There are seven parts to our spiritual defense: Truth, righteousness, preparation, faith, salvation, word of God, and prayer. *"Finally my brethren, be strong in the LORD, and in the power of His might. Put on the whole armour of God, that ye may be able to stand against the wiles of the devil. For we wrestle not against flesh and blood, but against principalities, against powers, against the rulers of the darkness of this world, against spiritual wickedness in high [places]. Wherefore take unto you the whole armour of God, that ye may be able to withstand in the evil day, and having done all, to stand. Stand therefore, having your loins girt about with truth, and having on the breastplate of righteousness; And your feet shod with the preparation of the gospel of peace; Above all, taking the shield of faith, wherewith ye shall be able to quench all the fiery darts of the wicked. And take the helmet of salvation, and the sword of the Spirit, which is the word of God: Praying always with all prayer and supplication in the Spirit, and watching thereunto with all perseverance and supplication for all saints."*—Ephesians 6:10-18.

Fight

There are five phases of the battle: Recognition of the enemy; amplification of defenses; the attack, the clean up; and warriors going home and a watch is set up.

The Hebrew word for fight means "to feed on, to consume, to battle. That is exactly how the immune system battles. II Chronicles 20 tells about a great battle and how it was fought. We do not have to consciously enter into the battle that goes on every day in our immune system. God set it up and He watches over it. If we cooperate in taking care of it, it fights for us. *"And, behold, God himself [is] with us for [our] captain, and his priests with sounding trumpets to cry alarm against you. O children of Israel, fight ye not against the LORD God of your fathers; for ye shall not prosper."*—II Chronicles 13:12. *"But thou, O man of God, flee these things; and follow after righteousness, godliness, faith, love, patience, meekness. Fight the good fight of faith, lay hold on eternal life, whereunto thou art also called, and hast professed a good profession before many witnesses."*—I Timothy 6:11,12. Every soul is required to fight the fight of faith. Every person has corrupt and sinful habits that must be overcome by vigorous warfare. Genuine faith always works by love. Faith that unites us to Christ helps us to fight against sin.

"And every man that striveth for the mastery is temperate in all things. Now they [do it] to obtain a corruptible crown; but we an incorruptible. I therefore so run, not as uncertainly; so fight I, not as one that beateth the air: But I keep under my body, and bring [it] into subjection: lest that by any means, when I have preached to others, I myself should be a castaway."—I Cor. 9:25-27. The warfare against self is the greatest battle that will ever be fought. The yielding of self, surrendering all to the will of God, requires a great struggle.

Enemies

Who is the enemy that we fight against? *"Be sober, be vigilant; because your adversary the devil, as a roaring lion, walketh about, seeking whom he may devour."*—I Peter 5:8.

The body's enemies are foreign invaders that come in to wreck havoc with the health, viruses, bacteria, foreign substances, such as smoke, and smog. Some we invite in by ingesting them, others come in uninvited.

Virus: A virus cannot reproduce itself. It must find a host cell and take it over, so that it can make more of the same virus. Satan has no creative abilities. He must take over people, so that he can make more corrupt people. The cold virus has more than 600,000 atoms, with a surface ridged with peaks and heavily corrugated with distinctive canyons. The virus roams the respiratory tract seeking a human cell with protuberances that precisely mirror the shape of its canyons. When canyon and peak meet, they lock together like pieces of a puzzle. The virus now has a foothold. Quickly it injects itself through the cell membrane and takes over the cell.

When Satan came to Jesus there was nothing in Him that responded. Now, while our great High Priest is making the atonement for us, we should seek to become perfect in Christ. Not even by a thought could our Saviour be brought to yield to the power of temptation. Satan finds in human hearts some point where he can gain a foothold; some sinful desire is cherished, by means of which his temptations assert their power. But Christ declared of Himself: *"the prince of this world cometh, and hath found nothing in Me."*—John 14:30. Satan could find nothing in the Son of God that would enable him to gain the victory. He had kept His Father's commandments, and there was no sin in Him that Satan could use to his advantage. This is the condition in which those must be found who shall stand in the time of trouble.

AIDS: Like Greeks hidden inside the Trojan horse, the AIDS virus enters the body concealed inside a helper T cell from an infected host. Almost always it arrives as a passenger in body fluids. In the invaded victim, helper T cells immediately detect the foreign T cell. But as the two T cells meet, the virus slips through the cell membrane into the defending cell. Before the defending T cell can mobilize the troops, the virus disables it. Some researchers believe the AIDS virus also may change the surface of the helper T cells so that they fuse together. That strategy makes it even easier for the virus to pass from cell to cell undetected. Once inside an inactive T cell, the virus may lie dormant for many months, even years. Then, perhaps when another unrelated infection triggers the invaded T cell to divide, the AIDS virus also begins to multiply. One by one, its clones emerge to infect nearby T cells. Slowly but inexorably, the body loses the very sentinels that should be alerting the rest of the immune system. Phagocytes and killer cells receive no call to arms. Because the B cells cannot recognize the clones as enemies, they are not alerted to produce antibodies. With no opposition, the AIDS clones can "run free" anywhere they want to go in the body. *"Beware of false prophets, which come to you in sheep's clothing, but inwardly they are ravening wolves."*—Matthew 7:15.

"While they promise them liberty, they themselves are servants of corruption: for of whom a man is overcome, of the same is he brought in bondage. For if after they have escaped the pollutions of the world through the knowledge

of the Lord and Saviour Jesus Christ, they are again entangled therein, and overcome, the latter end is worse with them than the beginning."—II Peter 2:19,20.

Autoimmune Disease

When the body starts attacking its own cells as if they were enemies, we have a renegade immune system and what is known as "autoimmune disease."

Rheumatoid arthritis is a good example of this. The immune system attacks the bones (principles and laws) and the joints (working together smoothly) of the body.

Early in the womb, the fetus develops lymphocytes that have receptors for antigens. These lymphocytes can differentiate between self and foreign substances so that they do not attack itself. Autoimmune diseases may result from two different mechanisms. First, a body cell may make a new antigen for which the body had not formed a tolerance. This happens because the antigen was not exposed to the fetal lymphocytes in the womb. The second mechanism happens when an alteration in the genetic program in a B or T cell produces a "forbidden clone" that produces a changed immunoglobulin that reacts with normal body antigens. The result is that the body is at war with itself. Some of these diseases are: rheumatoid arthritis, Addison disease, multiple sclerosis, Type 1 diabetes, lupus, etc.

Victory

Our immune system fights millions of enemies each day and overcomes most of them until it breaks down. We can overcome spiritually if we will stay connected to our Source of power. *"For whatsoever is born of God overcometh the world: and this is the victory that overcometh the world, [even] our faith. Who is he that overcometh the world, but he that believeth that Jesus is the Son of God?"*—I John 5:4,5. *"And I saw as it were a sea of glass mingled with fire: and them that had gotten the victory over the beast, and over his image, and over his mark, [and] over the number of his name, stand on the sea of glass, having the harps of God."*—Revelation 15:2.

It is in this life that we are to separate sin from us through faith in the atoning blood of Christ. Our precious Saviour invites us to join ourselves to Him, to unite our weakness to His strength, our ignorance to His wisdom, and our unworthiness to His merits. God's providence is the school in which we are to learn the meekness and lowliness of Jesus. The Lord is ever setting before us, not what we would choose, which seems easier and pleasanter to us, but the true lessons of life. It rests with us to co-operate with the agencies that Heaven employs in the work of conforming our characters to the divine model. None can neglect or defer this work but at the most fearful peril to their souls.

It is for a demonstration of what the Gospel can do for man's character that the world is looking for. Complete sanctification is the work of a lifetime as demonstrated symbolically in the lymph and immune systems. These two systems are in constant warfare against temptation. Every victory hastens the process. As victory is gained over a sin, one is immunized against that sin and it attracts him no more. As one is victorious over one besetment, so he is become victorious over every other sin. When this work is completed, when victory has been gained over pride, ambition, love of the world—over all evil, the victorious are ready for translation. Satan has no more temptations for him. He stands without fault before the throne of God. Christ places His seal upon him; He is safe. God is vindicated against the charges Satan made against Him; that His law was unfair and unable to be kept. *"Blessed [be] the God and Father of our Lord Jesus Christ, which according to his abundant mercy hath begotten us again unto a lively hope by the resurrection of Jesus Christ from the dead, to an inheritance incorruptible, and undefiled, and that fadeth not away, reserved in heaven for you, Who are kept by the power of God through faith unto salvation ready to be revealed in the last time. Wherein ye greatly rejoice, though now for a season, if need be, ye are in heaviness through manifold temptations: That the trial of your faith, being much more precious than of gold that perisheth, though it be tried with fire, might be found unto praise and honour and glory at the appearing of Jesus Christ: Whom having not seen, ye love; in whom, though now ye see [him] not, yet believing, ye rejoice with joy unspeakable and full of glory: Receiving the end of your faith, [even] the salvation of [your] souls."*—I Peter 1:3-9.

Chapter Twenty-Four

THE NERVOUS SYSTEM

Psalms 23:3

"He restoreth my soul: he leadeth me in the paths of righteousness for his name's sake."

The spiritual lessons in the nervous system are to teach us about the activity that is going on in Heaven and about God's plan to get rid of sin and restore in us His image. It is to show us Jesus, as He is, a Saviour. Let His hand draw aside the veil which conceals His glory from our eyes. It reveals Him in His high and holy place.

The numbers or divisions or shapes or functions of the organs of the nervous system all have their counter part in heaven just as the tabernacle that was built in the desert was patterned after the heavenly things. (Exodus 25:9.) That tabernacle was not God, but was a dwelling place for Him. We are not God, and will never be God but are only a dwelling place for Him to reflect His glory—His character. In the false religions that are becoming so popular today, there is an idea that God is in everything. He is especially in nature and in every person whether or not they love and obey Him. It is even popular to think that any one can be God by setting ones own standards for right and wrong. This is the same lie that Satan told Eve in the garden which caused man to worship the creatures instead of the CREATOR.

All the armies of Heaven are in the service of the Prince of Heaven exalting the Lamb of God, who taketh away the sins of the world. They are working for Christ under His commission to save to the uttermost all who look to Him and believe in Him. These heavenly intelligences are speeding on their mission. They confederate together to uphold the honor and glory of God. They are united in the holy alliance in a grand and sublime unity of purpose to show forth the power, compassion, love, and glory of the crucified and risen Saviour.

The nervous system is a copy of this intricate working of the spiritual forces in Heaven. It represents all that is taking place to save mankind from sin and a fallen nature. God would have us accept Him as a restoring force, so that we may be partakers of the joys of eternal life.

The Brain

The human brain is the most complex structure known, consisting of more than 100 billion nerve cells richly interconnected in networks that still defy analysis. Each neuron may have from 100 to 10,000 connections with other nerve cells. There may be as many as 100 trillion connections in the brain!

The brain is two halves of pinkish gray tissue, wrinkled like a walnut, with somewhat the consistency of porridge and weighs about three pounds. It can store more information than all the computers and the libraries can hold. In the brain countless nerve impulses constantly travel on fiber pathways recording the nature of the outside world, thinking, dreaming, feeling, and directing the muscles and organs of the body.

Basically the brain performs three types of activities:

1. It receives an input of sensory impulses from the eyes, ears, nose, and other sense structures;
2. It processes these impulses (thinking); and
3. It sends out impulses to direct the response to this information.

The Nervous System

Judah, the tribe from which Jesus came, represents the nervous system. It is the first system to begin to develop in the embryo. Judah was the lead tribe when the camp moved.

There are two kinds of cells in the nervous system: nerve cells and supporting cells. The supporting cells do not carry nervous impulses, but as their name implies, help to hold the nerve cells and their axons in place.

Neurons: Nerve Cells

Each neuron has three parts:

1. Cell body—This may be star-shaped, round, oval, or pyramid-shaped. It always has spreading branches that reach out to send or receive impulses to or from other cells. The cell body contains seven structures plus one nucleus. This makes 8, the number of super abundance.
2. Dendrite (tree)—Dendrites conduct nerve impulses toward the cell body. A neuron may have as many as 200 dendrites. We may bombard Heaven with multitudes of messages.
3. Axon—Axons carry impulses away from spinal cord as electrical impulses to the next neuron, muscle cell, or gland cell. There is generally only 1 axon. We should have only one LORD to give us directions.

There are 3 classes of neurons based on structure:

1. Unipolar,
2. Bipolar, and
3. Multipolar.

There are 3 messengers involved in messages from Heaven; the Father, the Son, and the angels. (Revelation 1:1.) To make the message heard, it takes two more components, man and the Spirit of God. *"Howbeit when he, the Spirit of truth, is come, he will guide you into all truth: for he shall not speak of himself; but whatsoever he shall hear, that shall he speak; and he will shew you things to come. He shall glorify me for he shall receive of mine, and shall shew it unto you."*—John 16:13,14.

If the cell body of a motor neuron were enlarged to the size of a baseball, the axon would extend about a mile and the dendrites would fill a large gymnasium.

The brain is the capital of the body, and the seat of all the nervous forces and mental action. All the organs of motion are governed by the communications they receive from the brain. The nerves carry the messages to and from the different parts of the body. In this way the vital action of every part of the body is controlled.

There are two different kinds of messengers. These are chemical and electrical messengers. There are two sides to the message that are sent from Heaven. These messages are: love and obedience, faith and works, mercy and justice. The message is sent along the nerve fiber as an impulse just like a message travels along a telephone wire. The nerve cells do not touch each other; there is a space between each neuron called a synapse. When the message reaches the space between the cells, it must be carried across by chemicals to the other side just like cars are carried on a ferry boat across a river.

Different regions of the plasma membrane are specialized to perform different roles. Some are channels and some are pumps. Channels act as pores through which ions can pass. (Ions are atoms that have acquired an electrical charge.) The pumps help to move the chemicals where they are needed to transport the messages. Some channels have electric gatekeepers and some have chemical gatekeepers. Receptor sites cause the channels to open or close. These gatekeepers decide whether to send the message on or not. If the nerves transmitted every message that came in from the senses, we would have an overload. So they decide what is important to process and act upon and what is not necessary. For instance, if you always had to be concentrating on how your socks felt on your feet or the sun shining on your head felt, your nervous system would be worn out so it rightly divides the messages. *"Study to show thyself approved unto God, a workman that needeth not to be ashamed, rightly dividing the word of truth."*—II Timothy 2:15.

The pumps require energy to drive the ions through the membrane. These pumps are powered by ATP, the usable energy formed in the mitochondria of the cells. The ATP represents the power of the Holy Spirit which is increased by faith and trust in God.

This spiritual lesson is that God allows room for individuality within His church as long as everyone obeys His laws. We can be different personalities, have different jobs and diverse means of accomplishing our work, but all must be guided and connected by the principles of the Bible. If a nerve refused to be guided by the message coming from the brain or spinal cord, it would short circuit the system it was connected with. *"Knowing this first, that no prophecy of the scripture is of any private interpretation."*—II Peter 1:20.

As you study the divisions of the nervous system, think of the different agencies that God is using to save man. The Central Nervous System (CNS) is composed of two divisions. First is the brain that is the command center and represents heaven. Next is the spinal cord that represents the communication between Heaven and Earth carried on by the angels.

The Peripheral Nervous System (PNS) represents God's children, who are the church. It consists of all the nerves outside the brain and spinal cord and has two types of systems that do two different jobs. One is the MOTOR (efferent away from) system, which carries orders *from* the CNS. This represents the responsibility of God's children to receive and distribute truth from God. The other is the SENSORY (afferent –toward) system which carries information *to* the CNS. This represents our other duty to pray for each other.

There are two divisions of the PNS. The somatic represents the relationship of God to us personally. This system has axons that individually go directly back and forth to the CNS. In the early life of David, we see a personal relationship with God in its fullness. In his life as a shepherd, God was teaching David lessons of trust. His experience in protecting his sheep taught him the love of the Great Shepherd and prepared him to defeat Goliath (I Samuel 17). His talents, as precious gifts from God, were employed to extol the glory of the Divine Giver. The visceral represents the relationship of God to us as a group. This system has nerve fibers of motor neurons that go from the CNS to interact within a ganglion (group of nerve cells) located outside of the CNS and then on to the destination and where the action occurs. The visceral system has two divisions: motor and sensory. Motor nerves carry messages from the CNS to the body, and sensory nerves that carry messages from the body to the CNS.

The motor division of the visceral PNS is called the peripheral autonomic nervous system, because it is concerned with internal functions that we usually don't have to be consciously concerned with. These functions are breathing, digestion, etc. The PNS is divided into two motor systems. (1) Parasympathetic which is active when the body is operating under normal conditions and (2) The sympathetic motor system helps the body to adjust to stressful situations.

A nerve is a bundle of fibers enclosed in a connective tissue sheath. Tracts are bundles of fibers and their sheaths found in the CNS. Nerves are the term for these bundles in the outlying PNS. A collection of cell bodies of neurons inside the CNS is called nuclei. A collection of cell bodies of neurons outside the CNS in the PNS is called a ganglion.

Coverings of the Brain and Spinal Cord

The brain is enclosed in three layers of cranial meninges (membranes). These coverings are represented by the three colors interwoven in the inner curtain of the sanctuary which was fine twined linen. These colors are blue, purple and scarlet.

BLUE is a symbol of obedience. *"Speak unto the children of Israel, and bid them that they make them fringes in the borders of their garments throughout their generations, and that they put upon the fringe of the borders a ribband of blue: And it shall be unto you for a fringe, that ye may look upon it, and remember all the commandments of the LORD, and do them; and that ye seek not after your own heart and your own eyes, after which ye use to go a whoring: That ye may remember, and do all my commandments, and be holy unto your God."*—Numbers 15:38-40. Blue is the color of the Dura mater (hard mother). It has two fused layers and divides the cranial cavity into three distinct compartments and adds support to the brain. Outer cranial dura mater contains veins and arteries that nourish the bones and is intimately attached to the bones (principles) of the cranial cavity. In certain locations the two layers separate to form channels (spaces - dural sinuses of the dura mater) that are lined with endothelium and contain and the drain venous blood from the brain. Between the inner dura mater and the arachnoid is a space. It is the subdural space that does not contain spinal fluid.

SCARLET represents the blood and sacrifice of Christ. The Hebrew word for scarlet is "worm," taken from the worm that the dye came from. The same word is used in Psalm 22:6, *"But I am a worm and no man; a reproach of men, and despised of the people."* This psalm is the prophecy of Christ on the cross. The same word is used in Job 25:4-6: *"How then can man be justified with God? or how can he be clean [that is] born of a woman? Behold even to the moon, and it shineth not; yea, the stars are not pure in his sight. How much less man, [that is] a worm? and the son of man, [which is] a worm?"* If you put this together with II Corinthians 5:21, you see that Christ came down to the depths of degradation to exchange places with us. *"For He hath made Him to be sin for us, who knew no sin; that we might be made the righteousness of God in Him."* Scarlet is the color of the arachnoid and is the next layer. It consists of delicate connective tissue which contains no blood vessels of its own. In the space under the arachnoid there is a network of bands of connective tissue (trabeculae) which contain cerebrospinal fluid and blood vessels. The fluid in the subarachnoid space provides a special environment on which the brain floats. This cushions it against hard blows and sudden movements.

PURPLE is the sign of obedience unto death (blue plus scarlet), even the death of the cross. (Philippians 2:8.) The altar of sacrifice was covered with this color when it was transported. (Numbers 4:13.) Purple is also the color of a king, and it was a purple robe the soldiers put on Jesus when they mocked Him and called Him "King of the Jews." (Mark 15:17,18.)

In Revelation 17, the Harlot is described as being decked with purple and scarlet, but no blue. This is because it is a fallen church that has only part of the truth. It talks of Christ's sacrifice and wants to wear the purple of royalty but not of death to self, and it has no obedience.

The scarlet third layer is the **pia mater** (tender mother), and this is the delicate innermost layer which directly covers and is attached to the surface of the brain. It dips down into the fissures between the raised ridges of the brain. The large number of small blood vessels in the pia matter supplies most of the blood to the brain.

The Brain

The brain has four major divisions. They are the brainstem, the cerebellum, the diencephalon, and the cerebrum. The brain is the organ and instrument of the mind, and controls the whole body.

In order for the other parts of the system to be healthy, the brain must be healthy; and in order for the brain to be healthy, the blood must be pure. If, by correct habits of eating and drinking, the blood is kept pure, the brain will be properly nourished. Nutrients can reach the brain only through the blood, and about one pint of blood is circulated to the brain every minute.

Blood which reaches the brain contains glucose (which the brain cannot store) and oxygen (20% of all used in the body is used by the brain). A lack of either oxygen or glucose will damage brain tissue faster than any other tissue. The brain has a built-in regulating device that makes it almost impossible to constrict blood vessels that would reduce the incoming blood supply. This subject will be covered further in the study of the Digestion & Cardiovascular Systems.

Cerebrum

All our conscious living depends on the cerebrum. It consists of two cerebral hemispheres, which are the same. "I and my Father are one," Jesus said in John 10:30. These hemispheres are each divided into five lobes: They are the frontal, parietal, temporal, occipital, and central (insula).

The LEFT HEMISPHERE is active in: Right hand control, spoken language, written language, scientific skills, numerical skills, reasoning, and sorting out parts.

The RIGHT HEMISPHERE is active in: Left hand control, music awareness, recognition of faces, recognition of three-dimensional shapes, art awareness, insight, imagination and grasping the whole subject.

Each hemisphere of the cerebrum is composed of the cortex, the white matter, and the basal ganglia.

Cortex

The cortex is made up of gray matter containing 50 billion neurons and 250 billion neuralgia cells. There are four types of neuralgia which are glue cells and non-conducting cells that protect and nurture and support the nervous system.

The raised ridges are called convolutions or gyri (jye-rye) which are separated by slit like grooves called sulci. Extremely deep grooves are called fissures. This rolling effect makes the surface area of the gray matter in the ratio of 3:1 to the white matter. Each person has a specific pattern of folds. The relations between God and each soul are as distinct and full as though there were not another soul upon the earth to share His watchcare. The cortex is divided on each side into six lobes: frontal, parietal, temporal, occipital, limbic, and insula (central).

White Matter

Beneath the cortex lies a thick layer of white matter which consists of interconnecting groups of axons projecting in two basic directions: (1) From the cortex to other cortical areas of both hemispheres, the thalamus, the basal ganglia, the brainstem, or the spinal cord. (2) From the thalamus to the cortex. The thalamus is functionally integrated with the cortex in the highest sensory and motor functions of the nervous system.

Three types of fibers are present in the white matter:

1. Association fibers link one area of the cortex to another area in the same hemisphere.
2. Commissural fibers are axons that project from a cortical area on one hemisphere to the opposite hemisphere. Anterior commissure and corpus callosum both connect the two hemispheres.
3. Projection fibers include the axons that project from the cortex to other structures of the brain, such as basal ganglia, thalamus and brainstem, and spinal cord.

Basal Nuclei

The basal nuclei are composed of gray matter deep within each cerebral hemisphere. They help coordinate muscle movements by relaying neural inputs from the cerebral cortex to the thalamus and finally back to the motor cortex of the cerebrum. The basal nuclei contain five parts: the caudate, the putamen, the globus pallidus, the subthalmic nucleus, and the substantia nigra. Five is the number of grace and redemption. Grace has quickening, assimilating power. "And God is able to make all grace abound toward you: that ye, always having all sufficiency in all things, may abound to every good work."—II Corinthians 9:8.

"For by grace are ye saved through faith; and that is not of yourselves: it is the gift of God: Not of works, lest any man should boast. For we are His workmanship, created in Christ Jesus unto good works, which God hath before ordained that we should walk in them."—Ephesians 2:5-8.

The basil nuclei are organized as an intricate network of neuronal circuits and processing centers. Along with the cerebellum, they act as the interface between the sensory systems and many motor responses.

The basic circuitry of a motor response is as follows: First the sensory information comes from all parts of the body. This information is processed in the thalamus. Then the information is sent to the cortex. The basal nuclei receives information from many areas of the cortex processes and integrates these inputs and relays them to the thalamus which in turn projects its output to the motor, premotor, and limbic cortical areas. These areas then exert their influences on the upper motor neuron pathways and other systems that affect motor, emotional, and cognitive behaviors. It is grace that works in every art of our life to control all of our behavior.

In all who will submit themselves to the Holy Spirit a new principle of life is to be implanted. This is the lost image of God that is to be restored in humanity. But man cannot transform himself by the exercise of his will. He possesses no power by which this change can be effected. Grace wholly from without must be put into the sinner before the desired change can take place. This must happen before one can be fitted for the kingdom of glory. It is by the renewing of the heart that the grace of God works to transform the life. No mere external change is sufficient to bring any one into harmony with God. The heart must be converted and sanctified. A profession of faith and the possession of truth in the soul are two different things. The mere knowledge of truth is not enough. True obedience is the outworking of a principle within. It springs form the love of righteousness, and essence of true righteousness is loyalty to our Redeemer. This will lead one to do right because it is right—because right doing is pleasing to God.

Ventral Thalamus

The ventral thalamus is located toward the front part of the diencephalon. It contains the subthalamic nucleus, one of the basal ganglia. This nucleus is the driving force that regulates and modulates the output of the basal ganglia and its influences on motor activity. This represents the regulation of grace. The knowledge of the law entered to show man his sin. Grace came to give mankind power to overcome sin and to walk in newness of life. It is regulated according to need. It is not given that one can keep on sinning, but that one might walk in harmony with God and his law. (Read Romans 5;20,21; 6:1,2.)

Malfunction of the basal nuclei results in abnormal involuntary movements or expressions of release phenomena in which certain inhibitory influences are reduced or lost like no brakes on a car. Parkinson's disease is characterized by rigidity in the skeletal muscles, tremor, and the tendency to be immobile. Without God's grace, no one is able to function spiritually.

Frontal Lobe

The frontal lobe is also called the "motor lobe" and is involved in two basic cerebral functions: They are in control of the voluntary movements, including those associated with speech, and the control of a variety of emotional expressions and moral and ethical behavior. From the motor cortex, nerve impulses are conveyed through the motor pathways to the brain, brainstem, and spinal cord to processing centers that stimulate the motor nerves of the skeletal muscles. Relatively large cortical areas are devoted to the face, larynx, tongue, lips and fingers especially the thumb. The thumb is so important for out dexterity that more of the brain's gray matter is devoted to manipulating the thumb than to controlling the thorax and abdomen. The arms and hands represent our works. Does that tell you something about how important our works are? The very front area of the brain has the control of emotional expressions and moral behavior. This is where we receive the seal of God.

Revelation 7:3 and Ezekiel 9 tell us about the seal or mark of God. It is a mark that cannot be seen by men, and will be placed upon those only who bear a likeness to Christ in character, and who are sighing and crying for the abominations that are done in the church. As wax takes the impression of the seal, so the soul is to take the impression of God's Spirit and retain the image of Christ. We are to copy no human being or let any human being stand between us and our Heavenly Father. The character of Christ is to be our model. Let us then take our minds off the perplexities of life and fix them on Christ, that by beholding we may be changed into His likeness. As we look to Him and think of Him, His character will be formed within us which is the hope of glory. Then the seal may be placed upon us.

The great mass of professing Christians in the church will meet with disappointment in the day of God because they do not have this seal in their foreheads. Like the foolish virgins who went out to meet the bridegroom in Matthew

25, they have neglected the work of perfecting their character and becoming like Christ. They have done only a superficial work and still cling to their selfish traits, never allowing God's Holy Spirit to transform them into the image of Christ. Even the wise virgins in the church went to sleep and had to be awakened by the "Loud Cry" (Isaiah 58:1) "Behold the Bridegroom Cometh." (Matthew 25:6.)

The olfactory nerve and bulb for the sense of smell are located beneath the frontal lobe. This is the only sense that does not go through the thalamus for interpretation.

Parietal Lobe

The parietal lobe evaluates general senses and tastes. It receives information from the skin, joints, muscles, and body organs. As the bone of the same name, it represents the righteousness of God and all of the information sent here should be chosen carefully. This lobe is divided into three parts: They are the postcentral gyrus, superior and inferior parietal lobes, and the supramarginal gyrus. The postcentral gyrus is the main sensory area of the cerebrum.

Temporal Lobe

This lobe is also divided into three convolutions. It has critical functions in hearing, equilibrium, emotion, memory, and understanding of the spoken word. It is involved in the promises of God.

Occipital Lobe

This lobe is also divided on its external surface into three convolutions. This is where visual images are assembled and made meaningful after being transmitted from the retina and the thalamus. This area is also involved in the understanding of the written and spoken word. Lesions on this part of the cortex can cause alexia (a failure to recognize written words).

It was to this part of the temple that God took Ezekiel and showed him what the religious leaders were doing in the chambers of the imagery. *"Then said he unto me, Son of man, hast thou seen what the ancients of the house of Israel do in the dark, every man in the chambers of his imagery? for they say, The LORD seeth us not; the LORD hath forsaken the earth."*—Ezekiel 8:12. The ancients are the ministers and leaders of the church who have forgotten that the LORD sees into the dark corners of the mind. *"Therefore judge nothing before the time, until the Lord come, who both will bring to light the hidden things of darkness, and will make manifest the counsels of the hearts: and then shall every man have praise of God."*—I Corinthians 4:5.

Central Lobe

The insula is buried deep in the central sulcus which is the fissure that separates the frontal and parietal lobes. It is an island in the middle of the brain. It is (along with the nuclei of the basal ganglia), the first part of the brain to be differentiated in the embryo according to Gray's Anatomy.

This part of the brain is represented by the laver in the courtyard of the sanctuary. Instead of being given with the instruction for the rest of the furniture in Exodus 25-26, the altar of incense and the laver were not mentioned until Exodus 30. This is after the instruction had been given for all the rest of the sanctuary, even the priests' clothing, the consecration service and the burnt offerings are mentioned before these two items. Then the instruction concerning the altar of incense and the laver are given. That is because this represents the part of sanctification that is left up to us.

The laver was for washing. The priests had to wash their hands and feet (works and ways) before they could offer sacrifices or go in before God. The LORD tells us what kind of washing this is in Isaiah 1:16,17: *"Wash you, make you clean; put away the evil of your doings from before mine eyes; cease to do evil; Learn to do well; seek judgment, relieve the oppressed, judge the fatherless, plead for the widow."* There was the washing of blood (Revelation 7:14) and the washing of water. (Acts 22:16 and Ephesians 5:16.)

Another washing is described in John 13:4-17, which is a mini-baptism that needs to be done as our ways get dusty walking through this world.

In all of these washings, we must put forth some effort to get clean. The central lobe is the area of the will and free choice where our part in the plan of salvation is housed. Heaven can carry on all the work for us but if we choose not to accept it and take salvation, it is all of no avail for us. Jesus died to preserve our power of choice and He will not violate it. We must do the choosing. This is why when the instructions were given for carrying the furniture in Numbers 4, there are no instructions for the laver administration.

The electric poser of the brain, promoted by mental activity, vitalizes the whole system. The will electrifies the nerve power. Pure religion has to do with the manifestations of one's own will. The will is the governing power in the nature of man. It brings all the other faculties under its sway. The will is not the taste or inclination, but it is the deciding power which works in the children of men unto obedience to God or unto disobedience. Man cannot transform himself by the exercise of his will. He possesses no power by which this change can be effected. The leaven—something wholly from without—must be put into the meal before the desired change can be wrought in it. So the sinner must receive the grace of God before he can be fitted for the kingdom of glory. The renewing energy must come from God. Only God's Holy Spirit can make the change. All who would be saved, high or low, rich or poor, must submit to the working of this power. It is up to us to yield our will to the will of Jesus and as we do this, God will immediately take possession and work in us to will and to do of His good pleasure. Our whole nature will then be brought under the control of the Spirit of Christ, and even our thoughts will be subject to Him. We cannot control our impulses, or our emotions as we may desire but we can submit our will to the One that made us. He can make an entire change in our life. By yielding up our will to Christ, our life will be hid with Christ in God and allied to the power from Heaven. We will have strength from God that will hold us fast to His strength and a new light even the light of living faith will be possible to us. But our will must cooperate with God's will. Our will is to be yielded to Him, that we may receive it again, purified and refined, and so linked in sympathy with the Divine that He can pour through us the tides of His love and power. The physiology books are not clear on what his part of the brain does. That could be because the doctors who study these areas do not relate to God's plan in building the body temple.

Limbic Lobe

This lobe is actually a subdivision of the cerebrum and is the ring of the cortex. It surrounds the central core of the cerebrum. *". . . there was a rainbow round about the throne, in sight like unto an emerald."*—Revelation 4:3.

"And God said, This is the token of the covenant which I make between me and you and every living creature that is with you, for perpetual generations: I do set my bow in the cloud, and it shall be for a token of a covenant between me and the earth."—Genesis 9:12,13. The bow represents Christ's love which encircles the earth and reaches unto the highest heavens. It connects men with God, and links Earth with Heaven. As we gaze upon the rainbow and upon the limbic lobe, we may be joyful in God and assured that He Himself is looking upon this token of His covenant. As He looks upon the rainbow He remembers the children of Earth to whom it was given. Their afflictions, perils, and trials are not hidden from Him. We may rejoice in hope for the bow of God's covenant is over us. He never will forget the children of His care.

The limbic lobe is composed of four parts: They are the cingulate gyrus, isthmus, parahippocampal gyrus, and uncus.

The limbic system includes the above named parts and the amygdala, the hippocampus, and parts of the thalamus and midbrain. The amygdala is a complex of nuclei located deep within the uncus. The hippocampus is in the floor of the lateral ventricle. The hippocampus and amygdala together look like a ram's horn or trumpet that was used to direct God's people. This trumpet was called a shofar.

Look in the Bible concordance under trumpet and see all the ways it was used, i.e. a call to worship, a call to war, a signal to march. The sons of Aaron followed the ark, each bearing trumpets. They were to receive directions from Moses which they translated to the people by speaking through the trumpets. These trumpets gave special sounds which the people understood and directed their movements accordingly.

A special signal was given by the trumpeters to call the people to attention; then all were to be attentive and obey the certain sound of the trumpets. There was no confusion of sound in the voices of the trumpets; therefore, there was no excuse for confusion in movements. The head officer of each company gave definite directions in regard to the movements they were required to make, and none who gave attention were left in ignorance of what they were to do.

If any failed to comply with the requirements given by the Lord to Moses and by Moses to the people, they were punished with death. It would be no excuse to plead that they knew not the nature of these requirements for they would only prove themselves willingly ignorant and would receive the just punishment for their transgression. If they did not know the will of God concerning them, it was their own fault. They had the same opportunities to obtain the knowledge imparted as others of the people had; therefore their sin of not knowing, or not understanding, was as great in the sight of God as if they had heard and then transgressed.

According to I Thessalonians 4:13-17, the first thing that dead Christians will hear is the voice of the archangel and the trump of God calling them forth from the grave when Jesus comes.

In the fall of Jericho, seven priests bearing seven trumpets led the march around the city. When the time came they blew a certain sound and the walls of Jericho fell. Likewise everyone who teaches the truth by precept and example will give the trumpet a certain sound.

People with lesions in the hippocampus have a loss of short term memory and lose their sense of time. They tend to become confused easily and forget the questions that were just asked and may reply to questions with irrelevant answers.

All expressions of the amygdala and hippocampus are associated with an interaction with other centers. The impact of the sensory systems from the sensory cortical areas upon these structures can trigger a variety of emotional and behavioral responses. They are actively involved in memory and emotions and the visceral and behavioral responses associated with them.

The limbic system and the hypothalamus control the emotional side of our lives. When we think gloomy, angry, unkind or selfish thoughts, the neurons in this area fire nervous impulses to the endocrine system which then produces pro-inflammatory corticoids. These substances were designed by God to equip the body to defend itself against disease. If we have an infection or toxin in the body, these substances generated help to fight it. But if these responses are called forth by negative thinking, their effect is harmful to the body, mind and soul. For example, if you are watching something on television that is exciting, the adrenal glands start to produce adrenaline. If this is not used for fight or flight as it was intended, it poisons the body. This is why watching television can be very harmful to bodily functions.

On the other hand, when our hearts are filled with love, tenderness, and sympathy, the limbic system causes us to produce A-C corticoids which tend to relax the body and cause a feeling of peace. Medical missionary work, properly motivated, was shown to be the greatest source of this production according to studies done.

"[Is] not this the fast that I have chosen? to loose the bands of wickedness, to undo the heavy burdens, and to let the oppressed go free, and that ye break every yoke? [Is it] not to deal thy bread to the hungry, and that thou bring the poor that are cast out to thy house? when thou seest the naked, that thou cover him; and that thou hide not thyself from thine own flesh? Then shall thy light break forth as the morning, and thine health shall spring forth speedily: and thy righteousness shall go before thee; the glory of the LORD shall be thy reward."—Isaiah 58:6-8.

Diencephalon

The diencephalon is the deep part of the brain that connects the midbrain with the cerebral hemispheres. It houses the third ventricle. It is composed of the thalamus, hypothalamus, epithalamus, and ventral thalamus. (The pituitary gland is connected to the hypothalamus.) This is the very center of the communication system, the command center as it were. This is a symbol of the **Most Holy Place** in the sanctuary that housed the Ark of the Covenant and the glory, the very presence of God. Into this chamber only the High Priest was allowed to go, and only once a year.

Thalamus (inner chamber)

The thalamus is located in the center of the cranial cavity directly beneath the cerebrum and above the hypothalamus. It is composed of two egg shaped masses of gray matter covered by a thin layer of white matter. It is the intermediate relay point and processing center for all sensory impulses that are coming from the spinal cord, brainstem, cerebellum, basal ganglia, and other sources. The only sense that is not interpreted here is the sense of smell.

There are four major areas of activity in the thalamus:

1. Sensory systems: fibers from the thalamic nuclei project into the sensory areas of the cortex, where input is decoded and translated into appropriate sensory reactions. For example, light is "seen" and sound is "heard." If we are under the control of God's Holy Spirit, He will interpret input for us.
2. Motor systems: The thalamus has a critical role in influencing the motor cortex. Some thalamic nuclei receive neural input from the cerebellum and basal ganglia and then project appropriate responses into the motor cortex. The motor pathways that regulate the skeletal muscles innervated by the cranial and spinal nerve originate in the motor cortex. *"Hold up my goings in Thy paths, that my footsteps slip not."*—Psalm 17:5. If we submit to God, He will guide all of our movements.
3. General neural background activity, such as sleep-wake cycles, and electrical brain waves: These cortical rhythms are generated and monitored by thalamic nuclei. When there are no more electrical brain waves the person is pronounced dead. This involves the electrical life force that comes from God, and is necessary for any life to exist.
4. Expressions of the cerebral cortex: Through its connections with the limbic system, it helps regulate many expressions of emotion and uniquely human behaviors. It is linked with the highest expressions of the nervous system, such as thought, creativity, interpretation and understanding of the written and spoken word and the identification of objects sensed by touch. Such accomplishments are possible because of the two way communication between the thalamus and the association area of the cortex.

Hypothalamus

This is a small region about the size of a lump of sugar, and only 1/300th of the brain's total volume. It has eight nuclei and extending from it is the pituitary gland. The hypothalamus is the highest integration center associated with the autonomic nervous system and is associated with eight functions:

1. It adjusts the activities of other regulatory centers; it modifies blood pressure, peristalsis and glandular secretions in the digestive system, secretion of sweat gland, control of the urinary bladder, and rate and force of the heartbeat.
2. It regulates the temperature of the body by monitoring the temperature of the blood.
3. It controls water and electrolyte balance. For example, when the concentration of water in the blood is reduced, causing the electrolyte concentration to rise, a hormone called anti-diuretic hormone (ADH) is produced by the hypothalamus and released into the blood stream by the posterior lobe of the pituitary gland. ADH stimulates the kidneys to absorb water from newly formed urine and return this water to the blood.
4. It controls sleep/wake patterns.
5. It regulates food intake by controlling the appestat, the area in the brain that is believed to regulate appetite and food intake.
6. It is associated with behavioral responses associated with emotion (pleasure, pain, anger, fear, love). The cerebral cortex activates the autonomic nervous system by way of the hypothalamus. In turn, the autonomic nervous system is responsible for changes in the heart rate and blood pressure, blushing, dryness of mouth, etc.

7. It helps with endocrine control, by producing oxytocin and an anti-diuretic hormone, and controls other hormones which are released by the posterior lobe of the pituitary gland.
8. It is associated with sexual responses. We need to be ever mindful of the will of God for our sexual activity.

There is not one area of activity in the body or any process that does not involve the hypothalamus and thalamus in some way, just as there is not one of our activities that God is not aware of and will regulate for us if we let Him.

Pituitary (or Hypophysis)

The pituitary gland is intimately connected to the hypothalamus. Even though it usually is classified as part of the endocrine system, we will discuss it here. It sits in the sella tursica of the spenoid bone (Ark of the Covenant bone) and is covered over with a tough membrane. The pituitary has two halves: (1) the adenohyophysis, which has an abundance of functional secretory cells which produce and secrete hormones and (2) the neurohypophysisis, which has a greater supply of large nerve endings and obtains its hormones from neurosecretory cells in the hypothalamus. These modified nerve cells project their axons down a stalk of nerve cells and blood vessels called the infundibular stalk, or infundibulum (funnel) into the pituitary gland. In this way a direct link exists between the nervous system and the endocrine system.

The job of the pituitary is to define the orders of the hypothalamus and send the messages on to the rest of the endocrine system (the regulators).

The job of the ten commandments is to define the character of God and His will in the terms that we understand. The two tables of stone on which the ten commandments were written were deposited in the Ark and had the mercy seat over top of them. The ten commandments are ten promises assured to us if we render obedience to the law governing the universe.

"If you love me, keep my commandments."—John 14:15e. Here is the sum and substance of the law of God. The lesson of the pituitary is the far-reaching influence of the law. The ten holy precepts spoken from Sinai's mount were the revelation of the character of God and made known to the world the fact that He had jurisdiction over the whole human heritage. That law of ten precepts of the greatest love is the transcript of the character of God: Loyalty, Fidelity, Reverence, Holiness, Respect for authority, Respect for life, Purity, Honesty, Truthfulness, and Contentment. The law is the definition of love.

Epithalamus

The epithalamus is in the back portion of the diencephalon, and located near the third ventricle. The pineal body is located here. It has a close connection with the thalamus and is called a neuroendocrine transducer which means that it converts a signal received through the nervous system into an endocrine signal which shifts concentrations of hormone secretion. Information about daily cycles of light and dark is detected by the eyes and conveyed via the optic nerve to the hypothalamus. From there, sympathetic nerves convey the signal to the superior cervical nerve ganglia in the upper part of the neck and then to the pineal gland.

This gland produces steady secretions of melatonin throughout the night; light inhibits the production of melatonin. This is why it should be dark during sleep. Melatonin has been isolated in the pineal gland and it definitely affects the sleep-wake cycle. Melatonin also affects depression, jet lag, shift work, insomnia, and Alzheimer's; and it can inhibit growth of tumors. It controls circadian rhythms, which are the timed cycles of the body, such as menstrual cycles, metabolism rate cycle, rest cycles, etc.

"To every thing there is a season, and a time to every purpose under heaven."—Ecclesiastes 3:1. God has set seven time cycles:

1. The day cycle was created in the first chapter of Genesis.
2. The weekly cycle was also created in Genesis 1, and is used to commemorate the Sabbath and to remember that God was the Creator.
3. The next timing was the month and year which is set by the movements of the Earth, the Moon and the Sun.
4. The sabbath year, or each seventh year, was established, in which the land was to rest. (Leviticus 25.)
5. The jubilee year was set every 49 years so that everything could be returned to its rightful owner.
6. Finally there was the 1000 year weekly cycle with 6000 of those years allotted to the earth as a probationary time, the seventh 1000 year period, the Sabbath millennium, to be spent with God in heaven.

The spiritual lesson of the pineal is the lesson of God's timing. It is an unpaired structure, which shows that there are not two sides to the timing of God. When He sets a time, that is what He means. Every person, family, city, nation, and world has a time to live and a time to die, a time to make a decision for God and when that probationary time is over, IT IS OVER!

Brainstem

The brainstem is the stalk of the brain, and looks like a stem holding up a flower. In *Grey's Anatomy,* the brain stem is likened to a trunk holding up a tree.

Messages are relayed between the spinal cord and the cerebrum through the brainstem. It narrows slightly as it leaves the skull and passes through the foramen magnum (the large hole in the center of the occipital bone) to merge with the spinal cord. There are long tracts of ascending and descending pathways between the spinal cord and parts of the cerebrum. A network of nerve cell bodies and fibers called the reticular formation is also located throughout the core of the entire brainstem. Ten cranial nerves emerge from the brain stem that send messages to the brain.

The brainstem is the last part of the nervous system in the head. It is the first thing as you see from the direction of the neck. In the sanctuary system, as the sinner came into the gate of the courtyard, the first thing seen was the altar of sacrifice. This is described in Exodus 27. It was a representation of the sacrifice that Christ made on the cross. As you study about the brainstem keep in mind the activity surrounding the altar of sacrifice.

This altar was a place where complete dedication took place. Christ was totally dedicated to the task of saving mankind. Jesus did not count Heaven as a place to be desired while we were lost. He left the heavenly courts for a life of reproach and insult and a death of shame. He who was rich in heaven's priceless treasure, became poor, that through His poverty we might be rich.

The altar of sacrifice was a type of the cross. The cross was associated with the power of Rome. It was the symbol of Tammuz the sun god, where human sacrifices were offered to him. Upon this instrument of torture the most cruel and humiliating form of death occurred. This is why we should never sing songs worshiping the cross. We must worship the One who died on the cross.

Without the death of Christ on the cross, man could have no connection with the Father. On Jesus Christ hangs our every hope of eternal life. In view of this relationship, the Christian may advance with the steps of a conqueror for Jesus paid the price that we deserved that we might receive the prize that He deserved.

When one exhibits love for souls for whom Christ died, it manifests as crucifixion of self. He who is a child of God would look upon himself as a link in the chain let down to save the world. We need to be one with Christ in His plan of mercy going forth with Him to seek and save the lost. The Christian is ever to realize that he has consecrated himself to God, and that in character he is to reveal Christ to the world. The self-sacrifice, the sympathy, the love, manifested in the life of Christ are to reappear in the life of the worker for God. *"Whosoever will save his life shall lose it; but whosoever shall lose his life for My sake, the same shall save it."*—Luke 9:24. Selfishness is death. No organ of the body could live should it confine its service to itself. The heart, failing to send its lifeblood to the hand and the head, would quickly lose its power. As our lifeblood, so is the love of Christ diffused through every part of His mystical body. We are members one of another, and the soul that refuses to impart will perish. Our privilege is to give ourselves wholly to Him who gave Himself for us. Then with the light of love that shines from His face on ours, we shall go forth to reflect this light to those in darkness.

What is the sacrifice that we are to bring to the altar*? "I beseech you therefore, brethren, by the mercies of God, that ye present your bodies a living sacrifice, holy, acceptable unto God, [which is] your reasonable service. And be not conformed to this world: but be ye transformed by the renewing of your mind, that ye may prove what [is] that good, and acceptable, and perfect, will of God."*—Romans 12: 1,2.

Reticular Formation (rf)

This network of nerve cell bodies and fibers plays a vital part in maintaining life. It runs through the entire length of the brainstem, with axons extending in both directions to communicate with the spinal cord and the diencephalon.

The reticular formation represents the myriad of heavenly beings that are at our disposal at any one moment. Angels from the courts above will attend the steps of those who come and go at God's command. (See Daniel 9:4-21) In the short time it took Daniel to pray his prayer an angel arrived from Heaven to answer. As swiftly as the messages go back and forth between the brain and the rest of the body, just that swiftly do the messages go back and forth between Heaven and Earth. Jesus would sooner send every angel out of Heaven to protect His people than to leave one soul that trusts in Him to be overcome by Satan. To us in the common walks of life, Heaven may be very near. *"Because thou hast kept the word of My patience, I also will keep thee."*—Revelation 3:10.

There are two specialized nuclear groups in the brainstem that are especially interesting:

1. The **raphenuclei**, which contain the neurotransmitter, serotonin; and
2. The **locus coeruleus**, which contain the norepinephrine.

They are different in that their axons project throughout the central nervous system and directly to the cerebral cortex and the cerebellum. They represent the two angels that are always in our presence, our guardian angel and our recording angel.

All the redeemed will understand the ministry of angels. When the saved arrive in Heaven, they shall be able to talk with their guardian angel. What a privilege, what a joy, what an awesome blessing it will be to converse with our leviathan protectorate and hear about the hour of our birth and the numerous times that we were saved from eminent danger. We will talk about the times that an Unseen Hand guided us in our upward walk. This beautiful celestial companion was there when we needed him.

The recording angel keeps an accurate record of our thoughts, words, and deeds. As the photographer records on film a true picture of the face, the recording angels writes upon the Books of Heaven an exact account of every deed affixing the character of the human being. We shall all be held accountable to God for these actions. The life He has given us is a sacred responsibility, and no moment of it is to be trifled with for we shall have to meet it again in the record of the judgment. The service given is there recorded if it be diligent or negligent. The temptations overcome or yielded to, the love bestowed or withheld, all is faithfully recorded by the angels. God judges every man according to his work. Not only does He judge but He sums up, day by day and hour by hour, our progress in well-doing as we become partakers of His divine nature by His grace.

Neurons within the "rf" are organized into several groups each having a specific and life-sustaining function. The functions are: breathing, heart rate, the diameter of blood vessels, level of awareness, and the brain's state of arousal or wakefulness. When stimulation comes through the brainstem, it produces a wakeup call to the cerebral cortex that in turn leads to the increased activity of the cortex. It helps the cerebellum coordinate selected motor units to produce smooth, coordinated contractions of skeletal muscles while maintaining muscle tone. The fibers conveying sensory input terminate in sensory nuclei. Fibers conveying motor output from the brainstem originate in motor nuclei. A lesion in the reticular formation may result in coma. A lack of self denial in our lives will put us in a spiritual coma.

The reticular formation is organized into three parts. They are:

1. The ascending sensory pathways (which are made up of three classes of neurons) from the spinal cord tracts and from the cerebellum. Sensory receptors convey information to the spinal cord and brain by input fibers, and this information is conveyed to processing centers (which contain 3 parts). Here the information is analyzed and acted upon in the different parts of the brain. Neurons of many sensory pathways cross over from one side of the cord or brainstem to the other side and because of this cross over, one side of the body communicates with the opposite side of the brain.
2. Descending motor (action-causing) pathways from the cerebral cortex and hypothalamus.
3. Ten cranial nerves that originate from the brainstem. The sensory and motor components of the cranial nerves are associated with the cranial nerve nuclei within the brainstem. (See section on cranial nerves).

Twelve Cranial Nerves

The first two cranial nerves are nerves of the cerebrum, the other ten are from the brainstem. They are concerned with the specialized senses of smell, taste, vision hearing, and balance, and with the general senses. They are also involved with the specialized motor activities of eye movement, chewing, swallowing, breathing, speaking and facial expressions. The tenth nerve (vagus) is an exception projecting fibers to organs in the abdomen and thorax.

1. OLFACTORY—Smell: This determines the atmosphere in the sanctuary. If the cribiform plate of the ethmoid bone is fractured, one loses the sense of smell. The olfactory is strictly a sensory nerve.
2. OPTIC—Sight: This nerve controls light (truth) entering. Lesions on this nerve causes loss of visual acuity. Matthew 13:11-16
3. OCULOMOTOR—Movement of eyes, eyelid, near vision and constriction of pupil: Lesions on this nerve causes double vision, pupil dilation, uncontrolled movement of eyeball, or loss of accommodation for near vision (self examination).
4. TROCHLEAR—Movement of eyes: In paralysis of this nerve, head is tilted to one side; diplopia (double vision) James 1:8; strabismus (squinting) can't stand light, John 1:19.
5. TRIGEMINAL—Facial movement: This nerve has three divisions and is concerned with the face area (spirit or character). (1) The opthalmic, sensory nerve which conveys general senses from the front of the scalp, forehead, upper eyelid, conjunctiva (inner membrane of the eyelid), cornea of the eyeball, and upper nasal cavity. (2) the maxillary, as sensory nerve, which conveys general sensations from the skin of the cheek, upper lip, upper teeth, mucosa of the nasal cavity, palate and parts of the pharynx. It covers the general area of the maxillary bone. (3) the mandibular is a mixed nerve which conveys general senses of the mouth (including the tongue but not sensations of taste), lower teeth, and skin around the lower jaw. The motor fibers innervate the chewing muscles. Injury of this nerve results in paralysis of muscles of mastication and loss of sensation of touch and temperature in structures supplied. (You can't tell if you are hot or cold in the spiritual area).
6. ABDUCENS—Movement of eye laterally. Controls lateral rectus muscle of the eyeball. With damage of this nerve the eyeball cannot move laterally beyond midpoint and eye is directed medially. (Can't see the broader vision or whole picture).
7. FACIAL—Secretion of saliva (the water of life) which mixes with food for digestion; tears (repentance); tongue receptors; the muscles of facial expression (face is the character or spirit); scalp movements. Injury to this nerve produces paralysis of the facial muscles, called Bell's Palsy, which is very painful and burns; other problems associated with this disorder are loss of taste; and eyes remain open, even in sleep.
8. VESTIBULOCOCHLEAR—Sound and balance. The vestibular branch conveys impulses associated with equilibrium. Injury to this nerve causes vertigo (a subjective feeling of rotation—going in circles) 2 Timothy 3:7. Ataxia nystagmus (involuntary rapid movements of the eyeball) is another problem associ-

ated with this nerve. The cochlear branch has control of the hearing. Injury to this nerve cause tinnitus (ringing in the ears) and deafness.

9. GLOSSOPHARYNGEAL—Motor: swallowing and secretion of saliva. Sensory: regulation of blood pressure and taste. The blood pressure is regulated by the carotid plexus located in the neck. Injury to this nerve cause pain during swallowing, reduced secretion of saliva, loss of sensation in the throat, loss of gag reflex and loss of taste on the posterior third of the tongue which is not noticeable except by testing.
10. VAGUS—Visceral muscle movement and swallowing: This nerve monitors oxygen and carbon monoxide concentrations in blood and senses blood pressure. This nerve is concerned with the three vital areas of maintaining life in the body (cardiovascular, respiratory, and digestive systems).
11. ACCESSORY—Joins vagus to larynx and controls neck muscles (attitude). This nerve has two divisions; the bulbar (swallowing) and the spinal (movement of head) If this nerve is damaged, you cannot turn the head or raise shoulders.
12. HYPOGLOSSAL—Motor to muscles of tongue, speech and swallowing. Injury to this nerve creates difficult in chewing, speaking, and swallowing. The tongue, when protruded, curls toward the affected side and the affected side becomes atrophied, (shrunken and deeply furrowed).

The lessons of the cranial nerves are the lessons of the vessels in the sanctuary. Besides the furniture in the sanctuary, there were vessels. (Exodus 25:38,39; 30: 26-28; 37:16; 40:9,10.) All the vessels were anointed.

When Hezekiah was trying to restore the true religion in Israel, everything had to be cleansed, including the vessels. (II Chronicles 29:18,19.) Notice also that some had been cast away in transgression. The vessels were also carried into captivity in Babylon. (Daniel 1:2.) Babylon means confusion. When Belshazzar drank wine out of these sacred vessels, it was the last blasphemous act he ever did (Daniel 5). There can be vessels of Baal in the Lord's Temple that must be removed and destroyed. (II Kings 23:4.) *"Nevertheless the foundation of God standeth sure, having this seal, The Lord knoweth them that are his. And, Let every one that nameth the name of Christ depart from iniquity. But in a great house there are not only vessels of gold and of silver, but also of wood and of earth; and some to honour, and some to dishonour. If a man therefore purge himself from these, he shall be a vessel unto honour, sanctified, and meet for the master's use, [and] prepared unto every good work."*—II Timothy 2:19-21.

In the scene in Revelation 4 and 5, you will see the 24 elders who are involved in the activity in Heaven. Revelation 5:9 says that they were redeemed from the Earth. When Jesus was resurrected, they were resurrected (Matthew 27:52,53) They went with Him when He returned to Heaven and are representatives of all the saved who will go to Heaven when He comes again. (I Corinthians 15:23.) There are 12 cranial nerves on each side of the body: 12 x 2 = 24. There are two representatives for each of the 12 tribes.

In Numbers 7:84-87 the head of each of the 12 tribes were to bring vessels. They brought 12 silver chargers, 12 silver bowls and 12 spoons of gold. The original Hebrew words for these vessels are enlightening. In Strong's Concordance the meanings are: <u>Charger</u>: A bowl, which comes from the root word to rend or cut out. *"the sacrifices of god are a broken spirit: a broken and a contrite heart, O God, thou wilt not despise."* Psalm 51:17. Bowl: A basin comes from a root word, to sprinkle. When you study "sprinkle" you find 2 words used in Hebrew; one means sprinkled; one means besprinkled, which means excessively. In Leviticus 14 the cleansing of the leper is described. He was besprinkled to be sure he was really clean. <u>Spoon</u>: A hollow hand or palm, and comes from the root word which means to bow down self. *"For thus saith the high and lofty One that inhabiteth eternity, whose name is Holy; I dwell in the high and holy place, with him also that is of a contrite and humble spirit, to revive the spirit of the humble, and to revive the heart of the contrite ones."*—Isaiah 57:15.

Spinal Cord

The spinal cord itself is about as big around as a pencil. The same three coverings that cover the brain cover it. The gray matter is composed of three pairs (3 x 2) of columns of neurons running up and down the length of the spinal cord from the upper cervical level to the sacral level. White matter is divided into three pairs (3 x 2) of columns funiculi (little ropes) of myelinated fibers that run the entire length of the cord. There are 12 columns altogether.

We read in Genesis 27 & 28 of a vision about the plan of redemption. The mystic ladder was referred to by Christ, *"Ye shall see heaven open and the angels of God ascending and descending upon the Son of man."*—John 1:51. Up to the time of man's rebellion, there had been free communion between God and man, but sin separated Earth from Heaven. Yet the world was not left in solitary hopelessness. Jesus is the appointed medium of communication. Had He not with His own life and merits bridged the gulf that sin had made, the ministering angels could have held no communion with fallen man. Christ connects man in his weakness and helplessness with the source of infinite power.

Spinal Nerves

There are 31 pairs of spinal nerves. Each one is divided in the vicinity of the cord into two: the ventral (front) ones convey motor information; the dorsal (back) ones convey sensory information.

The spinal nerves are named for the vertebrae they are associated with (cervical, thoracic, lumbar, sacral, or coccygeal) and are numbered accordingly. Most pass through a hole between the vertebrae and then are distributed to a specific segment of the body. The cervical spinal nerves take the number of the vertebra below each nerve. The rest take the number of the vertebra above. The first cervical nerve passes between the occipital bone and the first cervical vertebra and has practically no posterior root.

Sacral nerves are five in number on each side of the spinal column. The roots of the upper sacral nerves are the largest of all the spinal nerves while those of the lowest sacral and coccygeal nerves are the smallest.

Plexus

A network of nerves is called a "plexus." The plexus nerves accompany the arteries of the body. There are five major plexus in the autonomic nervous system, the number of grace and redemption. In every plexus, in all the varied systems that they affect, God's grace is at work if we will allow Him to control.

1. Cardiac Plexus –which has regulatory effect on heart.
2. Pulmonary Plexus –which has effect on bronchi. This controls the air passages. (3) Solar Plexus—the largest mass of nerve cells outside the CNS. It lies on the aorta behind the stomach. A sharp blow may cause unconsciousness by slowing the heart rate and reducing the blood supply to the brain. When people say that they have "butterflies in their stomach," this is the area from which that sensation comes. This is the feeling that is in "the pit of the stomach." From this plexus are derived nine more minor plexuses.
3. Hypogastric Plexus –connects the solar plexus to the pelvic and innervates organs and blood vessels of the pelvic region. This is our creative center.
4. Enteric Plexus receives both sympathetic and parasympathetic fibers, and it regulates peristalsis. This is the control that keeps nourishment moving so that it does not rot in the system.

There are five plexuses formed by the front (ventral) branches (rami) of spinal nerves. Note that the front branches of the T2 through the T12 vertebrae do not form plexuses. Instead each branch innervates a segment of the thoracic and abdominal walls.

1. Cervical plexus: This plexus is formed by the first four cervical nerves. It sets upon muscles of the scapula and is covered in by the sterno mastoid muscle. It is divided into 12 branches. It innervates skin and muscles of the neck and scalp and upper portions of the thorax and shoulder. It also controls motor innervation to the strap muscles attached to the hyoid bone and innervates the diaphragm. A fatal "broken neck" occurs when the spinal cord is severed above the fourth cervical nerve. The lesion severs the motor tracts from the brain. As a result, all muscles of respiration, including the diaphragm, and all thoracic and abdominal muscles are paralyzed. Death results because the victim cannot breathe. A blow to the back of the neck, if it is hard enough, can cause unconsciousness.

2. Brachial plexus: This plexus is formed by five nerves. It innervates the shoulder, arm, forearm, and hand.
3. Lumbar plexus: This plexus supplies the anterior and lateral abdominal wall, the external genitals, and the thigh.
4. Sacral plexus: This plexus innervates the thigh, leg, and foot muscles and controls our walking and the voluntary sphincters of the urethra and anus (which keeps us from being incontinent).
5. Coccygeal plexus: This plexus supply the skin in the coccyx region.

Reflex Action

In addition to linking the brain and most of the body, the spinal cord coordinates reflex action. A reflex is a predictable involuntary response to a stimulus such as quickly pulling your hand away from a hot stove. Spinal reflexes are carried out by neurons in the spinal cord alone, and do not immediately involve the brain. Reflex actions save time because the "message" being transmitted by the impulse does not have to travel from the stimulated receptor all the way to the brain. These reflexes remind us of our guardian angel who is constantly on guard to keep us from as much harm as possible.

In the lesson of the nervous system, we see the wonder of God's design as a marvel of communication that cannot be equaled. We have only scratched the surface. We look forward to the time when our classes will be continued in the life to come with Jesus Himself as our teacher.

Chapter Twenty-Five

THE ENDOCRINE SYSTEM

II Peter 1:4

"Whereby are given unto us exceeding great and precious promises: that by these ye might be partakers of the divine nature, having escaped the corruption that is in the world through lust."

The tribe of Asher represents the endocrine system. Review the traits of this tribe. The most prominent trait of this tribe is diplomacy which is defined by Webster as "the skill in handling affairs without arousing hostility." "Diplomat" comes from the root word meaning double. (Gr. *diplos*) This is a perfect description of the endocrine system. It has a feedback system that sends out hormones to start a reaction and still others to stop it. This system represents the promises of God which are only applicable if we conform to His conditions. Webster's definition of promise: "(1) a declaration that one will do or refrain from doing something specified; (2) a declaration that gives the person to whom it is made the right to expect or to claim the performance or forbearance of a specified act."

To every promise of God there are conditions. If we are willing to do His will, His strength is ours. Whatever gift He promises, is in the promise itself. *"the seed is the word of God."*—Luke 8:11. As surely as the oak is in the acorn, so surely is the gift of God in His promise. If we receive the promise we have the gift.

Most of the Israelites who left Egypt were not able to enter the promised land because they did not believe God's promises. *"Let us therefore fear, lest, a promise being left [us] of entering into his rest, any of you should seem to come short of it. For unto us was the gospel preached, as well as unto them: but the word preached did not profit them, not being mixed with faith in them that heard [it]."*—Hebrews 4:1,2.

Faith that enables us to receive God's gift is itself a gift of which some measure is imparted to every human being. It grows as it is exercised in appropriating the Word of God. In order to strengthen faith, we must often bring it in contact with the word. (This is why the blood with nourishment from the intestinal tract goes directly to the liver, the organ of faith).

There are nine endocrine glands in the endocrine system plus five structures that contain special endocrine tissues. Think of the significance here: nine, the number of judgment and finality, plus five, the number of grace, making a total of 14, spiritual perfection and completeness (seven) doubled! God's promises are so bountiful, who can comprehend it!

A hormone is a specialized chemical produced and secreted by an endocrine cell or tissue. They travel in the blood stream to all parts of the body with a specific message for specific target cells. Hormones are very effective in extremely small amounts and only a few molecules of a hormone may produce a dramatic response in a target cell. One hormone influences, depends on, and balances another in a controlling feedback system.

As we learn about the different structures, let us consider what types of promises are available to us.

Pituitary

The pituitary gland represents the Law of God and is located in the sphenoid bone, the representation of the ark of the covenant. *"For this [is] the covenant that I will make with the house of Israel after those days, saith the Lord; I will put my laws into their mind, and write them in their hearts: and I will be to them a God, and they shall be to me a people:"*—Hebrews 8:10. What a promise! To have the Law of God so imprinted in us that in obeying it, we will be doing what comes naturally. In every command and in every promise of the word of God is the POWER, the very life of God, by which the command may be fulfilled and the promise realized. So commands are promises as well!

Connected with the law are promises of blessings, for obedience and promises of curses for disobedience. Read Deuteronomy chapter 28.

The pituitary has three parts, two large sections, the Neurohypophysis and the Adenohypophysis, and a small one in the middle, the Pars Intermediate.

Pars Intermediate: They do not know the function of the pars intermediate, but it is possible that it may represent the law of Moses that was placed in the side of the ark. It may perform the function of defining the orders from the hypothalamus as the law of Moses laid out the fine details of the Law so that the people understood it.

Adenohypophysis: The adenohypophysis has three types of cells that synthesize seven separate hormones, that are all polypeptides (proteins). Five of these hormones go directly to target cells:

1. The thyroid stimulating hormone goes to the thyroid and stimulates the secretion of thyroxine which increases the oxygen consumption. These are promises of the help of God's Holy Spirit to come into harmony with God.
2. The adrenocorticotropic hormone stimulates the adrenal cortex to produce and secrete steroid hormones called glucocorticoids, which have five functions. Secretion of adrencorticotropic is controlled by the hypothalamus. Knowledge of God's Law gives us an understanding of His Holy Spirit which will then enable us to function as representatives of our Creator.
3. The luteinizing hormone goes to the ovaries and stimulates the monthly release of a mature egg. It also goes to the target cells in the male testes to stimulate the secretion of the male hormone, testosterone. The mechanism for the control of the luteinizing hormone depends on the hypothalamus.
4. The follicle stimulating hormone stimulates the cells of the testes that produce sperm and in the female follicle cells to secrete estrogen. This hormone is also controlled by the hypothalamus.
5. The melanocyte stimulating hormone is a precursor to an active hormone. It may be like the fifth ingredient of the incense that is never mentioned, salt, the sign of God's grace added to the creature. On the other hand, it may be utilized in generating eternal life in the Earth made new. The other two hormones, which affect all parts of the body instead of a target area are classified as somatropic.
6. These growth hormones are associated with growth and maintains the epiphyseal disks of the long bones. If these disks close, the body stops growing. The growth hormones stimulates the growth rate of the body cells by increasing the formation of RNA which speeds the rate of protein synthesis and decreases the breakdown of proteins. It also causes a shift from the use of carbohydrates for energy to the use of fats. The secretion of the growth hormone is controlled by two other hormones produced in the hypothalamus and transported to the adenohypophysis. If we stop growing in our knowledge of the will of God, we become spiritual dwarves. *"I will instruct thee and teach thee in the way which thou shalt go: I will guide thee with mine eye."* —Psalm 32:8.
7. The prolactin hormone has two functions in females, it stimulates the development of the duct system in the mammary glands during pregnancy, and it stimulates milk production after childbirth.

Neurohypophysis: The neurohypophysis does not manufacture any hormones but stores ADH (antidiuretic hormone) and oxytocin that is made in the hypothalamus. The neurohypophysis sends nerve impulses to the hypothalamus to stimulate the secretion of these hormones. ADH is composed of nine different amino acids and the main target organs are the kidneys. It increases the permeability of the kidney tubules to allow more water to be reabsorbed into the body. This shows control of the works by the law through the instruction of God's Holy Spirit. If ADH is deficient, it causes dehydration (a loss of water, truth) and as a constant spiritual thirst (ever learning and never coming to a knowledge of the truth). Oxytocin is a hormone that stimulates uterine contractions during childbirth. It also is a peptide composed of nine different amino acids. The spiritual side of this amino acid shows the promise of judgment and help to be reborn spiritually and nourished.

Pineal Gland

Head

This gland was discussed in the Nervous System. (Please review the lesson there). This gland represents the timing of God. Melatonin appears to be uniquely synthesized by the pineal gland. Melatonin is manufactured by the pineal gland during the nighttime hours and this hormone enables one to sleep at night. Serotonin is also manufactured during the night and enables the body to stay awake during the day. These two hormones are interdependent upon each other. Insufficient quantities of one will affect the production of the other.

The promises of prophecy's fulfillment right on time gives us the assurance that God is in charge. Like the stars in the vast circuit of their appointed path, God's purposes know no haste and no delay. *"When the fullness of the time was come, God sent forth His Son"*—Galatians 4:4. In Heaven's council, the hour for the coming of Christ had been determined. When the great clock of time pointed to that hour, Jesus was born in Bethlehem. Abraham's children were to sojourn 430 years beginning at the time when he was called out of Ur. They came out on the "self-same day." (Exodus 12:51.) Nebuchadnezzar was to eat grass of the field for seven years. At the end of the days that God had appointed, Nebuchadnezzar had learned his lesson. The books of Daniel and Revelation tell of a time period when the beast and its image will be in control, then Jesus will come. We can be sure that this will happen. Daniel 8 tells of a time prophecy that concerns the cleansing of the sanctuary, both the Heavenly and the bodily. The pineal gland assures us that these promises are all true.

Thyroid

The thyroid is a well-developed circulatory system through which amino acids, iodine, gland secretions and other substances are transported. It is composed of sacs which are filled with a gelatinous colloid in which thyroid hormones are stored. These sacs are make up of single layer of cells and follicular cells. There are two kinds of cells. They are: follicular, which synthesize and secrete the two thyroid hormones, and parafollicular cells, which synthesize and secrete calcitonin.

This important gland has many jobs. It has two lobes connected with a bridge (isthmus), which represents the two main pillars of God's government, Mercy and Justice, with our Bridge, Jesus, in between. *"Justice and judgment are the habitation of thy throne: mercy and truth shall go before thy face."*—Psalm 89:14.

Justice and mercy stood apart, in opposition to each other, separated by a wide gulf. Christ's object was to reconcile the prerogatives of justice and mercy, and let each stand separate in its dignity yet united. His mercy was not weakness, but a terrible power to punish sin because it is sin: yet a power to draw to it the love of humanity. Through Christ, justice is enabled to forgive without sacrificing one jot of its exalted holiness.

God's love has been expressed in His justice no less than in His mercy. Justice is the foundation of His throne and the fruit of His love. It had been Satan's purpose to divorce mercy from truth and justice. He sought to prove that the righteousness of God's law is an enemy to peace. But Christ shows that in God's plan they are indissolubly joined together; the one cannot exist with out the other. *"Mercy and truth are met together; righteousness and peace have kissed each other."*—Psalm 85:10.

There are eight functions of the thyroid gland:

1. It accelerates cellular reactions in most body cells.

2. It increases body metabolism (the rate at which cells use oxygen and organic molecules to produce energy and heat).
3. It causes the cardiovascular system to be more sensitive to sympathetic nervous activity, thus increasing the cardiac output and heart rate.
4. It affects the growth and stability of the skeletal system (principles) and the nervous system (communication with Heaven).
5. It stimulates cellular differentiation.
6. It stimulates protein synthesis.
7. It affects the growth rate.
8. It affects the water balance within the body.

Parathyroid

The parathyroid glands are tiny lentil-sized glands embedded in the posterior of the thyroid lobes. These are two glands in each lobe. Their removal causes death because they secrete parathormone, which regulates the concentrations of calcium and phosphate in the blood. Think how calcium affects everything in the body. Calcium affects muscles (judgment), bone (law), nerve tissue (control), heart rate, (desire, just to name a few of its functions). It represents love, and Romans 13:10 tells us that *"love is the fulfilling of the law."* An improper balance of calcium (love of the law) and phosphate (love of man) can cause faulty transmission of nerve impulses (messages to and from God), destruction of bone tissue (loss of God's principles), hampered bone growth (lack of growing in God's will), and muscle tetany (spasmodic judgment). If there is too much phosphorus, we become sentimental and excuse sin.

The wrong kind of love, the kind not balanced with mercy and justice, is called sentimentalism. Spiritualism has changed its form and is now assuming a Christian guise. While formerly it denounced Christ and the Bible, it now professes to accept both. But the Bible is interpreted in a manner that is pleasing to the unrenewed heart while its solemn and vital truths are made of no effect. Love is dwelt upon as the chief attribute of God, but it is degraded to a weak sentimentalism making little distinction between good and evil. God's justice, His denunciations of sin, the requirements of His holy law, are all kept out of sight. The people are taught that their own standards are to be their guide. Pleasing bewitching fables captivate their senses and lead men to reject the Bible as the foundation of their faith. Christ is as verily denied as before but Satan has so blinded the eyes of the people that deception is not discerned. They go on "self-esteem" and the idea that self is the main thing to be concerned with and taken care of. This eventually leads to creature worship instead of Creator worship. It goes back to the lie in the garden, *"ye shall be as gods."*

From the parathyroids, which represent the four bests around the throne, spiritually represents the promises of justice to be finally done. (Read Psalm 73.) David was expressing the thoughts of many of God's followers when he saw the prosperity of the wicked. He wondered why he was always in trouble and their lives seemed to run so smoothly. *"When I thought to know this, it was too painful for me; Until I went into the sanctuary of God; then understood I their end."*—Psalm 73:17. Then he was ashamed for doubting that justice would be served in the end.

"Dearly beloved, avenge not yourselves, but [rather] give place unto wrath: for it is written, Vengeance [is] mine; I will repay, saith the Lord. Therefore if thine enemy hunger, feed him; if he thirst, give him drink: for in so doing thou shalt heap coals of fire on his head."—Romans 12:19,20.

Read the description of the four beasts around the throne in Revelation 4:6-10. These are the same creatures that were seen in Ezekiel's vision by the river Chebar described by Ezekiel in chapter one. Read this chapter very carefully. Now read Ezekiel 10 and you will see that the same living creatures are there, standing at the east gate of the LORD's house (verse 19). The LORD, who had been standing at the threshold (verse 4), had now moved to a position above them. The eastern gate was the entrance that led to the altar of sacrifice (which is located in the brainstem). Now look at the picture of the thyroid and parathyroid on the previous page. Do you see the four (parathyroids), the wings, and the wheels? The thymus looks like more wings. Now read Ezekiel 9; this is the story of the last sealing of God's people and the destroying angels that follow. The same scene is being described in Revelation 6 when the judgments of God are announced by the four beasts.

Thymus

The thymus was discussed in the Lymphatic and Immune System study, please review it. This gland is well supplied with blood vessels but has only a few nerve fibers. It has an outer cortex and a medulla like the adrenal glands as well as clusters of cells called thymic corpuscles whose function is unknown. The thymus processes the T lymphocytes which are responsible for cellular immunity. The thymus produces thymic hormones that are responsible for developing and maintaining T cells. It represents the promises of victory in the battle against sin.

"But thanks [be] to God, which giveth us the victory through our Lord Jesus Christ."—I Corinthians 15:57. *"Call upon me, and I will answer thee, and shew thee great and mighty things, which thou knowest not."*—Jeremiah 33:3. *"When the enemy shall come in like a flood, the spirit of the LORD shall lift up a standard against him."*—Isaiah 59:19. *"He will swallow up death in victory; and the Lord GOD will wipe away tears from off all faces; and the rebuke of his people shall he take away from off all the earth: for the LORD hath spoken [it]. And it shall be said in that day, Lo, this [is] our God; we have waited for him, and he will save us: this [is] the LORD; we have waited for him, we will be glad and rejoice in his salvation."*—Isaiah 25:8, 9.

Read the story of Jacob wrestling with God in Genesis 32:24-30. He had mistrusted God's promise that he would have the birthright, and he sought to obtain it by deceit. That is what put him in the precarious position with his family.

There are three steps to gaining the victory over sin: humiliation, repentance, and self-surrender. We must be humble enough to admit that we have done the wrong thing. We must be sorry that we did it, and we must surrender to the Lord's way of doing things. After Jacob had done all three, his name was changed from Jacob (supplanter) to Israel (ruling with God). *"To him that overcometh will I grant to sit with me in my throne, even as I also overcame, and am set down with my Father in His throne."*—Revelation 3:21.

The power of God, combined with human effort, has wrought out a glorious victory for us. Shall we not appreciate this? All the riches of Heaven were given to us in Jesus. God would not have the evil forces say that He could do more than He has done. The worlds that He has created and the angels in Heaven could testify that He could do no more. God has resources of power of which we as yet know nothing, and from these, He will supply us in our time of need. But our effort is ever to combine with the divine. Our intellect, our perceptive powers, all the strength of our being, must be called into exercise. If we will work to overcome every defect in our characters, God will give us increased light and strength and help. *"There hath no temptation taken you but such as is common to man: but God [is] faithful, who will not suffer you to be tempted above that ye are able; but will with the temptation also make a way to escape, that ye may be able to bear [it]."*—I Corinthians 10:13. *"The Lord shall cause thine enemies that rise up against thee to be smitten before thy face: they shall come out against thee one way, and flee before thee seven ways."*—Deuteronomy 28:7.

Adrenal Glands

Adrenal means "upon the kidneys" in Latin, and this is exactly where they sit, one on each kidney. They are only about one to two inches long and they weigh only a fraction of an ounce each. Each gland is actually made up of two separate endocrine glands which produce different hormones and have different target organs: the medulla (L. marrow, meaning inside), which is reddish brown, and the outer cortex (L. bark, like the bark of a tree), which is yellow. These glands represent our trust in God's promises.

The cortex portion accounts for about 90% of the weight of the adrenal gland. It has three distinct zones. It secretes three classes of general steroid hormones. (1) Glucocorticoids (*gluco* refers to glucose and *corticoid* means it comes from the cortex). There are two types: cortisol, which does about 95% of the work and corticosterone. The cortisol must represent God's part of the work and the corticosterone represents man's part of the work.

These *glucocorticoids* have five jobs: (a) They affect the metabolism of all types of foods. (b) They act as anti inflammatory agents. (c) They affect growth rate. (d) They decrease the affects of physical or emotional stress. (e) They regulate the concentration of glucose from amino acids and fats. Which means that the cells do not need to remove glucose from the blood. So the blood sugar rises. The insulin from the pancreas keeps it from rising too high. Enzymes that help convert amino acids become plentiful in the liver in this process. It sounds complicated and it is complicated to keep our love for God balanced in all of the areas of our spiritual life. *"Who shall separate us from the love of Christ? [shall] tribulation, or distress, or persecution, or famine, or nakedness, or peril, or sword? As it is written, For thy sake we are killed all the day long; we are accounted as sheep for the slaughter. Nay, in all these things we are more than conquerors through him that loved us. For I am persuaded, that neither death, nor life, nor angels, nor principalities, nor powers, nor things present, nor things to come, nor height, nor depth, nor any other creature, shall be able to separate us from the love of God, which is in Christ Jesus our Lord."*—Romans 8:35-39.

Another effect of cortisol is to suppress allergic reactions and inflammatory responses. *"And we know that all things work together for good to them that love God, to them who are called according to His purpose."*—Romans 8:28.

Cortisol lowers the high temperature of fever. *"Remove thy stroke away from me: I am consumed by the blow of thine hand. When thou with rebukes dost correct man for iniquity, thou makest his beauty to consume away like a moth: surely every man is vanity."*—Psalm 39:10,11.

The secretion of cortisol is controlled by a negative feedback system between the cortex and the adenohypophysis and the hypothalamus.

The <u>mineralocorticoids</u> control the concentration of minerals: the main one being aldosterone, which controls the retention of sodium and the loss of potassium in urine and the reabsorption of sodium from sweat. Sodium spiritually represents service or works and potassium spiritually represents faith. It is important to keep these two balanced. The main target area of aldosterone is the kidneys, which represent works. It stimulates the kidneys, to reabsorb sodium. At the same time it stimulates the transport of potassium ions from the tubules into the urine. This keeps the acid/alkaline balance as well as a normal electrolyte balance in the body fluids.

The <u>gonadocorticoids</u> are the adrenal sex hormones. They are known as androgens, the male hormones, and estrogens, the female hormones. Disorders of the adrenals can cause masculine characteristics in females such as facial hair and deep voice and reduction in breast size. We need to trust that God knew best when He assigned the duties and positions and the distinction between male and female. We would not have so much divorce and abuse if we believed that God knows best.

The <u>adrenal medulla</u> secretes different hormones, and has a different tissue structure. It is derived from the same embryonic tissue as the ganglia of the sympathetic nervous system and has similar functions and effects. Granules in the cells in the medulla synthesize, store, and secrete two hormones, epinephrine and norepinephrine, which prepare the body for "fright, fight, or flight." This prepares the body to react quickly and strongly to emergencies. *"When thou passest through the waters, I [will be] with thee; and through the rivers, they shall not overflow thee: when thou walkest through the fire, thou shalt not be burned; neither shall the flame kindle upon thee. For I [am] the LORD thy God, the Holy One of Israel, thy Saviour: I gave Egypt [for] thy ransom, Ethiopia and Seba for thee. Since thou wast precious in my sight, thou hast been honourable, and I have loved thee: therefore will I give men for thee, and people for thy life. Fear not: for I [am] with thee: I will bring thy seed from the east, and gather thee from the west;"*—Isaiah 43:2-5. If we surrender our lives to His service, we can never be placed in a position for which God has not made provision. Whatever may be our situation, we have a Guide to direct our way. Whatever our sorrow, bereavement, or loneliness, we have a sympathizing Friend. If in our ignorance we make missteps, Christ does not leave us. *"He shall deliver the needy when he crieth; the poor also, and him that hath no helper."*—Psalm 72:12.

Abraham's life testifies to the fact that trust in God is a learning process. He got himself into some embarrassing situations by not trusting God, He even lied to the Pharaoh because he did not believe God would protect him. (Genesis 12.) He repeated the same mistake in Genesis 20. Ultimately He finally proved to the universe that he was able to be the father of the faithful as God had called him. He passed the test by offering Isaac, his son. (Genesis 22.)

Pancreas

The endocrine portion of the pancreas makes up only about 1% of the total weight of the gland and is called the pancreatic islets or isles of Langerhans. The other 99% acts as a digestive organ. This gland represents hope, which is one of three "greats" of I Corinthians 13—FAITH, HOPE, and CHARITY. It is absolutely necessary to all twelve systems. *"And now I stand and am judged for the hope of the promise made of God unto our fathers: Unto which promise our Twelve tribes, instantly serving God day and night, hope to come."*—Acts 26:6,7. Here we have 12 tribes serving day and night, just like the 12 systems.

Spiritually speaking, hope helps digest the word of God. *"They that fear thee will be glad when they see me; because I have hoped in thy word."*—Psalms 119:74.

The pancreatic islets are little islands of cells scattered throughout the gland. There are between 200,000 to 2,000,000 islets, and each contain four special groups of cells. They are: alpha, beta, delta, and F cells. Delta cells secrete somatostatin that inhibits the GH hormone of the hypothalamus. F cells secrete pancreatic polypeptide that is released into the bloodstream after a meal. Beta cells are the most common and are generally located near the center of the islet. They produce insulin. Alpha cells produce glucagon. Both of these hormones help regulate glucose metabolism. The pancreas works in close connection with the liver, (faith). When the concentration of blood glucose falls, glucagon stimulates the liver to convert glycogen into glucose to cause the blood sugar level to rise. Glucose is the fuel of the body and represents our love for God. This is what keeps us burning spiritually. Insulin's function is opposite to glucagon, and the two hormones work together to maintain normal blood sugar levels. Insulin's most important effect is to facilitate glucose transport across plasma membranes into the individual cells, or spiritual hope to each individual. Insulin causes the liver to convert glucose into glycogen so that it can be stored.

Ovaries and Testes

Review these glands in the reproductive system study. The promises of production and fruitfulness are seen in these glands. *"For I will pour water upon him that is thirsty, and floods upon the dry ground: I will pour my spirit upon thy seed, and my blessing upon thine offspring."*—Isaiah 44:3.

Endocrine Tissue

Endocrine tissue is located in the hypothalamus, heart, kidney, gastrointestinal tract and placenta. These are not a gland, but contain tissue that issues hormones. Each of these will be covered in other studies.

Study Questions

The Defense System

1. Give a short essay of your viewpoints concerning our ability to eradicate sin out of our lives.

The Lymphatic System

1. What are the four major functions of the lymphatic system?
2. Tiny capillaries called ____________________, which are most abundant near the innermost and outermost surfaces of the body, gather the lymph fluid for transport.
3. What are lacteals?
4. What three forces accomplish the actual movement of the lymph?
5. What is the function of the lymph nodes?
6. Medical doctors for many years have routinely removed the tonsils. What has been the result of this erroneous practice?
7. The tonsils are called the gatekeepers of the physical body. What are the gatekeepers of the spiritual body?
8. What is the function of the thymus?
9. What is the meaning of the phrase found in Matthew 22:14, *"Many are called but few are chosen."*?
10. If the spleen is removed from the body what could be the result?
11. What are Peyer's patches and what do they do?

The Immune System

1. What are the nonspecific defenses of the body and what do they accomplish?
2. What is an antigen?
3. Name the seven types of warriors in the immune reaction.
4. What are the five phases of battle against disease?
5. Why is it a good thing to be temperate? What is wrong with the statement *"And every man that striveth for the mastery is temperate in all things."*?
6. What happens when a cold virus enters the body?
7. In your opinion what is our best defense against the AIDS virus?
8. When the body starts attacking its own cells as if they were enemies, we have what is known as ______________________.
9. What are some autoimmune diseases?
10. How only can we gain the victory over sin?
11. Study chapters 8 & 9 of the book of Ezekiel and describe the sealing process.

The Nervous System

1. What are the spiritual lessons of the nervous system?
2. How many nerve cells are in the brain?
3. What are the three basic activities of the brain?
4. Give the three parts of the neuron?
5. Why is it important that we have two different kinds of messengers?
6. The __________ require _________ to drive the ions through the ____________.
7. Name some of the different agencies that God id using to save man.
8. What is the difference between the PNS and the CNS?
9. The PNS is divided into two motor systems. Name them and describe their function.
10. Describe the difference between the nuclei and ganglion.
11. What is the dura mater?

12. What function does the subarachnoid fluid perform?
13. The brain has four major divisions. What are they?
14. A lack of ___________ or _________ will damage brain tissue faster than any other tissue.
15. Give the functions of the brain's left and right hemispheres.
16. What are the three types of fibers in the white matter of the brain.
17. If man had not sinned would there be any need for grace? Explain your answer.
18. In all who will submit themselves to the Holy Spirit a new __________________________ is to implanted. Expand your answer.
19. Describe the ventral thalamus.
20. Why do you think we will be sealed in the frontal lobe?
21. The parietal lobe has three parts; name them. Do you think there is any connection between the word gyrus and the word gyroscope?
22. The Occipital lobe is involved in the understanding of the ___________ and _____________ word.
23. What is the first part of the brain to be differentiated in the embryo?
24. We cannot control our _____________ or our ____________ as we may desire but we can _____________ our will to _____________.
25. Describe the functions of the limbic lobe.
26. What is the diencephalon?
27. Give the four major areas of activity in the thalamus.
28. There is not one activity in the body or any process that does not involve the _______________ and ____________ in some way. In your own words what do these two areas have to do with the area of the brain that is called "God's throne room."
29. What is the job of the pituitary gland?
30. Describe what is meant by the term "neuroendocrine transducer."
31. What is the spiritual lesson of the pineal gland?
32. Messages are relayed between the spinal cord and the cerebrum through the ________________.
33. Write a short essay on the phrase, "the soul that refuses to impart will perish.
34. Neurons within the "rf" are organized into several groups each having a specific and life-sustaining function. What are these functions?
35. Name the twelve cranial nerves.
36. What are the spiritual lessons of the cranial nerves?
37. Name the five major plexus in the autonomic nervous system.
38. What is a reflex action and of what use are they?

✓The Endocrine System

1. The pituitary gland has three parts. Name them and give their function.
2. Discuss the functions of the thyroid gland.
3. The parathyroid gland secretes ___________________ which regulates the concentrations of ____________ and ___________.
4. The thymus gland spiritually represents what?
5. Give the functions of the glucocorticoids found in the adrenals.
6. The pancreatic islets contain four special groups of cells. What are they and what do they do?

Chapter Twenty-Six

THE SKELETAL SYSTEM

<u>**Acts 20:35**</u>

"I have showed you all things, how that so labouring ye ought to support the weak, and to remember the words of the Lord Jesus, how he said, It is more blessed to give than to receive."

The skeletal system is the support of the body and represents laws and principles. *"And thou shalt teach them ordinances and laws, and shalt show them the way wherein they must walk, and the work that they must do."*—Genesis 18:20. *"For when for the time ye ought to be teachers, ye have need that one teach you again which [be] the first principles of the oracles of God; . . ."*—Hebrews 5:12.

Vertebral Column

As you look at the spinal vertebrae sideways, you can definitely see the form of a serpent. Satan always takes the truths of God and counterfeits and confuses the meaning behind them. We know that the Devil was represented by a serpent all through the Bible, but so also was Jesus. In John 3:14 & 15 we read: *"And as Moses lifted up the serpent in the wilderness, even so must the Son of man be lifted up: That whosoever believeth in Him should not perish, but have eternal life."* He was referring to the incident in the desert when the remedy for snake bite was looking at the serpent on the pole. (Numbers 21.) We know that this serpent represented Christ, because there is never any real healing in Satan.

In Exodus 7: 10-12 the story of the two serpents is told again; Aaron's rod represented Christ and the magicians' rods represented Satan and his evil angels.

Also when Moses' rod became a serpent, it became his rod of authority. (Exodus 4:1-5) He was commanded to pick it up by the tail. As we take hold of Christ, He becomes our rod of authority as He was for His disciples. *"And Jesus came and spake unto them, saying, All power is given unto me in heaven and in earth. Go ye therefore, and teach all nations, baptizing them in the name of the Father, and of the Son, and of the Holy Ghost: Teaching them to observe all things whatsoever I have commanded you: and, lo, I am with you alway, [even] unto the end of the world. Amen."*—Matthew 28:18-20.

The whole body represents Jesus and His bride. The plan of salvation that Our Heavenly Father has laid out for us gives us the authority to carry on the work for the saving of mankind. We would expect to find the attributes of His character revealed especially in the spine, vertebra by vertebra, or the spiritual rod of authority, ending with the tail bone.

Cervical Vertebrae

Each of the seven cervical vertebrae has a pair of openings called *transverse foramina,* which are found only in these vertebrae. The neck is the area in which wisdom is revealed and is the connecting link between the head and the body, or Heaven and Earth. (See Chapter 20.) Seven means spiritual perfection or completeness. The neck contains the seven cervical vertebrae which represents the principle of wisdom. *"Wisdom hath builded her house, she hath hewn out her seven pillars."*—Proverbs 9:1. James 3:17 tells us what the seven principles of true wisdom are: *"But the wisdom that is from above is first pure, then peaceable, gentle, and easy to be entreated, full of mercy and good fruits, without partiality, and without hypocrisy."*

The reason that man does not understand spiritual things is that he does not have that connecting link between the Head and the body, WISDOM. *"But of him are ye in Christ Jesus, who of God is made unto us wisdom and righteousness, and sanctification and redemption."*—I Corinthians 1:30. *"Happy is the man that findeth wisdom, and the man that getteth understanding."* Proverbs 3:13. *"The fear of the LORD is the beginning of wisdom: a good understanding have all they that do His commandments: His praise endureth forever."*—Psalm 111:10.

The Atlas

The first vertebra is called the *atlas* and supports the head. It is ringlike and has no spinous process or body. Look at a picture of this vertebra in the "Anatomy Coloring Book;" this structure makes it possible to nod "yes" to the Great Physician when we want to be healed. The first character trait found in the book of James is pure. The definition of "pure" is very revealing: *"Having a uniform composition, not mixed; Free from adulterants or impurities, full of strength; free from dirt, defilement, or pollution; clean; free from foreign elements. Containing nothing inappropriate or extraneous; complete, thorough, perfect sinless; Chaste, virgin; free from discordant qualities."* This is a perfect description of Christ and should be a description of His followers when they reflect His image.

The Axis

The axis vertebra has a protrusion (Dens) that fits up inside of the Atlas. It forms a pivot point for the Atlas to move the skull in a "no" motion. It is the responsibility of the man in his home (High Priest) to say "NO" to any worldliness. When the transverse ligament of the atlas snaps, the dens crushes the lower part of the brainstem and the adjacent spinal cord and death occurs. The principle of wisdom found here is *Peaceable*, which means undisturbed or calm. Read Luke 8:23-25; Romans 12:18; John 14:27.

Vertebrae 3, 4, 5, and 6 are similar in shape and their facets are always positioned in the same way.

Gentle.—Considerate or kindly in disposition; tender and patient; read Isaiah 42:3; John 10:11-16; Hebrews 13:20; Matthew 18:12-14. The shepherd's gentle care of his flock is untiring. As the earthly shepherd knows his sheep, so the Divine Shepherd knows His flock that are scattered throughout the world. Jesus knows us individually, and is touched with the feeling of our infirmities. He knows us all by name. He knows the very house in which we live, and the name of each occupant. Every soul is as fully known to Jesus as if he were the only one for whom the Saviour died. The distress of everyone touches His heart.

Easy to be entreated.—Gracious; read Matthew 7:7-11; John 11:22; Jeremiah 33:3; Isaiah 65:24

Full of mercy and good fruits.—Merciful; read James 1:17; John 4:10; Isaiah 55:1-3

Without partiality.—Impartial; read Acts 10:34; Deuteronomy 10:17; II Chronicles 19:7

Without hypocrisy.—The seventh vertebra is different from the rest in that it has an exceptionally long un-forked spinous appendage with a tubercle at its tip that can be felt as a lump at the back portion of the bottom of the neck. A ligament attaches here, with the other end attached to the occipital protuberance. This vertebra represents without hypocrisy. (Sincere, Real.) *"Blessed is the man unto whom the LORD imputeth not iniquity, and in whose spirit there is no guile."*—Psalm 32:2. Guile comes from the word *decoy* or *bait.* Christ's severest rebukes were given to hypocrites. (Matthew 16:6.) As He would speak impressive truth, the Scribes and Pharisees, under the pretense of being interested, would assemble around the disciples and Christ and divert the minds of the disciples by asking questions to create controversy. They pretended that they wanted to know the truth. Christ wished His disciples to listen to the words He had to say, and He would not allow anything to attract and hold their attention. Therefore He warned them, *"Beware of the leaven of the Pharisees, which is hypocrisy."* They feigned a desire to get as close as possible to the inner circle. As Jesus presented truth in contrast to error, the Pharisees pretended to be desirous of understanding the truth; yet they were trying to lead His mind in other channels.

Hypocrisy is like leaven or yeast that is hidden in the flour and its presence not known until it produces its effect. By penetration, it soon pervades the whole mass. Hypocrisy works secretly and, if indulged, it will fill the mind with pride and vanity. The hypocrisy of the Pharisees was the product of self-seeking. The glorification of themselves was the object of their lives. When the Saviour gave this caution, it was to warn all that believe in Him to be on guard. Watch against imbibing this spirit and becoming like those who tried to ensnare Him. He was transparent. There was nothing hidden in His character; not one thread of selfishness or self-seeking.

Thoracic Vertebrae

There are 12 thoracic (or dorsal) vertebrae to which the 12 ribs are attached. The ribs spiritually represent the commandments, and you will see that the principles of the vertebrae provide the basis for the commands. (Read the description of the boards and their tenons of the sanctuary given in Exodus 26:15-30). There were 48 boards in the building and there are 24 ribs; 12 on each side, and 24 regular vertebrae. Compare the description of the corner boards and the way they are coupled together with the way the bottom ribs are coupled together. As you contemplate the sternum, think of the middle bar that reached from end to end in verse 28 of Exodus 26.

The ribs provide the cage that protects the vital organs of respiration and circulation, the lungs, and the heart. The commandments provide the protection of God against the wiles of the Devil. These are ten promises like the ribs that protect us IF we render obedience to the law governing the universe. *"IF ye love me, keep my commandments"* is the sum and substance of the law of God. The ten holy precepts spoken by Christ upon Sinai's mount were the revelation of the character of God, and made known to the world the fact that He had jurisdiction over the whole human heritage.

That law of ten precepts of the greatest love that can be presented to man is the voice of God from heaven speaking to the soul in promise, *"This do, and you will not come under the dominion and control of Satan."* There is not a negative in that law. It is DO, and live. When some insist that the commandments were nailed to the cross, which of these precepts or principles should we do away with? Which rib would you like to be without?

The Law of God existed before the creation of man, or else Adam could not have sinned. After the transgression of Adam the principles of the law were not changed, but were definitely arranged and expressed to meet man in his fallen condition. The principles of love to God and love to your neighbor were more explicitly stated to man after the fall and worded to meet the case of fallen intelligencies.

Loyal—"Thou shalt have no other gods before me."—Exodus 20:3. There is only ONE GOD. There are not two or three Gods; Our God reins supreme over all. He is loyal to us, and we should be loyal to Him. Jehovah is the eternal, self-existent, uncreated One. He is the Source and Sustainer of all, and He is *alone* entitled to supreme reverence and worship. Man is forbidden to give to any other object the first place in his affections or his service. Whatever we cherish that tends to lessen our love for God or to interfere with the service due Him, of that do we make a god.

Faithful—"Thou shalt not make unto thee any graven image, or any likeness of anything that is in heaven above, or that is in the earth beneath, or that is in the water under the earth: thou shalt not bow down thyself to tem, nor serve them."—Exodus 20:4,5. The second commandment forbids the worship of the true God by images or similitudes. Many heathen nations claimed that their images were mere figures or symbols by which the Deity was worshiped, but God has declared such worship to be sin. The attempt to represent the eternal One by material objects would lower man's conception of God. The mind, turned away from the infinite perfection of Jehovah, would be attracted to the creature rather than to the Creator. And as his conceptions of God were lowered, so would man be come degraded. *"I the Lord thy God am a jealous God."* The close and sacred relation of God to His people is represented under the figure of marriage. (Isaiah 54:5.) Idolatry is spiritual adultery; the displeasure of God against it is fitly called jealousy. *"Visiting the iniquity of the fathers upon the children unto the third and fourth generation of them that hate Me."* The consequences of disobedience to this commandment not only effects the one who disobeys, but it also effects future generations. We need to take into consideration the dire results of going contrary to God's commands. By inheritance and example the sons become partakers of the father's sins. Wrong tendencies, perverted appetites, and debase morals, as well as physical disease and degeneracy, are transmitted as a legacy from father to son to the third and fourth generation. This fearful truth should have a solemn power to restrain men from following the course of sin. *"Showing mercy unto thousands of them that love Me, and keep My commandments."* God offers to those that become converted - GRACE. Future generations do not have to suffer for evil done in the previous generation. God has given us the privilege to stop the results of sin upon our children.

Reverent— "Thou shalt not take the name of the Lord thy God in vain: for the Lord will not hold him guiltless that taketh His name in vain." This commandment not only prohibits false oaths and common swearing, but it forbids us to use the name of God and call our selves "Christians" in a light or careless manner. By the thoughtless mention of God in common conversation, by appeals to Him in trivial matters, and by frequent and thoughtless repetition of His name (even in our prayers), we dishonor Him. *"Holy and reverend is His name."*—Psalm 111:9. All should meditate upon His majesty and His purity and holiness, will impress the heart with a sense of His exalted character. His holy name should be uttered with reverence and solemnity.

Holy— "Remember the Sabbath day, to keep it holy. Six days shalt thou labor, and do all thy work: but the seventh day is the Sabbath of the Lord thy God: in it thou shalt not do any work, thou nor thy son, nor thy daughter, thy manservant, nor thy maidservant, nor thy cattle, nor thy stranger that is within thy gates: for in six days the Lord made heaven and earth, the sea, and all that in them is, and rested the seventh day: wherefore the Lord blessed the Sabbath day, and hallowed it." The Sabbath was founded at creation when the seven-day week was established. It is to be remembered and observed as the memorial of the Creator's work. Being the only commandment that points to God as the Maker of the Heavens and the Earth, it distinguishes the true God from all false gods. All who keep this day holy acknowledge the fact that God is *The Creator.* Thus, the Sabbath is the sign of man's allegiance to God. The fourth commandment is the only one of all the ten in which are found both the *Name* and the *Title* of the *Lawgiver*. It is the only one that shows by whose authority the law is given. Thus it contains the seal (goldenseal) of God, affixed to His law as evidence of its authenticity and binding force. God has given men six days to labor and tend to their secular business, and other than acts of necessity and mercy, unnecessary labor is to be avoided on the Sabbath Day. Everyone within your gates should unite to honor God in willing service to Him upon His holy day. The Sabbath is a sign of His *creative* and *redemptive* power to make us holy. *"If thou turn away thy foot from the Sabbath, [from] doing thy pleasure on my holy day; and call the Sabbath a delight, the holy of the LORD, honourable; and shalt honour him, not doing thine own ways, nor finding thine own pleasure, nor speaking [thine own] words: Then shalt thou delight thyself in the LORD; and I will cause thee to ride upon the high places of the earth, and feed thee with the heritage of Jacob thy father: for the mouth of the LORD hath spoken [it]."*—Isaiah 58:13-14.

Meek— "Honor thy father and thy mother: that thy days may be long upon the land which the LORD thy God giveth thee." Parents are entitled to a degree of love and respect which is due to no other person. God, Himself, who has placed upon them a responsibility for the souls committed to their charge, has ordained that during the earlier years of life, parents shall stand in the place of God to their children. And he who rejects the rightful authority of his parents is rejecting the authority of God. This commandment requires children not only to yield respect, submission, and obedience, to their parents, but also to give them love and tenderness. The children are also to lighten their cares, to guard their reputation, and to comfort them in old age. Jesus had been the Commander of Heaven, and His angels had delighted to fulfill His word. But when on earth, He was a willing servant and an obedient son. He lived in a peasant's home and faithfully and cheerfully acted His part in bearing the burdens of the household. Until He was thirty years old. He waited submissively for the appointed time to enter His work. He always performed the duties of a faithful Son. As Jesus was dying upon the cross, He had a thoughtful care for His mother and entrusted her care to John.

Respectful— "Thou shalt not kill." As the Creator, Jesus had a great respect for life and for the creatures He had made. His whole mission was to preserve life and to try to improve the quality of life for all with which He came in contact. *"I am come that they might have life, and that they might have it more abundantly."*—John 10:10. All acts of injustice tend to shorten life: This includes the spirit of hatred or revenge, or the indulgence of any passion that leads too injurious acts toward others. When one wishes to harm someone else (for "whosoever hateth his brother is a murderer,") it is a violation of God's will. A selfish neglect of caring for the needy or suffering; self indulgence or unnecessary deprivation; or excessive labor that tends to injure health—are, to a greater or lesser degree, violation of this command. "If any man defile the temple of God, him shall God destroy; for the temple of God is holy, which [temple] ye are."—I Corinthians 3:17

Clean— "Thou shalt not commit adultery." It is fitting that this command be the "seventh" commandment for it encompasses all the other nine. In the physical sense this commandment forbids not only acts of impurity, improper sexual relations, but sensual thoughts and desires or any practice that tends to excite them. Purity is demanded not only in the outward life but in the secret intents and emotions of the heart. Christ, who taught the far-reaching obligation of the law of God, declared that if lust was in the heart, that the evil thought or look was as truly sin as is the unlawful deed. In the *mental* sense this commandment it involves mixing the true and the false. We are not to go to Babylon to receive our education—this is mental adultery. When we become Christians and send our children to, or we ourselves knowingly study at, colleges and universities that teach falsehoods in education, are in danger of violation of the seventh commandment. In a *spiritual* sense this involves all that call themselves Christians, and yet hold on to perverted baggage and merchandise of this world. The loud cry of Revelation 18 commands that God's people *"come out from among them and be ye separate."* All who allow beliefs and practices that are not according to the Biblical teachings into their lives or homes are also in violation of the seventh commandment.

Honest—"Thou shalt not steal." Both public and private sins are included in this prohibition. It forbids wars of conquest. It condemns theft and robbery. It demands strict integrity in minutest details of the affairs of life. It forbids cheating in business, and requires the payment of just debts and wages. It declares that every attempt to advantage oneself by the ignorance, weakness, or misfortune of another, is registered as fraud in the books of Heaven. This commandment forbids the stealing of our time and talents that could and should be used in the service of God. It forbids the waste of our finances in worldly interests.

Truthful—"Thou shalt not bear false witness against thy neighbor." False speaking in any matter and every attempt or purpose to deceive our neighbor is here included. An intention to deceive is what constitutes falsehood. By a glance of the eye, a motion of the hand, an expression of the face, a falsehood may be told as effectually as by words. All intentional overstatements, every hint or insinuation calculated to convey an erroneous or exaggerated impression, even the statement of facts in such a manner as to mislead, is falsehood. This precept forbids every effort to injure our neighbor's reputation by misrepresentation or evil surmising or by slander or tale bearing. Even the intentional suppression of truth, by which injury may result to others, is a violation of this principle. Can you imagine how difficult it would be to be a lawyer and obey this command? Could you imagine Heaven with deceitful people living next door? Not only are the streets there transparent gold, but so are the inhabitants transparent and perfectly truthful.

Content—"Thou shalt not covet thy neighbor's house, thou shalt not covet thy neighbor's wife, nor his manservant, nor his maidservant, nor his ox, nor his ass, nor anything that is thy neighbor's." The tenth commandment strikes at the very root of all sins. It prohibits the selfish desire from which springs all sinful acts. He who is obedience to God's law refrains from indulging even a sinful desire. That which belongs to another and will not be guilty of a wrong act toward his fellow creatures. It was when Satan became discontented with his position in Heaven and wanted God's place that sin manifested (Isaiah 14:13,14.) Paul said he had learned to be content (Philippians 4:11), and that we should be content with such things as we have. (Hebrews 13:5.) Jesus was always content because He was always in the will of God.

The eleventh and twelfth ribs are floating ribs, not hooked to the sternum in the front. They represent the two great commandments that Jesus talked about in Matthew 22:37-39. All the commandments are summed up in these two.

Devoted—"Thou shalt love the Lord thy God with all thy heart, and with all thy soul, and with all thy mind." Jesus' life on Earth was one of total devotion to His Father. God takes men as they are, and educates them for His service if they will yield themselves to Him. The Spirit of God, received into the soul quickens all its faculties. Under the guidance of God's Holy Spirit, the mind that is devoted unreservedly to God develops harmoniously and is strengthened to comprehend and fulfill the requirements of God. The weak vacillating character becomes changed to one of strength and steadfastness. Continual devotion establishes so close a relation between Jesus and His disciples that the Christian becomes like his Master in character.

Unselfish—"Thou shalt love thy neighbor as thyself." Unselfishness, the principle of God's kingdom, is the principle that Satan hates. He denies its very existence. From the beginning of the great controversy, he has endeavored to prove God's principles of action to be selfish, and he deals selfishly with all that serve God. To disprove Satan's claim is the work of Christ and of all that bear His name. It was to illustrate by His own life of unselfishness that Jesus came in the form of humanity, and all that accept this principle are to be workers together with Him in demonstrating it as a living epistle.

The Sternum (breastbone)

The breastbone is made of three parts: Manubrium (handle), Body and Xiphoid process (sword-shaped). Spiritually speaking, three is the number of divine perfection.

The manubrium is united with the body of the sternum at the movable manubriosternal joint, which acts as a hinge to allow the sternum to move forward during inhalation. This joint forms a slight angle that can be felt through the skin as opposite the second rib. The xiphoid process is the smallest, thinnest, and most variable part of the sternum, and it has the diaphragm attached to it. It does not totally harden until about the age of forty, and it is very fragile. There is a white line of connective tissue that attaches to it, and the other end is attached to the pubic bone.

There are 9 pairs of muscles attached to the sternum. Nine is the number of finality or judgment. The sternum represents the righteousness of God: *"For He put on righteousness as a breastplate"*—Isaiah 59:17.

The 10 ribs (Ten Commandments) attach to this bone with cartilage. When we study connective tissue, which represents love, we will study cartilage. There are different kinds of cartilage and connective tissue just as there are different manifestations of love. The costal cartilage that is connected with the ribs is hyaline cartilage. This is a bluish transparent tissue that supports and allows movement in the rib cage. The love here represented is the love of the law, and is connected with righteousness. *"Great peace have they which love Thy law, and nothing shall offend them."* —Psalms 119:165. When we understand that all the law is a manifestation of the character of God, why would we not love it?

Lumbar Vertebrae

There are five lumbar vertebrae, and they are the largest and strongest, giving support for all the other vertebrae.

The Bible's firm supports for truth are the first 5 Books of the Bible. When Jesus taught, He always began in these books. *"And beginning at Moses and all the prophets, He expounded unto them in all the scriptures the things concerning Himself."*—Luke 24:27.

These vertebrae have short, blunt, four-sided spinous processes, which are adapted for attachment to the lower back muscles. The arrangement of the facets on the articular processes maximizes forward and backward bending. Lateral bending is limited and rotation is practically eliminated. *"Only be thou strong and very courageous, that thou mayest observe to do according to all the law, which Moses my servant commanded thee: turn not from it [to] the right hand or [to] the left, that thou mayest prosper whithersoever thou goest. This book of the law shall not depart out of thy mouth; but thou shalt meditate therein day and night, that thou mayest observe to do according to all that is written therein: for then thou shalt make thy way prosperous, and then thou shalt have good success."*—Joshua 1:7,8. We should take heed to these words of the Lord to Joshua, for they were spoken as the children of God were getting ready to go into the Promised Land. It is time now for His children to go into the real *Promised Land,* and so we need the instruction in these first five books of the Bible.

"Search the scriptures; for in them ye think ye have eternal life: and they are they that testify of me."—John 5:39. The first five books of the Bible give us a clear picture of the merits in the many roles of Christ.

Sacrum (Sacred)

The sacrum is five bones fused together. All the vertebrae rest upon it, and it joins the two bones of the hips. The sacrum cradles the private or sacred parts of the reproductive system. These parts are to be kept sacred for the marriage relationship.

Christ is the sure foundation to those who believe, just as the sacrum is the bone that gives support to the vertebral column, and provides strength and stability to the pelvis.

Like bone, stones in the Bible stand for laws and principles. God instructed His followers to use stones just as they were. The Hebrews were not to form them to their own liking (Deuteronomy 27:5.) Christ is the Rock of our salvation.

He is all the law and principle rolled up into One. *"And whosever shall fall on this stone shall be broken: but on whomsoever it shall fall, it will grind him to powder."*—Matthew 21:44.

When we come to Jesus and allow our old carnal nature to be broken up, He can put His spiritual nature within us and make us a new creature. He says to build our house upon the Rock and not upon the sand. When a person builds upon the world's policies instead of God's principles, they will be among those crying for the rocks and mountains to fall on them (Revelation 6:16.)

Coccyx

From the back view, the coccyx looks like a ram's head with horns. This bone represents the principle of humility. It is often referred to as the "tail bone," and is spoken of as something left over from evolution that is really not needed. This is the farthest thing from the truth.

There are one single and four pairs of muscles of major importance that attach to the coccyx. This is a total of nine muscles, the Biblical number of finality or judgment. The muscles anchored to the coccyx support pelvic viscera. They help maintain intra-abdominal pressure and acts as a sphincter to constrict the anus. Without this anchor, it would be impossible to reproduce or control waste elimination. *"Nay, much more those members of the body, which seem to be more feeble, are necessary: And those [members] of the body, which we think to be less honourable, upon these we bestow more abundant honour; and our uncomely [parts] have more abundant comeliness. For our comely [parts] have no need: but God hath tempered the body together, having given more abundant honour to that [part] which lacked:"*—I Corinthians 12:22.

The ordinance of feet washing, described in John 13, most forcibly illustrates the necessity of true humility. While the disciples were contending for the highest place in the promised kingdom, Christ girded Himself and performed the office of a servant by washing the feet of those who called Him Lord. He, the pure spotless Lamb of God, was presenting Himself as a sin offering. And as He ate the Passover with His disciples, He put an end to the sacrifices that for four thousand years had been offered. In the place of the Passover, He instituted a memorial service in the ceremony of feet washing. This sacramental supper was to be observed by His followers through all time. All Christians should ever repeat Christ's act, too see that true service calls for unselfish ministry.

Humility is an active principle growing out of a thorough consciousness of God's great love, and will always show itself by the way in which it works. Real humility fulfills god's purposes by depending upon His strength. Remember that the disciple is not above his Master, or the servant greater that his Lord. We need to cherish that humility and humbleness of mind that dwelt in Christ.

The Spine

The spine represents the ladder that Jacob saw in Genesis 28 and 29. Jacob thought to gain the right to the birthright through deception, but he found himself disappointed. He thought he had lost everything, his connection with God, his home, and all, and there he was a disappointed fugitive. But what did God do? He looked upon him in his hopeless condition. He saw his disappointment, and He saw there was material in him that would render back glory to God. No sooner does Jesus see Jacob's condition than He presents the mystic ladder, which represents Him. Here is a man, who had lost all connection with God. The God of Heaven looks upon Jacob and consents that Christ shall bridge the gulf which sin has made. We might have looked and said: I long for Heaven but how can I reach it? I see no way. That is what Jacob thought, and so God shows him the vision of the ladder. That ladder connects Earth with Heaven. A man can climb it, for the base rests upon the Earth and the topmost round reaches into Heaven. Through Jesus Christ,—whose long human arm encircles the race, and with His divine arm that grasps the throne of the Infinite—the gulf is bridged with His own body. This atom of a world that was separated from the continent of Heaven by sin became an island again reinstated because Christ bridged the gulf.

We cannot climb this ladder while we load ourselves down with earthly treasures. We wrong ourselves when we place our convenience and personal advantages before the things of God. If we gain a genuine experience in climbing, we shall learn that as we ascend we must leave every hindrance behind. Those who will mount higher must place their feet firmly on every round of the ladder (II Peter 1:3-11). There is continual advancement in every stage of the knowledge of Christ It is a knowledge of the perfection of the divine character manifested to us in Him that opens up to us communion with God. It is by appropriating the great and precious promises that we are to become partakers of the divine nature having escaped the corruption that is in the world through lust. Scarcely can the human mind comprehend what is the breadth and depth and height of the spiritual attainments that can be reached by becoming partakers of the divine nature. The human agent who daily yields obedience to God, and becomes a partaker of the divine nature, finds pleasure daily in keeping the Commandments of God.

Pectoral Girdle

The pectoral girdle is designed more for mobility than stability. It consists of the clavicle and scapula. It is held in place and surrounded by muscles and ligaments. These muscles and ligaments provide the stability of the shoulder. The shoulder is often the site of injuries such as shoulder dislocations.

These yoking bones give mobility to the arms, which represent service. *"Take my yoke upon you,"* Jesus says. The yoke is an instrument of service. Cattle are yoked for labor, and the yoke is essential that they may labor effectively. By this illustration Christ teaches us that we are called to service as long as life shall last. We are to take upon us His yoke that we may be coworkers with Him. This is how His rest is to be found. It is the love of self that brings unrest. By His yoke God confines us to His will which is high and noble and elevating. He desires that we shall patiently and wisely take up the duties of His service. The yoke of service Christ, Himself, has borne in humanity when He came down from heaven not to do His own will but the will of Him that sent Him (John 6:38).

Love for God, zeal for His glory, and love for fallen humanity brought Jesus to Earth to suffer and to die. This was the controlling power of His life. This principle He bids us adopt.

As you look at a picture of the clavicle, notice that it is attached to the sternum. Remember that the sternum represents God's righteousness. The clavicle represents the principle of *responsibility*. It acts as a strut to hold the shoulder joint and arm away from the rib cage, allowing the upper limb much freedom of movement. Because of its vulnerable position and relative thinness, the clavicle is broken more often that any other bone in the body. When it is broken, the whole shoulder is likely to collapse. When responsibility fails, whole governments collapse.

In this structure there are two bones, one on each side. There are always two sides of a coin. We can either refuse responsibility or take on responsibility that is not ours.

The scapula (shoulder blade) connects with the clavicle, and they form the shoulder. The muscles of the arm and chest attach to this bone, and it is free moving across the upper back and ribs like wings. Notice that these bones have many ridges, projections, and connections. This represents the help that we receive from Heaven in our service for the Lord. It comes from all directions, and is attached to the responsibility bone and the large bone of our service, the *humerus,* and to the muscles of our chest and arm.

There are rights that belong to every individual. We have an individuality and identity that is our own. No one can submerge his identity in that of any other. All must act for themselves according to the dictates of their own conscience. As regards our responsibility and influence, we are amenable to God by deriving our life from Him. This we do not obtain from humanity but from God only. We are His by creation and by redemption. Our very bodies are not our own. We can not treat our body temple as we please. We cannot cripple it by habits that lead to decay, making it impossible to render to God perfect service. Our lives and all our faculties belong to Him. He is caring for us every moment; He keeps the living machinery in action. If we were left to run it for one moment, we should die. We are absolutely dependent upon God.

Christ is our only hope. We may look to Him for He is our Saviour. We may take Him at His word, and make Him our dependence. He knows just the help we need, and we can safely put our trust in Him. If we depend on merely human wisdom to guide us, we shall find ourselves on the losing side. If we are yoked to Him, we may be sure of divine help. Let us carry our questions and difficulties to God humbling us before Him. There is a great work to be done, and while it is our privilege to counsel together, we must be sure in every matter to counsel with God for He will never mislead us. We are not to make flesh our arm.

On the other hand, we need to be sure that it is God who is laying a responsibility upon us. We should beware of taking upon ourselves responsibilities that God has not authorized us to bear. Men frequently have too high an estimate of their own character or abilities. They may feel competent to undertake the most important work when God sees that they are not prepared to perform aright the smallest and humblest duties.

The high priest of the sanctuary carried the names of the tribes of Israel graven in stone upon his shoulders. (Exodus 28:9-12.) This was a symbol of his responsibility for the people, and was a lesson about the responsibility that Jesus has for His people. He carries us on His shoulders, just like He carried the lost sheep in Luke 15:5.

Upper Limbs

There are three large bones in each arm. There were three tribes camped on each side of the sanctuary. There are thirty bones in each limb that is the number of order times divine perfection.

We need two arms and hands to do complete service. If we are missing one limb, we have a hard time doing everything. There were two parts of Christ's work as He ministered to the people. His work consisted of teaching and healing. Two is the number of division and support. This is how the two arms work separately and together. Christ first ministered to the sick, healing whole villages, and then He taught them the Truth. He was a Medical Missionary. Physical healing is bound up with the Gospel commission (spiritual healing). In the work of the Gospel, teaching and healing are NEVER to be separated. God often reaches hearts through the efforts made to relieve physical suffering.

The *humerus bone* of the teaching arm represents the principles and methods of Christ's teaching. In the Saviour's parable, teaching is an indication of what constitutes the true "higher education." Divine wisdom and infinite grace were made plain by the things of God's creation. Through nature and the experiences of life, men were taught of God. *"The invisible things of Him since the creation of the world"* were *"perceived through the things that are made, even His everlasting power and divinity."*—Romans 1:20.

Christ might have opened to men the deepest truths of science. He might have unlocked the mysteries that have required many centuries of toil and study to penetrate. He might have made suggestions in scientific lines that would have afforded food for thought and stimulus for invention to the close of time. But He did not do this. He said nothing to gratify curiosity or to satisfy man's ambition by opening doors to worldly greatness. In all His teaching, Christ brought the mind of man in contact with the Infinite Mind. He did not direct the people to study man's theories about God, His Word, or His works. He taught them to behold Him as manifested in His works, in His Word, and by His providences. Christ did not deal in abstract theories, but in that which is essential to the development of character and enlarge man's capacity for knowing God and increase his efficiency to do good. He spoke to men of those truths that relate to the conduct of life and that take hold upon eternity.

There are two bones in the forearm, the *ulna* and the *radius*. Both are connected to the same bone but having different functions. A spiritual application for this is that there are two branches of teaching: the ministry and the laymen. Both uses the same principles but bearing different responsibilities.

See how both the ulna and radius are connected to the elbow and the wrist and how these connections provide the mobility for the hand. Both bones are needed to work together or the limb would not be movable enough to be of much use. The ordained ministers alone are not equal to the task of warning the world. God is calling upon the laymen to help to finish the Gospel. *"Go ye therefore, and teach all nations, baptizing them in the name of the Father, and of the Son, and of the Holy Ghost: Teaching them to observe all things whatsoever I have commanded you: and, lo, I am with you alway, [even] unto the end of the world. Amen."*—Matthew 28:19-20. The Saviour's commission to go includes all believers in Christ to the end of time. It is a fatal mistake to suppose that the work of saving should depends alone on the ordained minister. All to whom the heavenly Inspiration has converted are put in trust of the Gospel. All who receive the life of Christ are ordained to work for the salvation of their fellow men.

There are eight carpal bones in the wrist. Let's look at another spiritual application. God gave us eight ministries in the teaching branch of service: *"And God hath set some in the church, first apostles, secondarily prophets, thirdly teachers, after that miracles, then gifts of healings, helps, governments, diversities of tongues."*—I Corinthians 12:28.

These gifts make the work complete and effective. All are joined together by love but all distinctly doing their own job. Eight is the umber of superabundance and regeneration, which is the result of the Gospel.

There are five metacarpal bones. The spiritual application of Five is the number of GRACE. These bones represent the steps of redemption in the teaching of the Gospel. The principles of the everlasting Gospel from Genesis to Revelation are the same:

1. Repentance - *"Therefore I will judge you, O house of Israel, every one according to his ways, saith the Lord GOD. Repent, and turn [yourselves] from all your transgressions; so iniquity shall not be your ruin. Cast away from you all your transgressions, whereby ye have transgressed; and make you a new heart and a new spirit:"*—Ezekiel 18:30.
2. Confession - *"He that covereth his sins shall not prosper: but whosoever confesseth and forsaketh them shall have mercy."*—Proverbs 28:13.
3. Conversion - *"Jesus answered and said unto him, verily, verily, I say unto thee, Except a man be born again, he cannot see the kingdom of God."*—John 3:3.
4. Sanctification - *"Sanctify them through they truth: thy word is truth." "And for their sakes I sanctify myself, that they also might be sanctified through the truth."*—John 18:17,19.
5. Glorification - "But we all, with open face beholding as in a glass the glory of the Lord, are changed into the same image from glory to glory, even as by the Spirit of the Lord."—II Corinthians 3:18.

The palm of the hand is divided into eight *carpals.* We have two healing hands. Eight plus two equals ten, the number of completeness. These represent the simple remedies of nature that God has provided:

1. *Nutrition* is of great importance to health. God's ideal diet for man is given in Genesis 1:29: grains, nuts and fruit. After man sinned, vegetables and herbs were added because of their high mineral content, so that the broken body could be mended (Genesis 3:18). After flesh food was added, in less than only ten generations, man's life span decreased from nine centuries to less than two. One of the causes for failure of the Israelites to enter the Promised Land was their insistence for flesh food—perversion of taste. (Numbers 11.)
2. *Exercise* is a vital factor of good health. You lose what you don't use. In order to keep all the body fluids flowing, it is necessary to exercise. It is during movement that the circulation is at its best. A sedentary lifestyle is detrimental to all systems of the body just as a lack of spiritual exercise ruins the Christian's experience.
3. Water is necessary inside and outside the body. Every cell is bathed in fluid. The body is 70% water, and the supply needs to be renewed daily by drinking sufficient quantities of pure water. A daily bath helps to cleanse the skin, improves digestion and circulation, and soothes the nerves. Simple hydrotherapy treatments have been used for centuries to restore health.
4. *Sunshine* was given by God for the benefit of plants and animals. It causes the body to manufacture vitamin D, and this lowers blood pressure. Sunshine is one of the blessings that God has given us to heal the body of cancer and many other diseases. Jesus is the *"Sun of Righteousness with healing in his wings"* mentioned in Malachi 4:2.
5. *Trust* in God is absolutely necessary for good health. God must be seen as the Father who wants only the best for His children. Then it is easy to follow His commandments and statues that were given for our health. It brings peace to the soul.
6. *Pure air* is the most immediate need of the body and, if it is not available death occurs in minutes. Most people are shallow breathers and do not take in full respiration's of air when they breathe. This deprives the cells of oxygen and they cannot burn fuel efficiently. Everyone needs to sit and stand properly so that the lungs can work most efficiently. The word for "breath" in the Bible is usually translated "spirit." If we do not partake largely of Christ's spirit, we become weak.

7. *Rest* in the proper quantity and at the proper time is important to maintain perfect health. Burning the candle at both ends wears out the nervous system. Jesus said if we come to Him, He will give us rest.
8. *Temperance* means abstaining from all harmful things, and using those things that are *good* in moderation. It is one of the most important lessons to learn if health is to be restored. In this day of excess, temperance is very seldom mentioned. This is a fruit of the Spirit (Galatians 5:22), and if we have Jesus, we will have His Spirit.
9. *Benevolence* toward our fellow men and to God is essential for our well being. A giving heart is not selfish. Isaiah 58 tells us that this is the fast that God has chosen.
10. *An Attitude of Gratitude* should be shown by All of God's children for the wonderful blessings that God has given. This will bring peace to the soul and promote healing within the body as nothing else can.

NOTE: The last two health remedies or health laws are added to the eight laws of health to balance them with the laws of the Decalogue.

There are five metacarpals in the hand. Spiritually these represent the steps in restoration of health:

1. Recognize the problem.
2. Be willing to change.
3. Revise your health plan.
4. Progress step by step.
5. Restoration of health.

There are fourteen *phalanges* in the palm of the hand. Spiritually speaking: The twelve bones of the fingers represent the twelve tribes working together. The two bones of the thumb represent the help of the angels and God's Holy Spirit, which gives us dexterity and efficiency. The joints spiritually represent the closeness with which we must work with each other in order to get the work done for Christ.

Lower Limbs

The *pelvic girdle* yokes the lower limbs together. Spiritually speaking this represents the lower throne room of the body or the creative throne room. The body reflects the head (the northern part), the dividing point being the neck, making this area the "north." Lucifer said that he would sit in the sides of the north indicating that he wanted the throne of God for himself. (Isaiah 14:12-14.) Satan has, especially in this day and age, accomplished his goal of sitting in this creative throne room. Satan is using this as the center of everything in the entertainment world. He also uses television, music, advertisements, computers, internet, electronics, books, etc, and thereby gained control of the masses. God invented sex and He made it beautiful and holy. Satan has turned sex into the basest form of self-indulgence.

Notice that the pelvic girdle forms an arch around the sacrum like wings. These bones provide the stability and foundation that affords walking and standing. Less freedom of movement is displayed here than in the upper yoking bones. The feet and legs function best when walking straight ahead and not sideways or backwards. Similarly, it is the same with the way in Christ. *"I am the way."*—John 14:6.

The ball and socket joint cause the leg to swing so that it falls in a direct line under the spine when walking, thereby making for strength and stability. If we stay in line with the merits of Christ, and claim them for our own, we will walk straight.

The *femur* spiritually represents the way of Christ. It is also called the thighbone. The Bible indicates that the thigh was used to swear by an oath thereby indicating trustworthiness.

When Jacob was wrestling with Jesus (Genesis 32:24-32), he was crippled in the hollow of the thigh. Biblically, this area of the body was used as a sign of authority as submission was recognizing a higher authority, when one placed the hand under their thigh. God manifested His authority in Jacob's life and Jacob, The Deceiver, became Israel, The Overcomer. When we submit to God's authority, we too will become overcomers. The Lord is willing to show us the way to walk in if we are willing to listen to Him. But when we chose to walk in our own way, trouble surely follows. *"There is a way that seemeth right unto a man, but the end thereof are the ways of death."*—Proverbs 14:12.

Many are losing their way in consequence of thinking that they must climb to Heaven. These people believe that they must do something to merit the favor of God. They seek to make themselves better by their own unaided efforts. This they can never accomplish! Christ has made the way by dying as our sacrifice, by living as our example, and by becoming our Great High Priest. He declares, *"I am the way, the truth, and the life"*—John 14:6. If by any effort of our own we could advance one step up the ladder, the words of Christ would not be true. But when we accept Christ, good works will appear as fruitful evidence that we are in the way of life, that Christ is our way, and that we are treading the true path that leads to Heaven. *"Ye shall observe to do therefore as the LORD your God hath commanded you: ye shall not turn aside to the right hand or to the left. Ye shall walk in all the ways which the LORD your God hath commanded you, that ye may live, and [that it may be] well with you, and [that] ye may prolong [your] days in the land which ye shall possess."*—Deuteronomy 5: 32, 33.

There are two bones in the lower leg called the *tibia* and *fibula*. The only way to stay in the way of Christ is to <u>trust</u> and <u>obey</u>. *"Trust in the LORD and do good; so shalt thou dwell in the land, and verily thou shalt be fed."*—Psalm 37:3. These two bones work together, and if we try to do without one or the other, we cannot walk. The tibia and fibula are joined to the femur by a joint that allows the leg to bend.

The *patella* or kneecap helps to take the stress as the leg bends. The patella represents the principle of submission. If we are submitted to the Lord, our walk will be easy. When an injury occurs in this area, the walk is stiff-legged and very painful.

Someday everyone will bow before the Lord. Some will bow gladly, some too late, but everyone will finally acknowledge the goodness and mercy of God in the plan of salvation. *"That at the name of Jesus every knee should bow, of [things] in heaven, and [things] in earth, and [things] under the earth; And [that] every tongue should confess that Jesus Christ [is] Lord, to the glory of God the Father."*—Philippians 2: 10, 11.

The *tarsal bones* in the foot represent the perfect completeness in the Way of Christ. They join with the trust (faith) principle and the obey principle to form the basis for our walk. *"And beside this, giving all diligence, add to your faith virtue; and to virtue knowledge; And to knowledge temperance; and to temperance patience; and to patience godliness; And to godliness brotherly kindness; and to brotherly kindness charity. For if these things be in you, and abound, they make [you that ye shall] neither [be] barren nor unfruitful in the knowledge of our Lord Jesus Christ. But he that lacketh these things is blind, and cannot see afar off, and hath forgotten that he was purged from his old sins. Wherefore the rather, brethren, give diligence to make your calling and election sure: for if ye do these things, ye shall never fall."*—II Peter 1:5-10. Here is the secret of keeping your feet under you and never falling down. *"Ponder the path of thy feet and let thy ways be established. Turn not to the right hand nor to the left: remove thy foot from evil."*—Proverbs 4:26, 27.

The five metatarsal bones are in the soles of our feet, and remember we have already recognized that the number five means grace and redemption. *"How beautiful upon the mountains are the feet of him that bringeth good tidings, that publisheth peace; that bringeth good tidings of good, that publisheth salvation; that saith unto Zion, Thy God reigneth!"*—Isaiah 52:7. There are five divisions of this message of grace connected with the feet. The meanings of the phrases according to Strong's concordance are:

1. Bringeth good tidings—"to be fresh, rosy, full, cheerful, announce glad news, tell good tidings."
2. Publisheth peace—(shama) "to hear intelligently (often with implication of attention, obedience, etc.; cause to tell etc.) attentively, call together carefully;" (shalom) "safe; well, happy, friendly; also welfare; health; prosperity."
3. Bringeth good tidings—"the news gets better; to be in favor; loving; precious prosperity; wealth; be well favored"

4. Publisheth salvation—"salvation means deliverance, aid, victory prosperity, health, help, save, salvation, welfare."
5. Thy God reigneth—"means to ascend the throne, to induct into royalty, set up a king, begin to reign, rule."

To understand why the above passage is good news, the verses just before Isaiah 52: 7 need to be studied. *"Now therefore, what have I here, saith the Lord, that the people is taken away for nought? They that rule over them make them to howl, saith the Lord; and my name continually every day is blasphemed. Therefore my people shall know my name: therefore they shall know in that day that I am HE that doth speak: behold, it is I."*—Isaiah 52:5,6.

"But we see Jesus, who was made a little lower than the angels for the suffering of death, crowned with glory and honour; that he by the grace of God should taste death for every man. For it became him, for whom [are] all things, and by whom [are] all things, in bringing many sons unto glory, to make the captain of their salvation perfect through sufferings."—Hebrews 2:9,10.

Now we can see what the good news, the Gospel, is all about Jesus became a human and joined humanity to divinity, and in this form, He reigns and invites His fellow sanctified human beings to join Him on the throne. Wonder, O Heavens; be astonished, O Earth!

All this is the message of the redemption represented in the sole of the foot. In the commission to His disciples given in the last verses of Matthew, Christ not only outlined our work but gave us our message. Teach the people, He said, *"TO OBSERVE ALL THINGS WHATSOEVER I HAVE COMMANDED YOU."* We are to teach what Christ taught. That which He had spoken, not only in person but through all the prophets and teachers of the Old Testament, is here included. Human teaching is shut out. There is no place for tradition, for man's theories and conclusions, or for church legislation. No laws ordained by ecclesiastical authority are included in the commission. No church manual was to be written. Christ ordains none of these. *"The law and the prophets,"* with the record of His own words and deeds are the treasure committed to His disciples to be given to the world. Christ's name is their watchword, their badge of distinction, their bond of union, the authority for their course of action, and the source of their success. Nothing that does not bear His superscription is to be recognized in His kingdom.

In the phalanges of the foot we again have the twelve tribes of Israel represented. Showing that all, no matter what their temperament or talent is have a part in keeping the body balanced and spreading the good news. Of course in the great toe, as in the thumb, we are promised the help of God's Holy Spirit and the help of His angels.

Jesus told us that the way we walk here on this earth is narrow, and most of the time it is strewn with stumbling blocks and pitfalls that we must be watching for constantly. Finally, when we reach the promised land, there will be no rugged paths to wound our feet. We will only have the way of holiness to walk in.

"And an highway shall be there, and a way, and it shall be called The way of holiness; the unclean shall not pass over it; but it [shall be] for those: the wayfaring men, though fools, shall not err [therein]. No lion shall be there, nor [any] ravenous beast shall go up thereon, it shall not be found there; but the redeemed shall walk [there]: And the ransomed of the LORD shall return, and come to Zion with songs and everlasting joy upon their heads: they shall obtain joy and gladness, and sorrow and sighing shall flee away."—Isaiah 35: 8-10.

See how wonderfully the principles all fit together to restore in us the image of God. May they come together as in Ezekiel 37 and may we become a mighty army to vindicate the character and principles of our LORD.

The things Jesus suffered for us become even worse when you know the significance of this afflicted body. They whipped His back, thereby trying to destroy His merits. They put a crown of thorns on Him, thereby ridiculing His right to be their KING. They nailed His hands and feet to show the rejection of His grace. They pulled His hair out of His cheeks thereby ridiculing His experience in the truth. He suffered what we deserve, that we might receive the reward that He deserves. Let us not disappoint Him. Let us receive His gift; His free gift that was given to us by our Father at such an awesome price.

Chapter Twenty-Seven

THE CARDIOVASCULAR SYSTEM

Jeremiah 17: 9, 10

"The heart [is] deceitful above all [things], and desperately wicked: who can know it? I the LORD search the heart, [I] try the reins, even to give every man according to his ways, [and] according to the fruit of his doings."

The mechanism of the human body cannot be fully understood. It presents mysteries that baffle the most intelligent. It is not as the result of a mechanism, which, once set in motion continues its work. Consider now that the pulse beats and breath follows breath. The physical organism of man is under the supervision of God. It is not like a clock, which is set in operation, and must go by perpetual motion. The heartbeats, pulse succeeds pulse, breath follows breath, and the entire being system and movement is under the supervision of God. *"Ye are God's husbandry, ye are God's building."*—I Corinthians 3:9. In God we live and move and have our being. Each heartbeat, each breath, is the inspiration of Him who breathed into the nostrils of Adam the breath of life, the inspiration of the ever-present God, the Great I AM—LOVE personified!

Every system of the body tells the story of the redemption of man. Some *parts* of the plan show more clearly in one system than in another. The Bible uses the word for heart to describe the innermost feelings and desires because this organ is located in an innermost protected place, surrounded by ribs, spiritually called the commandments of God. The breastbone, spiritually representing the righteousness of God covers them. As you see the role of the heart in the body, then the reason for this action becomes clear. Here is a symbolic representation of the church, both in its collective sense and in an individual member sense. Remember that "God has a church. It is not the great cathedral, neither is it the national establishment, neither is it the various denominations; it is the people who love God and keep His commandments. Where two or three are gathered together in My name, there I am in the midst of them. Where Christ is even among the humble few, *this is Christ's church*. "For the presence of the High and Holy One who inhabiteth eternity can ALONE CONSTITUTE A CHURCH." (Upward Look, p. 315.) God's church is the **heart** of the work here on Earth.

When we recognize the condition of the heart (church) and allow the LORD to give us a new heart, His promise is, *"For I will take you from among the heathen, and gather you out of all countries, and will bring you into your own land. Then will I sprinkle clean water upon you, and ye shall be clean: from all your filthiness, and from all your idols, will I cleanse you. A new heart also will I give you, and a new spirit will I put within you: and I will take away the stony heart out of your flesh, and I will give you an heart of flesh. And I will put my spirit within you, and cause you to walk in my statutes, and ye shall keep my judgments, and do [them]. And ye shall dwell in the land that I gave to your fathers; and ye shall be my people, and I will be your God."*—Ezekiel 36:24-28. This is the everlasting covenant that was given to man from the Garden of Eden. *". . . I will be your God, and ye shall be my people . . ."*—Jeremiah 7:23. When our innermost being comes into line with this concept, then we will walk according to His law, and represent His character.

The cardiovascular system is composed of the heart and the blood vessels. It is the main transport system of the body and carries the blood to every cell. The tribe of Rueben represents this system. Review the characteristics of Rueben before beginning this study. This system can be *". . . my firstborn, my might, and the beginning of my strength, the excellency of dignity, and the excellency of power."*— Genesis 49:3. This is the first system to be completely developed in the fetus, and is fulfilling the nutritional need of the embryo by the end of the third week or gestation. The heart actually begins to pump blood in one direction by day 28.

If we do not take care of this system, it can be like Rueben, *". . . unstable as water, and thou shalt not excel . . ."* —Genesis 49:4. It is up to each one of us to make the choices that will strengthen the heart and vessels.

It is the heart that pumps the blood to every part of the body. The blood spiritually represents the Word. "And the Word was made flesh and dwelt among us, and we beheld His glory, the glory as of the only begotten of the Father, full of grace and truth."—John 1:14. (The details of the blood were seen in the study on the fluids of the body.)

"Then Jesus said unto them, 'Verily, verily, I say unto you, Except ye eat the flesh of the Son of man, and drink his blood, ye have no life in you. Whoso eateth my flesh, and drinketh my blood, hath eternal life; and I will raise him up at the last day. For my flesh is meat indeed, and my blood is drink indeed. He that eateth my flesh, and drinketh my blood, dwelleth in me, and I in him. As the living Father hath sent me, and I live by the Father: so he that eateth me, even he shall live by me. This is that bread which came down from heaven: not as your fathers did eat manna, and are dead: he that eateth of this bread shall live for ever... It is the spirit that quickeneth; the flesh profiteth nothing: the words that I speak unto you, [they] are spirit, and [they] are life.'" John 6:53-58, 63.

The nutrients in the blood come from the digestive system. *"Thy words were found, and I did eat them; and Thy word was unto me the joy and rejoicing of my heart: for I am called by Thy name, O LORD God of hosts."*—Jeremiah 15:16 *"My son, attend to my* ***words****; incline thine ear unto my sayings. Let them not depart from thine eyes; keep them in the midst of thine heart. For they are life unto those that find them, and health to all their flesh. Keep thy heart with all diligence; for out of it are the issues of life."*—Proverbs 4:20-23. *"For the life of the flesh is in the blood: and I have given it to you upon the altar to make an atonement for your souls: for it is the blood that maketh an atonement for the soul."*—Leviticus 17:11.

The job of God's Church is to provide the Word to all the world. The job of the heart is to provide the blood to all parts of the body. From the beginning God has wrought through His people to bring blessings to all the world. To the ancient Egyptian nation God made Joseph a fountain of life. Through the integrity of Joseph, the lives of that nation were preserved. Through Daniel God saved the life of all the wise men of Babylon. And these deliverances are as object lessons offered to the world through connection with the God whom Joseph and Daniel worshiped. Everyone in whose heart Christ abides, and everyone who will show forth His love to the world is a worker together with God for the blessings of humanity. As he receives from the Saviour grace to impart to others, from his whole being flows forth the tide of spiritual life. *"And to make all [men] see what [is] the fellowship of the mystery, which from the beginning of the world hath been hid in God, who created all things by Jesus Christ: To the intent that now unto the principalities and powers in heavenly [places] might be known* ***by the church*** *the manifold wisdom of God, According to the eternal purpose which he purposed in Christ Jesus our Lord:"*—Ephesians 3:9-11.

The cardiovascular system is created with the number 4, (the spiritual number of created works). There are four heart chambers; four heart sounds; four pulmonary veins; four stages in cardiac cycle; and four main blood types; four heart valves. There are eight (2 x 4) forms of valve malfunction, and there are eight events recorded in an electrocardiogram. The heart of the embryo beats four weeks after fertilization. Review the number four in the study "God's Design in Numbers." Four represents man's weakness and his utter inability to save himself. Keep this in mind as this system is studied.

In order to have good health, we must have good blood for the blood is the current of life. Blood repairs waste and nourishes the body. When supplied with the proper food elements and when cleansed and vitalized by contact with pure air, it carries life and vigor to every part of the system. The more perfect the circulation, the better will this work be accomplished.

At every pulsation of the heart, the blood should make its way quickly and easily to all parts of the body. Its circulation should not be hindered by tight clothing or bands or by insufficient clothing of the extremities. Whatever hinders the circulation forces the blood back to the vital organs, and produces congestion.

Think about the tribe of Rueben, which represents the cardiovascular system. This tribe took the path of least resistance just like the blood. When the arteries or the capillaries are clogged, the blood will take another path and not completely nourish that clogged part of the body. If the extremities are cold for a period of time, the blood will back up into the trunk of the body causing congestion rather than continuing on to the cold portion of the extremities. That is why the fingers and toes are the first to freeze if exposed to prolonged cold. When the tribe of Rueben became connected to God, they were mighty warriors. Like the tribe of Rueben, the heart works as a mighty warrior and as long as it can.

In order to have good blood we must breathe well and correctly. Full, deep respiration's of pure air, fill the lungs with oxygen, and purify the blood. The lungs impart to the blood a bright red color and send it, (a life giving) current, to every part of the body. A good respiration soothes the nerves, stimulates the appetite, renders the digestion more perfect, and induces sound refreshing sleep.

When an insufficient supply of oxygen is received, the blood moves sluggishly. The waste or poisonous matter that should be thrown off in the exhalations from the lungs is retained and the blood becomes impure. Not only the lungs, but the stomach, liver, and brain are affected. The skin becomes sallow; digestion is retarded; the heart is depressed; the brain is clouded; the thoughts are confused; gloom settles upon the spirits; the whole system becomes depressed and inactive and peculiarly susceptible to disease.

Oxygen represents God's Holy Spirit that must be present so that we can understand and apply the word of God to our life.

The heart, together with its vessels, takes shape and begins to function long before any other major organ. It begins beating in a human embryo during the fourth week of development, and continues throughout the life of a person at a rate of about 70 times a minute, 100,000 times a day, or about 2.5 billion times during a 70 year lifetime. The specialized cardiac muscle tissue that performs this extraordinary feat is found only in the wall of the heart.

The heart is about the size of the clenched fist of its owner. It is located in the center of the chest cavity. A sac called the pericardium protects the heart. This sac spiritually represents the protective care of Christ for His church. *"Husbands, love your wives, even as Christ also loved the church, and gave Himself for it."*—Ephesians 5:25.

The heart is held in place by connective tissue (love) that binds the pericardium to the sternum (the principles of the righteousness of God) and the vertebral column (the principles of the character of Christ), and other parts of the thoracic cavity. The pericardium, with its three layers, is tough and inelastic yet loose-fitting enough to allow the heart to move in a limited way. The serous pericardial fluid moistens the sac and minimizes friction between the membranes as the heart moves during its contraction-relaxation phases.

Deep breathing that moves the diaphragm up and down and assists the heart in its work. Earnest prayer for God's Holy Spirit brings Him into our hearts to assist with our work.

The heart is a hollow organ and has four chambers. Each side of the heart contains an elastic upper chamber called an atrium where blood enters the heart and a lower pumping chamber called a ventricle, where blood leaves the heart. This heart is actually a double pump. The circulation between the heart and the lungs is called the pulmonary circulation. The circulation between the heart and the rest of the body is called the systemic circulation.

Cardiac muscle cells in the atria or ventricles function as a coordinated unit in response to physiological stimulation rather than as a group of separate units as skeletal muscle does. Cardiac muscle cells act in this way because they are connected end to end by intercalated disks which are especially designed to cause the muscle cells in the entire chamber to contract almost in unison.

The wall of the heart is made up of three layers. This number represents solid completeness. The outer shiny membrane is called the epicardium (epi-kard-I um). (*Kardia* is the Greek word for heart.) Inside the epicardium, and often surrounded with fat, are the main coronary blood vessels that supply blood to the heart and drain blood from the heart. The middle layer, the myocardium (my-o-KARD-I um), is a thick layer of cardiac muscle that gives the heart its special pumping ability. This myocardium has three layers of cardiac muscle which are attached to a fibrous ring that forms the cardiac skeleton. The spiral is the most effective arrangement for squeezing blood out of the heart's chambers. When blood is pumped, it is wrung out of the ventricles like water from a wet cloth.

The cardiac skeleton is made up of a fibrous *trigone* (triangular) and four rings or cuffs One surrounds each of the heart's four openings. Valves regulate these heart openings. Four rings of the cardiac skeleton support the bases of the valves and prevents them from stretching. This arrangement prevents blood from flowing backward from the ventricles into the atria and from the arteries back into the ventricles.

There are three cusps (valves) on the right atrioventricular valve and two cusps on the left one. The cusps are like panels of a parachute that are secured to the papillary muscles in the ventricles by tendinous chordae tendineae. The cusps overflap as ventricular blood bulges into them during ventricular contraction. There are three cusps on each semilunar valve, making 11 cusps in all. This marks the inner heart with the number signifying imperfection and disintegration. This makes sense if you follow the history of the inner working part of the church down through the ages.

The human body contains about one to one and a half gallons of blood. The heart takes only a little more that a minute to pump a complete cycle of blood throughout the body. In times of strenuous exercise, the heart can quintuple this output. On the average, the adult heart pumps about 2000 gallons of blood throughout the body every day.

Blood Path

Blood enters the atria. Oxygen poor blood from the body flows into the right atrium at about the same time as newly oxygenated blood from the lung flows into the left atrium.

Blood is pumped into the ventricles. The heart's pacemaker (sinoatrial node) fires an electrical impulse that coordinates the contractions of both atria. Blood is forced through the one-way atrioventricular valves into the relaxed ventricles.

The ventricles, filled with blood, hesitate for an instant.

The ventricles contract sending blood to the body. The left ventricle pumps the newly oxygenated blood through the aortic semilunar valve into the aorta. The right ventricle forces blood low in oxygen out through the right and left pulmonary arteries to the lungs. The left and right ventricles pump almost simultaneously so that equal amounts of blood enter and leave the heart. By this time, the atria have already started to refill preparing for anther cardiac cycle.

The Pacemaker

If you put separate pieces of heart muscle on a plate, all pulsating at a different rate, soon they will come together and all pulsate at the same beat. They are connected end to end by intercalated disks. This allows action to be transmitted from one cardiac cell to another so that they will be coordinated even though they are not connected.

Cardiac muscle has a built-in pacemaker that initiates a beat independently of the central nervous system. This pacemaker is called the *sinoatrial node* (SA). Impulses from the SA sweep across the cardiac muscle of the atria causing the atria to contract. These contractions slow as they pass through the atrioventricular node (AV), and then descend via the atrioventricular bundle and spread throughout the myocardium of the ventricles. Ventricular contraction results. This all happens within 2/10 of a second and is much like spreading ripples on a pond.

The SA node is a mass of specialized heart muscle tissue. The muscles of the atria are not continuous with the muscles of the ventricles, so they must be coordinated at every beat of the heart. The impulse travels from the SA node through the AV node and through hundreds of tiny specialized cardiac muscle fibers called Purkinje fibers. Such an arrangement directs the electrical impulse in a definite pathway so that it can make contact with all areas of the muscle of the ventricles.

The SA node represents love for God; and the AV node, love for man. *"Jesus said unto him, Thou shalt love the Lord thy God with all thy heart, and with all thy soul, and with all thy mind. This is the first and great commandment. And the second [is] like unto it, Thou shalt love thy neighbour as thyself."*—Matthew 22:37-39. Christ bound men to His heart by the ties of love and devotion; and by the same ties He bound them to their fellow men. With Him, love was life, and life was service. *"Freely ye have received,"* He said, *"freely give."* (Matthew 10:8.) It was not on the cross only that Christ sacrificed Himself for humanity. As He *"went about doing good"* (Acts 10:38), every day's experience was an outpouring of His life. In one way only could such a life be sustained. Jesus lived in dependence upon God and communion with Him. To the secret place of the Most High, under the shadow of the Almighty, men now and then repair. They abide for a season, and the result is manifest in noble deeds. Then their faith fails; the communion is interrupted; and the lifework marred. But the life of Jesus was a life of constant trust and sustained by continual communion. His service for Heaven and Earth was without failure or faltering.

As a man, Jesus supplicated the throne of God until His humanity was charged with a heavenly current that connected humanity with divinity. Receiving life from The Father, Jesus imparted life to men. When we lose this love, the church ceases to function properly.

Vessels

Arteries carry blood away from the heart to the organs and tissues of the body. There are two different kinds of arteries, *elastic* and *muscular* arteries. The walls of elastic arteries expand slightly with each heartbeat. The muscular arteries branch into smaller arterioles, which play important roles in determining the amount of blood going to any organ or tissue and in regulating blood pressure. Arterioles branch into smaller vessels called *metarterioles,* which carry blood to the smallest vessels in the body called the *capillaries*.

Capillaries are microscopic vessels with walls mostly one cell thick. The thin capillary wall is porous and allows the passage of water and small particles of dissolved material. Capillaries are distributed throughout the body except in the dead outer layers of skin and in such special places as the lenses of the eyes.

Capillaries converge into larger vessels called *venules,* which merge to form even larger vessels called veins. Veins carry blood toward the heart. They are generally more flexible than arteries, and they collapse if blood pressure is not maintained.

The central canal of all blood vessels is called the lumen. Surrounding the lumen of a large elastic artery is a thick wall composed of three layers. The veins have the same three layers, but the middle layer is much thinner and has less elastic tissue which make veins able to distend and compress easily. Veins contain paired valves that keep the blood from flowing backward. Exercise helps the veins to return the blood to the heart.

The Marvelous Heart Muscle

The heart muscle needs more oxygen (which spiritually represents God's Holy Spirit in this study) than any other organ except the brain. To obtain this oxygen, the heart must have a generous supply of blood. Like other organs, the heart receives its blood supply from arterial branches that arise from the aorta. Blood is supplied to the heart itself by the right and left coronary arteries which are the first branches off the aorta. The branching of the coronary arteries varies from person to person. The flow of blood that supplies the heart tissue itself is the coronary circulation. The heart needs its own separate blood supply because blood in the pulmonary and systemic circulations cannot seep through the lining of the heart. The heart pumps about 5% (100 gallons) of all the blood pumped every day to its own muscle tissue.

When these arteries become clogged as a result of eating sugar, oils and animal fat, then the worldly physician may perform bypass surgery. Veins are stripped from the legs (the ways), and are routed around the old clogged arteries (angioplasty). If the lifestyle is not changed, these new vessels soon become clogged, and the heart has the same old problem again.

Spiritual Application

The members of the church have become like the clogged arteries because they have been fed too many smooth and sugared sermons (Babylon). This has caused a build up of cholesterol (Laodiceanism) in the blood stream (church). They have become "*rich and increased with goods and have need of nothing.*" (Revelation 3:17.) They do not know that they are *"wretched, and miserable, and poor and blind and naked."* This is why they need the "Loud Cry" (Isaiah 58) to come out of Babylon. They need a series of spiritual chelation treatments (straight testimonies) to ream out their cholesterol.

"For my people have committed two evils; they have forsaken me the fountain of living waters, [and] hewed them out cisterns, broken cisterns, that can hold no water."—Jeremiah 2:13 *"He that turneth away his ear from hearing the*

law, even his prayer [shall be] abomination. Whoso causeth the righteous to go astray in an evil way, he shall fall himself into his own pit: but the upright shall have good [things] in possession."—Proverbs 28:9, 10. The main cause of death in this country is cardiovascular disease, just as spiritual death is caused by spiritual problems of the heart. As we look at some of the problems of the physical heart, think of how these can apply spiritually.

Aneurysm

An aneurysm is a balloon-like, blood-filled dilation of a blood vessel. This happens when the muscle cells of the middle layer become lengthened and the wall weakens. The wall may also be weakened by arteriosclerosis (hardening of the arteries). An aneurysm can burst, and if it is a cerebral vessel, a stroke may result.

Arteriosclerosis

This is the leading cause of coronary disease. It is characterized by deposits of fat, fibrin, cellular debris, and calcium on the inside of arterial walls and in the underlying tissues. These built-up materials, called atheromatous plaques, adhere to the inner lining, narrowing the lumen and reducing the elasticity of the vessel. This is caused first by the damage of the interior wall of the vessel making it rough instead of smooth. Tobacco smoke and animal fat and cholesterol in the diet can cause this. Once the damage occurs, the rough walls become a place where fat substances can adhere making the passage narrow or making a place where a blood clot can form. Then the blood cannot travel freely and parts of the body die. It is written: *"And thou shalt take all the fat, that covereth the inwards . . . and burn them upon the altar."*—Exodus 29:13.

In the Bible fat represents sin. See Leviticus 4:35. In Leviticus 7:23, the children of Israel were commanded to eat no fat. God wanted the fat all burned to show that sin is to be destroyed completely. "*The proud have forged a lie against me: but I will keep thy precepts with my whole heart. Their heart is a fat as grease; but I delight in Thy law."*—Psalm 119:69,70.

Let's equate this plaque to selfishness which slows down the blood of Christ. which is able to cleanse us from all sin. Selfishness is death to the soul, as plaque is death to the body. One little bit tolerated grows bigger and bigger until it finally plugs up the channels of grace. The lover of self is "*a transgressor*" of the law. The promise of obedience from a person appears to be fulfilled when it requires no sacrifice. When self-denial and self-sacrifice are required, and the cross needs to be lifted, one draws back. Conviction of duty wears away and known transgression of God's commandments becomes a habit. The ear may hear God's word but the spiritual perceptive powers have departed. The heart is hardened, and the conscience seared. Do not think that one does manifest decided hostility to Christ and still abide in His service. This is deception. When we withhold that which God has given us to use in His service, be it time of means or any other of His entrusted gifts, we are at enmity with God. The hardening of the arteries is not noticed until the artery is so closed that the heart attack is suffered because of lack of blood and oxygen. The spiritual condition is not noticed until a crises comes, and it is realized that the connection with Christ is gone. Remember the parable of the foolish virgins in Matthew 25, God's Holy Spirit is not present and the light has gone out.

The cardiovascular system has no filtering mechanism of its own. It relies upon the lymphatic system, the liver and the kidneys to purify the blood. Whatever enters the blood is allowed to be circulated through the body and if it were not for the other systems removing and purifying the blood we would soon die. For example only two teaspoons of sugar can be in the blood at one time without causing major problems. Think of the work the other systems must do when we ingest eight teaspoons of sugar in a soda or thirty-two teaspoons of sugar in a banana split. The fat from homogenized milk enters directly into the blood stream instead of going into the lymphatics first. This causes hardening of the arteries worse than smoking cigarettes. So you see that if we ingest things that have been altered by refining processes or are toxic, the body adjusts if possible. If not, it deposits the harmful material in the vessels, in the lungs, in fat cells, in any place to get it out of the blood stream. Without perfect circulation it is impossible to have perfect health.

"It shall be a perpetual statute for your generations throughout all your dwellings, that ye eat neither fat nor blood."—Leviticus 3:17.

Our Creator knows what fuel the body machine needs to operate best. We would do well TO TAKE HEED!

Exercise is a key in keeping the cardiovascular system health. Over and over in the Bible we are enjoined to work - it is a blessing. It is even commanded in the Ten Commandments. *"Six days shalt thou labor and do all thy work . . ."* —Exodus 20:9. *"Go to the ant, thou sluggard: consider her ways, and be wise: which having no guide, overseer, or ruler, provideth her meat in the summer, and gathereth her food in the harvest."*—Proverbs 6:6-8.

Have you ever seen how ants scurry around? No couch potato time for them. None of them are overweight and none die of heart disease!

Portal Systems

Most veins transport blood directly back to the heart from a capillary network, but in the case of a portal system, the blood passes through two sets of capillaries on its way to the venous system.

The human body has two portal systems:

1. The *hypothalamic-hypophyseal* portal system moves blood from the capillary bed of the hypothalamus directly, by way of veins, to the sinusoidal bed of the pituitary gland.
2. The *hepatic* portal system varies considerably from one person to the next but follows a basic pattern. (A) The hepatic portal vein is formed behind the neck of the pancreas by the union of the splenic and superior mesenteric veins. The splenic mesenteric vein returns blood from the spleen, and the superior mesenteric vein returns blood from the small intestine. The hepatic portal (pertaining to liver) vein also receives coronary, cystic, and pyloric branches. (B) The splenic vein receives the left short gastric vein from the greater curvature of the stomach and the inferior mesenteric vein which receives blood from the veins of the large intestine. (C) The systic vein brings blood from the gallbladder to the hepatic portal vein. The right gastric (pyloric) vein arises from the stomach and then enters the hepatic portal vein. The left gastric (coronary vein also empties into the hepatic portal vein. Because the veins of the hepatic portal system lack valves, the veins are vulnerable to the excessive pressures of venous blood. This situation can produce hemorrhoids. Hemorrhoids make the elimination of waste becomes painful. A spiritual application would be if we get too much sin backed up, it makes its elimination much more painful. This is known as spiritual constipation.

Much of the venous blood returning to the heart from the capillaries of the spleen, stomach, pancreas, gallbladder, and intestines contains products of digestion. This nutrient-rich blood is carried by the hepatic portal vein to the sinusoids of the liver. After leaving the sinusoids of the liver, the blood is collected in the hepatic veins and drained into the inferior vena cava, which returns it to the right atrium of the heart.

In our study on the endocrine and digestive systems the student will see the spiritual lessons of the portal systems. The liver represents faith, and the pituitary represents law of God. Unless our spiritual nourishment goes through these two filters, we are confused and debilitated.

Spiritual Challenge

There are hundreds of verses in the Bible that mention heart. Each one has a lesson for us. To every household and every school, to every parent, teacher, and student upon whom the light of the Gospel has shone, comes the question asked of Esther the queen by her Uncle Mordecai. *"Who knoweth whether thou art come to the kingdom for such a time as this?"*

Few give thought to the suffering that sin has caused our creator. All heaven suffered in Christ's agony; but that suffering did not begin or end with His manifestation in humanity. The cross is a revelation to our dull senses of the pain that, from its very inception, sin was brought to the heart of God. Every departure from the right, every deed of cruelty, every failure of humanity to reach the ideal, brings grief to our Heavenly Father. There came upon Israel the calamities that were the sure result of separation from God, subjugation by their enemies, cruelty, and death; it is written: *"His soul was grieved for the misery of Israel."*—Judges 10: 16.

To all, great or small, learned or ignorant, old or young the command is given. *"Go ye therefore, and teach all nations, baptizing them in the name of the Father, and of the Son, and of the Holy Ghost: Teaching them to observe all things whatsoever I have commanded you: and, lo, I am with you alway, [even] unto the end of the world. Amen."* —Matthew 28:19 –20.

Chapter Twenty-Eight

THE MUSCULAR SYSTEM

Psalm 37:5

"Commit thy way unto the LORD; trust also in Him; and He shall bring [it] to pass. And He shall bring forth thy righteousness as the light, and thy judgment as the noonday."

The muscular system is represented by the tribe of Dan. *"Dan shall judge his people."*—Genesis 49:16. The lesson of the muscular system is a lesson of judgment. In the New Testament the word that judgment is translated from is "krisis," from which we get our word, crisis (decision; to decide).

The tribe of Dan is not represented in the Holy City. Instead of exercising good judgment, he exercised a critical back-biting character, and this cause him to be lost forever. Both the tribe of Dan and the tribe of Judah were called the "lion's whelp" but look at the difference. From the tribe of Judah came the great JUDGE of all the universe. Read Revelation 5; in this scene we have the Lion and the Lamb, the same Person. The One who is our Judge (John 5:22) is also our Sacrifice (John 1:36.)

There are three types of muscle tissue: skeletal, cardiac, and smooth. Each muscle type is designed to perform different functions.

There are three classes of judgment that matter: human judgment, God's judgment and the judgment of the universe:

1. Human judgment should be made after all the facts have been considered. We all have to make a judgment about our relationship to God; our moment of truth, as it were. Over and over we see this demonstrated in the Bible. *"Who is on the LORD's side? Let him come unto me."*—Exodus 32:26. *". . . choose you this day whom ye will serve." Joshua 24:15. "How long halt ye between two opinions? If the LORD be God, follow Him; but if Baal, then follow him."*—I Kings 18:21.
2. God's judgment is made after our lives are reviewed. Every man's work passes in review before God and is registered for faithfulness or unfaithfulness. Every deed of kindness, every encouraging word spoken, every evil overcome, every temptation resisted, every act of sacrifice, every suffering and sorrow endured for Christ's sake—all are entered in the Book of Remembrance. *"Then they that feared the LORD spake often one to another: and the LORD hearkened, and heard [it], and a Book of Remembrance was written before him for them that feared the LORD, and that thought upon his name. And they shall be mine, saith the LORD of hosts, in that day when I make up my jewels; and I will spare them, as a man spareth his own son that serveth him. Then shall ye return, and discern between the righteous and the wicked, between him that serveth God and him that serveth him not."*—Malachi 3:16. Opposite each name in the Books of Heaven is entered with terrible exactness every wrong word, every selfish act, every unfulfilled duty, and every secret sin, with every artful dissembling. Heaven-sent warnings or reproofs neglected, wasted moments, unimproved opportunities, the influence exerted for good or for evil, with its far-reaching results, all are chronicled by the recording angel. *"Let us hear the conclusion of the whole matter: Fear God, and keep his commandments: for this [is] the whole [duty] of man. For God shall bring every work into judgment, with every secret thing, whether [it be] good, or whether [it be] evil."*—Ecclesiastes 12:13.
3. The judgment of all the universe is proclaimed after this Earth's probation is over and the results of sin are

> seen in their completeness. The inhabitants of Heaven and of other worlds have had to decide whether Lucifer lied about God. This was not easy for them, for Satan was the covering cherub and greatly beloved of all before his fall. Satan boasted that this Earth is the demonstration of his better government with freedom from God's law. For the good of the entire universe, through ceaseless ages, Satan must more fully develop his principles when this happens, charges against the divine government will be seen in their true light by all created beings. The justice and mercy of God and the immutability of His law might forever be placed beyond all question. Since the service of love can alone be acceptable to God, the allegiance of His creatures must rest upon a conviction of His justice and benevolence. *"That at the name of Jesus every knee should bow, of [things] in heaven, and [things] in earth, and [things] under the earth; And [that] every tongue should confess that Jesus Christ [is] Lord, to the glory of God the Father."*—Philippians 2:10,11.

As we have said, the muscular system is a lesson of judgment.

The driving force behind movement is muscular tissue. The word muscle is based on the Latin musculus, which means "little mouse," because the movement under the skin was thought to resemble a running mouse.

There are three properties of muscle tissue that make muscles do their job properly:

1. The ability to contract or shorten;
2. The capacity to receive and respond to a stimulus; and
3. The ability to be stretched and return to its original shape after being stretched or contracted.

Contraction is the phase in which work is done. For instance, skeletal muscles contract to move lower limbs when walking or running. In order to sit up straight, the muscles in the back must contract. Food is passed along the digestive tract as the smooth muscle contracts. The contraction of heart muscles propels the blood out of the heart. The contraction of skeletal muscle produces heat to help maintain a stable body temperature.

The three basic works of muscle tissue are movement, posture, and heat production. There are three phases of judgment: Investigation, Declaration of sentence, and execution of sentence.

Become familiar with the following terms:

Myo (MIE-o; Gr. *Mys,* muscle)—any word with this beginning pertains to the muscles. For instance, *myotherapy* means a treatment on the muscles.
Myofilbril (*myo* + *fibrilla*)—a small fiber.
Sarco (Gr. *sarkos,* flesh)—muscle is considered flesh.
Sarcolemma (flesh + *lemma,* husk)—the membrane surrounding each skeletal muscle fiber.
Sarcoplasm (sarko–plas im)- a special type of cytoplasm in the cell of the muscle.
Sarcomere—(sarko-mer) Greek for *meros,* part.
Striated (STRI-ate-ed) Latin –channeled or striped.
Fila (L. thread)—filament (fine thread).
Fascicle (FAS-eh-kul)—little bundle.
Articular (ar-tick-u-lar)—relating to a joint.

Skeletal Muscle

The skeletal muscle is attached to the skeleton and makes it move. It is called striated muscle because its fibers or cells are composed of alternating light and dark stripes. It is also called voluntary muscle because it can contract it when it wants to. Skeletal muscle is composed of individual cells that have a long cylindrical shape and more than one nuclei. These cells are called muscle fibers and contain still smaller fibers called myofibrils which are made up of thick and thin treads called myofilaments. There are thick myofilaments (which are composed of a fairly large protein called myosin) and thin myofilaments (composed mostly of a smaller protein called actin). Each fiber contains several nuclei and many mitochondria and a large number of myofibrils. Sarcolemma, a specialized membrane, encloses each muscle cell (fiber).

When seen with a microscope, skeletal tissue shows a pattern of red and white bands which give the fiber its layered look. These bands are caused by the arrangement of the myosin strands and the actin strands. The red is an overlapping of the thick myosin strands and the thin actin strands. The white is the thin actin strands alone. The intermediate color is found in the myosin strands only. The sarcomere is the section of the muscle fiber that contracts. The color red spiritually represents sacrifice, while the color white represents the righteousness and the intermediate color represents the mercy needed when we are judged. God's judgment can be either a vindication or a condemnation, depending upon the attitude and life lived by the one being judged. "For we must all appear before the judgment seat of Christ; that every one may receive the things [done] in [his] body, according to that he hath done, whether [it be] good or bad."—II Corinthians 5:10.

David knew that it was a thing to be desired to have God pronounce a decision about him. *"Judge me, O LORD; for I have walked in mine integrity: I have trusted also in the LORD; [therefore] I shall not slide. Examine me, O LORD, and prove me; try my reins and my heart. For Thy lovingkindness [is] before mine eyes: and I have walked in Thy truth."*—Psalm 26:1 He knew it would vindicate him before his enemies. "Judge me, O LORD my God, according to thy righteousness; and let them not rejoice over me."—Psalm 35:24. *"Judge me, O God, and plead my cause against an ungodly nation: O deliver me from the deceitful and unjust man."*—Psalm 43:1.

On the other hand, there are some who are not looking forward to judgment. *"The king shall mourn, and the prince shall be clothed with desolation, and the hands of the people of the land shall be troubled: I will do unto them after their way, and according to their deserts will I judge them; and they shall know that I [am] the LORD."*—Ezekiel 7:27. Ezekiel is the Bible Book of Judgment. The Hebrew word *shaphat*—to judge or pronounce sentence, for or against—is translated as "judge" twenty-one times in Ezekiel.

Muscle fibers can contract individually as the eye muscles do. But they usually contract in groups. In this they are like individuals or corporate judgments from God. They are packed together in fascicles averaging about 150 fibers. The number of bundles of fibers determines the size of muscles. The largest muscles are found where large, forceful movements are needed, such as the back, arms and legs. In people with larger muscles, their muscles have the same amount of fibers as the smaller-muscled person, but the fibers are increased in diameter.

Dan will not be found among the redeemed. In this pronouncement about Dan (Deuteronomy 33:22), scripture says that he shall leap from Bashan." The word leap comes from a word that means sort (as when you put a difference between something), so scripture is actually saying that he will sort from Bashan. Since the tribe of Dan did not live in Bashan, he was coming from Bashan in attitude not as from a dwelling place. Bashan was the land of the giants (Deuteronomy 3). Giant muscles obtained from weightlifting, are a distortion of the muscular system. It is a form of hero worship connected with appearance. Dan was a false judge and had this kind of judgment.

Skeletal muscles vary in length and size from the tiniest, like the muscle that moves the stapes bone in the ear which is about 0.04 inch long; to the longest, which is 12 inches long and known as the sartorius muscle of the thigh.

Coverings

Skeletal muscles are held together and covered with a fibrous connective tissue that is part of a network called fascia (FASH–ee–uh; Latin, *bandage*). There are two forms of fascia, superficial and deep.

The superficial fascia is located in the dermis of the skin and connects the skin with the deep fascia. It is generally composed of loose connective tissue containing blood vessels, nerves, lymphatic vessels, and many fat cells. In the face, it connects the muscles of facial expression. It provides a protective layer of insulation and allows the skin to move freely over deeper structures.

The deep fascia is made up of several layers of dense connective tissue. It extends between muscles and groups of muscles. It is inelastic.

Wrapped around each muscle fiber (cell) is a protective covering of connective tissue called an endomysium (*endo,* Gr. within + *mysium*). Many fibers are bundled together and called fascicles (L. little bundle). And they are wrapped with a layer of connective tissue called perimysium (*peri*, Gr. around or about). The connective tissue covering that surrounds the whole muscle is called epimysium (epi, Gr. upon, over, in addition to). These sheaths of connective tissue contain blood vessels, lymphatic vessels, and nerves. In our muscles, oxygen, nutrition, the protection of lymph, and messages from the control center, are carried in the connective tissue. In the spiritual judgment, God's Holy Spirit, the Word of God, the cleansing from sin and the protection against it, and directions from God are all carried in a covering of love. *"He that hath my commandments, and keepeth them, he it is that loveth me: and he that loveth me shall be loved of my Father, and I will love him, and will manifest myself to him."*—John 14:21.

Muscle Attachments

Tendons attach muscle to bone or to a cartilage. Muscle attachments show how spiritually, judgment is related to principle (bone). Tendons are an extension of the deep fascia and/or the epimysium surrounding the muscle. It can extend into the outside covering of the bone and into the bone as perforating fibers. Tendons are white glistening fibrous cords of connective tissue that are devoid of elasticity. The color white spiritually represents righteousness, and righteousness is not negotiable. One is either righteous or he is not. Tendons are sparingly supplied with blood vessels and nerves and the smaller tendons have neither. The largest tendon is the calcaneal (Achilles) tendon.

Aponeuroses (AP-o-nue-roe-sis; Gr. *apo,* from + *neuro,* neuron, nerve, tendon) is a tendon which is spread out like a broad, flat sheet. It is present in the abdominal wall and beneath the skin in the palm of the hand. It has no nerves.

A muscle moves a bone by contracting, and becoming shorter. The part of the muscle that contracts is called the belly. When a muscle contracts, one of the bones attached to it stays in place while the other bone moves with the muscle. The end of the muscle attached to the moving bone is called the insertion. The other end is called the origin. This is only a general rule because in different actions the diversity of movement can change which bone moves.

Motor Units

Each skeletal muscle fiber is contacted by at least one nerve ending. One motor nerve fiber can stimulate several muscle fibers at the same time. A motor neuron, together with the muscle fiber it innervates, is called a motor unit. A motor unit can contain from about 6 to 30 fibers (in the small eye muscles) to over 1000 fibers (in the leg muscles).

When a strong enough signal is received by a motor neuron, the impulse is transmitted to the muscle fibers in a motor unit. The amount of stimulus that is necessary to contract the muscle is called the threshold stimulus. The intensity of the stimulation has to reach this point for the muscle to contract. If this intensity is increased, it doesn't make any difference. The contraction will not increase because when a muscle contracts it contracts all the way or not at all. This is called the all or none principle. *"Whatsoever thy hand findeth to do, do [it] with thy might; for [there is] no work, nor device, nor knowledge, nor wisdom, in the grave, whither thou goest."*—Ecclesiastes 9:10.

The smaller the muscle, the more refined can be the movement.

Muscle Action

Bones give the body its support and muscles give it its power, but neither support or power be possible without moveable joints. Joints provide the mechanism that allows us to move freely. Articulation (L. *articulus,* small joint) is the place where two bones or cartilages meet. Most joints are movable. Try to walk or sit without bending your legs and you will see how important joints are to the mechanics of movement.

There are three different types of joints: Fibrous, Cartilaginous, and Synovial.

Fibrous joints lack a cavity, and fibrous connective tissue unites the bones. They are joined together tightly, so they generally are immovable. This kind of joint includes (1) sutures that hold the skull bones together, like the sealed joints to protect the brain; (2) bones that are held together by ligaments, like the tibia and fibula, and also the radius and ulna; and (3) peg and socket joints that hold the teeth to the bones.

A plate of hyaline cartilage and a disk that unite bones are called cartilaginous joints. These kinds of joints permit growth not movement, and are present in the growing long bones. These usually become bone in an adult, except in a few places, such as the connection of the first pair of costal cartilages (a bar of hyaline cartilage that attaches a rib to the sternum) and the manubrium (the handle-like part of the sternum or malleus). (Remember that the first rib spiritually represents the first commandment. It is an ever-growing knowledge of the character of God that causes us never to put anyone or anything before Him.)

Most of the joints in the body are synovial joints. These joints allow the greatest range of movement. The ends of the bones are covered with a smooth hyaline articular cartilage and the joint is lubricated by a thick fluid called synovial fluid (Gr. *syn,* together + *ovum,* egg). It is called this because its consistency is like the white of an egg. These joints are enclosed in a flexible articular capsule (articular—pertaining to a joint—articular capsule: a sac like envelope enclosing the cavity of a synovial joint).

If articular capsules are reinforced with collagenous fibers they are called fibrous capsules. A ligament is the portion of this capsule that joins one bone to its adjoining bone. They are considered to be inelastic, yet they do allow movement at the joints. They will tear rather than stretch under too much stress.

Cardiac Muscle

This tissue contains the same type of myofibrils and protein components found in skeletal muscle, and has similar striations. These cells are closely packed, but they are separate, each with its own nucleus, joined end to end by intercalated disks (L. *intercalatus,* to insert between) that attach one cell to another, connect the myofibril filaments of adjacent cells, and contain gap junctions that help synchronize the contractions of cardiac muscle cells by allowing electrical impulses to spread rapidly from one cell to the next.

This can be used as an object lesson to show us that individual church members are responsible for their own actions and judgment, but, at the same time are closely bound to the other members of the church. All have to all work together for a common purpose. *"That there should be no schism in the body; but [that] the members should have the same care one for another. And whether one member suffer, all the members suffer with it; or one member be honoured, all the members rejoice with it."*—I Corinthians 12:25-26.

Cardiac muscle cells contain huge numbers of mitochondria, 35% of the cell, compared to 2% in the skeletal muscle cell. One can understand the extreme dependence of cardiac muscle on aerobic metabolism (metabolism requiring oxygen). From the Cardiovascular System study we learned that the heart spiritually represents the church, and it requires a great measure of God's Holy Spirit in order to function properly. Any healing creates scar tissue, not muscle tissue. If the heart is deprived of oxygen any longer than 30 seconds, the cardiac muscle cells may stop contracting; resulting in heart failure.

The heart has a built-in safety feature that protects it from tetanic contraction, which would be fatal in cardiac muscle. Tetanization is when a muscle is in a steady state of contraction with no relaxation at all between stimuli (like "lockjaw"). This is caused by repeated stimuli at a rapid rate so that the muscle cannot relax. The skeletal muscle contacts for 20 to 100msec. and the resting period is about 1 to 2 msec. In cardiac muscle the resting period is about 200 msec., which is almost as long as the contraction, which is between 200 and 250 msec. This pace allows it to have time for blood to leave the chambers of the heart. If the members of the church are not receiving and spreading the good news, the church is spiritually constipated. Then the church stagnates and dies.

Calcium is necessary in the function of muscles, and the cardiac muscle is affected by calcium imbalances sooner than any other muscle. This is because the excitation contraction (coupling of the heart muscle) becomes calcium dependent. In the nutrients of the body, calcium spiritually represents love for the law of God. If we have too little love for the law or if we have the wrong attitude toward it, we will be unbalanced. Wrong kinds of calcium cannot be assimilated and utilized and causes calcium deposits in the tissues and joints. This causes one to experience pain and discomfort—(legalism). If we have too little calcium in our bodies we get holes in our bones (principles) and our heart stops. If we do not love the law, a manifestation of God's character, one can uselessly try to work one's way to heaven. Thus the Jews were led to keep the letter of the law without the spirit of the law. *"Therefore judge nothing before the time, until the Lord come, who both will bring to light the hidden things of darkness, and will make manifest the counsels of the hearts: and then shall every man have praise of God."*—I Corinthians 4:5.

Those whom Christ commends in the judgment may have known little of theology, but still they have cherished His principles. Through the influence of God's Holy Spirit, they have been a blessing to those about them. Even among the heathen are those who have cherished the spirit of kindness. Among the heathen are those who worship God ignorantly, those to whom the light is never brought by human instrumentality; yet they will not perish. Though ignorant of the written law of God, they have heard His voice speaking to them in nature and have done the things that the law required. Their works are evidence that God's Holy Spirit has touched their hearts, and they are recognized as the children of God.

Cardiac muscle is involuntary. When we love Jesus, it follows automatically, like the cardiac muscle, that we will work for Him. *"For the love of Christ constraineth us . . ."*—II Corinthians 5:14.

Smooth Muscle

Smooth muscle cells contain only one nucleus, which is usually located near the center of the fiber.

Smooth muscle is not connected to bones. It generally forms sheets in the walls of large hollow organs such as the stomach and urinary bladder. Smooth muscle fibers are often formed in parallel layers, and the exact arrangement of the fibers varies from one location to another. In the walls of the intestines, smooth muscle fibers are arranged at right angles to each other One of these layers running long ways, and the next is wrapped around the circumference of the tubular intestine. These two layers working together are necessary to supply the constrictions that move the intestinal contents along.

In the bladder and uterus, the layers are arranged in several different directions. Connective tissue in the empty spaces binds them into bundles.

In the walls of hollow organs, smooth muscle fibers are wrapped around some blood vessels like tape around a rubber hose. They function to change the diameter of the vessels.

Smooth muscle action is rhythmical, and its contraction and relaxation periods are slower than those of any other type of muscle. It does not tire easily.

In the digestive system, after food is swallowed, all muscular contractions are involuntary until the body initiates the process of defecation.

"I beheld till the thrones were cast down, and the Ancient of days did sit, whose garment [was] white as snow, and the hair of his head like the pure wool: his throne [was like] the fiery flame, [and] his wheels [as] burning fire. A fiery stream issued and came forth from before him: thousand thousands ministered unto him, and ten thousand times ten thousand stood before him: the judgment was set, and the books were opened."—Daniel 7:9, 10.

In the judgment the use made of every talent will be scrutinized. How have we employed the capital lent us of Heaven? Will the Lord at His coming receive His own with usury? Have we improved the powers entrusted us, in hand, heart, and brain, to the glory of God and for blessing the world? No value is attached to a mere profession of faith in Christ. Only the love that is shown by works is counted genuine. Yet it is love alone which in the sight of Heaven makes any act of value. Whatever is done from love, however small it may appear in the estimation of men, is accepted and rewarded of God.

This investigation must take place before the Lord returns. On the day of atonement the sins of the people were blotted out before the High Priest came out of the sanctuary. Before Christ will return to gather up His own, the seal of God will be on His faithful children. So the investigation must take place to see who is faithful. *"He that is unjust, let him be unjust still: and he which is filthy, let him be filthy still: and he that is righteous, let him be righteous still: and he that is holy, let him be holy still. And, behold, I come quickly; and my reward [is] with me, to give every man according as his work shall be."*—Revelation 22:11,12.

As this event takes place the righteous and the wicked will still be living upon the Earth in their mortal state. Men will be planting and building, eating and drinking, all unconscious that the final irrevocable decision has been pronounced in the sanctuary above. Before the flood, after Noah entered the ark, God shut Noah in and shut the ungodly out. For seven days, the people, did not know that their doom was fixed, and they continued their careless pleasure-loving life and even mocked the warning of impending judgment. *"So shall also the coming of the Son of Man be."* —Matthew 24:39. Silently, un-noticed as the midnight thief, will come the decisive hour that marks the fixing of every man's destiny. This will be the final withdrawal of mercy's offer to guilty man.

The final phase of the judgment is to carry out the sentence pronounced by God. *"And I saw heaven opened, and behold a white horse; and he that sat upon him [was] called Faithful and True, and in righteousness he doth judge and make war."*—Revelation 19:10. *"For, behold, the day cometh, that shall burn as an oven; and all the proud, yea, and all that do wickedly, shall be stubble: and the day that cometh shall burn them up, saith the LORD of hosts, that it shall leave them neither root nor branch. But unto you that fear my name shall the Sun of righteousness arise with healing in his wings; and ye shall go forth, and grow up as calves of the stall. And ye shall tread down the wicked; for they shall be ashes under the soles of your feet in the day that I shall do [this], saith the LORD of hosts."*—Malachi 4:1-3.

"Behold the righteous shall be recompensed in the earth: much more the wicked and the sinner."—Proverbs 11:31.

"But the heavens and the earth, which are now, by the same word are kept in store, reserved unto fire against the day of judgment and perdition of ungodly men."—II Peter 3:7.

It is written that even Satan will be brought to ashes and never will exist again to tempt and destroy. *"Thine heart was lifted up because of thy beauty, thou hast corrupted thy wisdom by reason of thy brightness: I will cast thee to the ground, I will lay thee before kings, that they may behold thee. Thou hast defiled thy sanctuaries by the multitude of thine iniquities, by the iniquity of thy traffic; therefore will I bring forth a fire from the midst of thee, it shall devour thee, and I will bring thee to ashes upon the earth in the sight of all them that behold thee. All they that know thee among the people shall be astonished at thee: thou shalt be a terror, and never [shalt] thou [be] any more."*—Ezekiel 28:17-19.

"And I saw a new heaven and a new earth: for the first heaven and the first earth were passed away and there was no more sea."—Revelation 21:1. The great controversy will be ended. Sin and sinners will be no more. The entire universe will be clean with one pulse of harmony and gladness beating through the vast creation. *"And I heard a great voice out of heaven saying, Behold, the tabernacle of God [is] with men, and he will dwell with them, and they shall be his people, and God himself shall be with them, [and be] their God. And God shall wipe away all tears from their eyes; and there shall be no more death, neither sorrow, nor crying, neither shall there be any more pain: for the former things are passed away."*—Revelation 21:3,4. The Creator is finally united with His family from the renewed Earth. *"One pulse of harmony and gladness beats through the vast creation. From Him who created all, flow life and light and gladness, throughout the realms of illimitable space. From the minutest atom to the greatest world, all things, animate and inanimate, in their unshadowed beauty and perfect joy, declare that God is love."*—E. G. White, Faith to Live By, p. 371.5.

Chapter Twenty-Nine

THE RESPIRATORY SYSTEM

Genesis 2:7

"And the LORD God formed man [of] the dust of the ground, and breathed into his nostrils the breath of life; and man became a living soul."

"Though I speak with the tongues of men and of angels, and have not charity, I am become [as] sounding brass, or a tinkling cymbal. And though I have [the gift of] prophecy, and understand all mysteries, and all knowledge; and though I have all faith, so that I could remove mountains, and have not charity, I am nothing. And though I bestow all my goods to feed [the poor], and though I give my body to be burned, and have not love, it profiteth me nothing. Charity suffereth long, [and] is kind; charity envieth not; charity vaunteth not itself, is not puffed up, Doth not behave itself unseemly, seeketh not her own, is not easily provoked, thinketh no evil; Rejoiceth not in iniquity, but rejoiceth in the truth; Beareth all things, believeth all things, hopeth all things, endureth all things. Charity never faileth: but whether [there be] prophecies, they shall fail; whether [there be] tongues, they shall cease; whether [there be] knowledge, it shall vanish away. For we know in part, and we prophesy in part. But when that which is perfect is come, then that which is in part shall be done away. When I was a child, I spake as a child, I understood as a child, I thought as a child: but when I became a man, I put away childish things. For now we see through a glass, darkly; but then face to face: now I know in part; but then shall I know even as also I am known. And now abideth faith, hope, charity, these three; but the greatest of these [is] charity."—I Corinthians 13.

The tribe of Simeon spiritually represents the respiratory system. The people composing the tribe of Simeon were people characterized by focused life; zealous nature; and with dedicated purpose. Like Simeon, our lungs must be focused on a life-delivering purpose to our bodies. Simeon has a name on one of the gates in the New Jerusalem indicating that he overcame the negative traits of passion and put his zealousness to work for God. The more objectionable our traits, the more we need our Heavenly Father. All through the systems of the body is the lesson of God's Spirit of LOVE, but especially so in the respiratory system. Oxygen is so necessary for every process of life and it is a symbol of God's presence with us. *"And the Lord God formed man of the dust of the ground, and breathed into his nostrils the breath of life and man became a living soul."*—Genesis 2:7.

Breath in the Bible is synonymous with life. Most of the words that are translated into the English words, soul, spirit, breath, come from the Hebrew and Greek words that mean *breath* or *breathing.* (ncshamah = puff; rawach = wind; nephesh = breathing creature, "soul") *"And, behold, I, even I, do bring a flood of waters upon the earth, to destroy all flesh, wherein [is] the breath (rawach) of life, from under heaven; [and] every thing that [is] in the earth shall die."*—Genesis 6:17.

"All in whose nostrils was the breath (ncshamah) of life, of all that was in the dry land, died."—Genesis 7:22.

When a person dies *"His breath (rawach, wind) goeth forth, he returneth to his earth; in that very day his thoughts perish."*—Psalm 146:4. *"Then shall the dust return to the earth as it was: and the spirit (rawach, wind) shall return unto God who gave it."*—Ecclesiastes 12:7.

"For that which befalleth the sons of men befalleth beasts; even one thing befalleth them: as the one dieth, so dieth the other; yea, they have all one breath; so that a man hath no preeminence above a beast: for all [is] vanity. All go unto one place; all are of the dust, and all turn to dust again. Who knoweth the spirit of man that goeth upward, and the spirit of the beast that goeth downward to the earth?"—Ecclesiastes 3:19-21.

In the story of the widow's son being resurrected (I Kings 17), Elijah asked that his soul (nephesh) come into him again. It is from verses like this that people get the idea that there is an entity separate from the body when in reality, Elijah was only asking that the life force be returned. The idea that the soul is separate from the body is a Greek concept that was carried over from pagan beliefs.

All the references in the Old Testament to the Holy Spirit or Spirit of God are translated from rawach. In the New Testament soul is translated from *psuche; pneuma,* meaning breath or breeze, is translated *spirit. "God that made the world and all things therein, seeing that he is Lord of heaven and earth, dwelleth not in temples made with hands; Neither is worshipped with men's hands, as though he needed any thing, seeing he giveth to all life, and breath, and all things."*—Acts 17:24, 25

God has given the assurance that though our breath leaves us at death, the body alone turns to dust, He will resurrect us in the last day God will restore unto us the breath of life. He calls this death a sleep but not a death in the final sense. Read the story of Lazarus' death and resurrection in John 11. Jesus awoke him out of sleep after his body had begun to rot. He was still in the tomb; ALL OF HIM. The only thing missing was the breath of life that Jesus gave back to him. *"For I know [that] my redeemer liveth, and [that] he shall stand at the latter [day] upon the earth: And [though] after my skin [worms] destroy this [body], yet in my flesh shall I see God:"*—Job 19:25, 26. *"And this is the will of Him that sent me, that every one that seeth the Son, and believeth on Him, may have everlasting life: and I will raise him up at the last day."*—John 6:40.

The human body can survive without food for several weeks and without water for several days; but if breathing stops for three to six minutes, death is likely. Body tissues, especially the heart and brain, require a constant supply of oxygen. The respiratory system delivers air containing oxygen to the blood and removes carbon dioxide, which is the gaseous waste product of metabolism. The respiratory system includes the lungs, the several passageways leading from outside to the lungs, and the muscles that move air into and out of the lungs.

For our study of the respiratory system, it would be well to become familiar with the following terms:

1. Pharynx (FAIR-inks) meaning *throat* in Greek.
2. Larynx (LAR-inks) is the voice box.
3. Naso—is related to the nose.
4. Epithelium (epp-uh-THEE-lee-um Latin. *epi,* on, over, upon + Greek. *thele,* nipple) are cells are those that cover something.
5. Cilia (SIL-ee-uh) are hair-like processes projecting from the epithelium, which propel mucus and foreign matter.
6. Mucus is the goblet cell, or one-celled gland, that produces a carbohydrate-rich glycoprotein called mucin (MYOO-cin). This is secreted in the form of mucus, a thick lubricating gel. This cell is found in the lining of the respiratory tract and other linings of the body.
7. Mucosa is the membrane that contains goblet cells.

Respiratory Tract

Air flows through respiratory passages because of differences in pressure produced by chest and trunk muscles during breathing. Except for the beating of cilia in the respiratory lining, the passageways are simply a series of openings through which air is forced. These openings make up the respiratory tract.

The respiratory tract is a combination of bone and cartilage. Cartilage is classified as connective tissue that has been covered in the Study of the Tissues. The spiritual significance of the connective tissue is love. Cartilage is tough love and bone is love in principles and is even tougher love. In order to understand love we must understand that love is a principle, not the sentimentalism that some people in today's world consider as love.

Nose

The external nose is supported at the bridge by nasal bones. Along with the vomer and perpendicular plate of the ethmoid bone, the septal cartilage forms the nasal septum that divides the nasal cavity and separates the two openings of the nostrils.

The nasal cavity fills the space between the base of the skull and the roof of the mouth. It is supported above by the ethmoid bones and on the sides by the ethmoid, maxillary, and inferior conchae bones. (There are three bones in the inferior conchae: superior, middle, and inferior, which may be confusing because they are called inferior conchae.) These bones are covered with mucosa, which contain many blood vessels. These blood vessels bring a large quantity of blood to this area which warms the air inhaled through the nose. Since warm air holds more water than cold air does, the air is also moistened as it moves through the nasal cavity. Between the ridges of the conchae are folds called superior, middle, and inferior meatuses (Latin, *passages*).

The surface of the nasal cavity is lined with two types of epithelium (cellular covering):

1. Nasal epithelium, which warms and moistens inhaled air. It has millions of cilia, which also help capture minute particles and microorganisms.
2. Olfactory epithelium contains sensory nerve endings for smell.

Hard Palate

The hard palate, strengthened by the palatine and parts of the maxillary bones, forms the floor of the nasal cavity. (If these bones do not grow together normally by the third month of development, the condition is known as *cleft palate,* which makes it impossible for the baby to suck.)

Soft Palate

The soft palate is a flexible muscular sheet that separates the nasal cavity from the middle pharynx. It joins the hard palate and consists of several skeletal muscles covered by mucous membrane. It extends between the oral and nasal portions of the pharynx with a small fleshy cone called the *uvula* (YOO-vyoo-luh; L. small grapes). Depression of this area during breathing enlarges the air passage into the pharynx. The soft palate is elevated during swallowing and the uvula prevents food from entering the nasal pharynx.

Sinuses

The paranasal sinuses include the frontal, sphenoidal, maxillary, and ethmoid sinuses. They are blind sacs that open through a small foramen (hole) into the nasal cavity. These are named after the bones in which they are located. They give the voice a full, rich tone. When these become infected and full of debris, it can interfere with breathing. When our spiritual life becomes infected with the debris of sin, and selfishness, we have trouble staying in touch with God. *"In all their affliction he was afflicted, and the angel of his presence saved them: in his love and in his pity he redeemed them; and he bare them, and carried them all the days of old. But they rebelled, and vexed his holy Spirit: therefore he was turned to be their enemy, [and] he fought against them."*—Isaiah 63:9, 10.

The *nasolacrimal* (Latin nose and tear) duct leads from each eye to the nasal cavity through which excessive tears from the surface of the eye are drained into the nose. That is why weeping may be accompanied by a runny nose.

Pharynx

The pharynx leads to the respiratory passage and the esophagus. It is connected to both the nasal cavity and the mouth. It is divided into three portions (*Naso,* related to the nose, *Oro*, related to the mouth, and *Laryngo,* related to the larynx.) The *nasopharynx* is above the soft palate. The auditory tubes open into it. The *oropharynx* at the back part of the mouth is separated from the mouth by a pair of membranous narrow passageways called *fauces* (FAW-seez) and extends from the soft palate to the epiglottis. The *laryngopharynx* extends from the level of the hyoid bone to the larynx.

Spiritual food may only enter the body where there is an open passage for the breath of life to enter. The spiritual definition for the following three words are:

naso, which is related to the altar of incense and prayer, oro, which determines what kind of food is admitted, and laryngo, which lets words be returned that show our spiritual condition.

Tonsils

There are three types of tonsils: (1) the pharyngeal tonsils, (2) the palatine tonsils, and the (3) lingual tonsils. Tonsils are clusters of lymphatic nodules enclosed in a capsule of connective tissue. They are part of the lymphatic system and are discussed in that study. The pharyngeal tonsils (adenoids) can interfere with breathing through the nose if they become swollen, and may block the opening of the auditory tube.

Larynx

The larynx (voice box) is the dividing point of the air tube and the food tube. It is an air passage. The larynx is supported by nine cartilage. The most prominent is the thyroid cartilage (Adam's apple), which consists of two plates that fuse to form an acute angle at the midline. The thyroid cartilage is larger in males than in females.

The larynx is supported above the thyroid cartilage by ligaments connected to the hyoid bone. Below the thyroid cartilage is the cricoid cartilage, a ring of cartilage that connects the thyroid cartilage with the trachea (the wind pipe). The *cricoid cartilage* (KRI-koid; Gr. *kridos,* ring + *eidos,* form) is shaped like a signet ring.

Sound is produced by a complex coordination of muscles. The immediate source of most human sound is the vibration of the vocal cords that are pared (2) strips of mucosa-lined ligaments at the base of the larynx. These cords are made of layered, scaly, epithelium. They are at the base of the larynx and have a front to back slit between them called the glottis. A flap of cartilage folds down over the glottis during swallowing and swings back up when the act of swallowing ceases. This flap is called the epiglottis. The vocal cords are held in place and regulated by a pair of arytenoid cartilages. Above and beside the vocal cords is a pair of vestibular folds (false vocal cords) that protrude into the space of the larynx.

Connected with the speech is the support from the hyoid (spiritually represented as the principle of separation), nine cartilages (the spiritual number of judgment), an Adam's apple, the number two (2) prevalent also (spiritually support and division), a sealing instrument, a set of true and a set of false vocal cords. *"But I say unto you, that every idle word that men shall speak, they shall give account thereof in the day of judgment. For by thy words thou shalt be justified, and by thy words thou shalt be condemned."*—Matthew 12:36, 37.

The variable shapes of the nasal cavity and sinuses give voices their individual qualities. In fact, each person has such a distinct set of vocal overtones that every human voice is as unique as a set of fingerprints. The intensity, volume, or loudness of vocal sounds is regulated by the amount of air passing over the vocal cords, and that in turn is regulated by the pressure applied to the lungs mainly by the thoracic and abdominal muscles.

As the air from the lungs passes over the cords, it makes them vibrate. The tension on the cords is controlled by the forward motion of the thyroid cartilage relative to the fixed location of the arytenoid cartilages. Because the male thyroid cartilage is larger, his voice is deeper. Women have a higher and softer voice which is only one of the many differences in male and female. (They also have different roles to fulfill. A wife is to support and ennoble her husband.

His role is to lead and protect.) The actual size of the cord also contributes to pitch. Most men have longer, thicker cords.

Tone from the cord and changes in the positions of lips, tongue, and soft palate make the final quality of the sound. The tongue that is controlled by the brain via cranial nerve XII shapes the sounds coming from the larynx. The vocal cords are much like harp strings. Could this be why we will all know how to play the harp in heaven? *"The mouth of the just bringeth forth wisdom: but the froward tongue shall be cut out. The lips of the righteous know what is acceptable: but the mouth of the wicked [speaketh] forwardness."*—Proverbs 10:31,32.

Jesus said that out of the abundance of the heart, the mouth speaketh (Matthew 12:34. Think of words as the seed of the creative (reproductive) part of the brain (Mark 4:14). This seed is received into the ear, and joins with the information we have stored there and brings forth some kind of creation (ideas, thoughts, or words) either good or evil depending upon the spiritual condition. *"Hear; for I will speak of excellent things; and the opening of my lips [shall be] right things. For my mouth shall speak truth; and wickedness [is] an abomination to my lips. All the words of my mouth [are] in righteousness; [there is] nothing froward or perverse in them. They [are] all plain to him that understandeth, and right to them that find knowledge."*—Proverbs 8:6-9.

There are three veils or curtains involved with breathing and swallowing. The outer veil is the lips; the second veil is the soft palate with the uvula; the third veil is the epiglottis. Since air must pass from the pharynx with is behind the mouth to the larynx that is in front of the esophagus (the food tube), there is a crossover between the respiratory and digestive tracts. Nothing but air is allowed to enter the air chamber known as the lungs. If this rule is broken, death can occur. The epiglottis is the gate keeper to keep this law safe from law breakers.

There are three veils in the sanctuary that represents Jesus. As you breathe and swallow, keep in mind that if it were not for Jesus' life and death as our substitute, this would not be possible. *"Having therefore, brethren, boldness to enter into the holiest by the blood of Jesus, By a new and living way, which He hath consecrated for us, through the veil, that is to say, His flesh; And [having] an High Priest over the house of God; Let us draw near with a true heart in full assurance of faith, having our hearts sprinkled from an evil conscience, and our bodies washed with pure water."*—Hebrews 10:19–22.

There are five muscles that are connected with the vocal cords and rima glottis, denoting grace and judgment (available from God) in our speech. *"The LORD God hath given me the tongue of the learned, that I should know how to speak a word in season to him that is weary; He wakeneth morning by morning, He wakeneth mine ear to hear as the learned."*—Isaiah 50:4.

Trachea

The *trachea* (TRAY-kee-uh), or windpipe, is an open tube extending from the base of the larynx to the top of the lungs where it forks into two branches known as the right and left *bronchi* (BRON-kee). It is kept open by 16–20 C-shaped cartilaginous rings that are open on the backside next to the esophagus. It is both flexible and can be extended because the rings are connected by fibroelastic connective tissue and longitudinal smooth muscle. It is lined by epithelium that produces moist mucus and upward beating *cilia* (SIHL-ee-uh; L., eyelash). Dust particles and microorganisms that are not caught in the nose and pharynx may be trapped in the trachea and carried up to the pharynx by the cilia to be swallowed or spit out.

Bronchi and Lungs

The right and left bronchi that enter the right and left lungs divide into smaller secondary bronchi that divide into tertiary (segmental) bronchi. These bronchi continue to branch into smaller and smaller tubes called bronchioles and then terminal bronchioles. The whole system inside the lung looks so much like an upside-down tree that is it is commonly called the "respiratory tree." There are 10 branches on each side of the tree. The last branches, respiratory bronchioles, spread out into *alveolar* (al-VEE-uh-lur; L. small hollow or cavity) ducts. These ducts lead into microscopic air sacs called *alveoli* (al-VEE-oh-lye) where gas exchange takes place.

The right and left lungs are different sizes because the other structures in their neighborhood, the heart, the esophagus, the thoracic duct, nerves and major blood vessels, have to fit in the thoracic cavity also. The right lung has three main lobes and the left has two making a total of five (the spiritual number of grace).

There are ten subdivisions (lobules) in each lobe that are fed by the 10 branches of the bronchial tree. Each lobule functions somewhat independently from the others because they are surrounded by elastic connective tissue. Each lobule is served by a bronchiole carrying gases, a lymphatic vessel, a small vein, and an artery.

The number ten in this system spiritually alludes to the complete sufficiency of God's Holy Spirit of LOVE. The number five shows God's grace in giving us this wonderful gift, and the elastic cartilage alludes to His control over changing circumstances.

There are many things being cleansed out of this air chamber; in particular debris from pollution and mucous. Coughing and sneezing carries on this process. Foreign particles cause much trouble if they enter there and are not ejected.

Pleurae

The lungs have a covering called the inner visceral *pleura* (PLOOR-uh, Gr. side) directly on the outside surface. Next to it is a small space called the pleural cavity. Then there is the lining that lines the whole thoracic cavity known as the *parietal* (Gr. walls) pleura. In between these two membranes is the *pleural cavity,* which contains fluid to enable the lubrication of the lungs as they expand and contract.

Alveoli

Each lung contains over 350 million alveoli, each surrounded by many capillaries. They are clustered in bunches like grapes, and provide enough surface area to allow ample gas exchange. The total interior area of adult lungs provides an area almost as big as a tennis court. This is twenty times greater than the surface area of the skin. If your lungs are healthy, every time you inhale you expose this much area to fresh air.

At the surface of water, where air and water meet, the water molecules are more attracted to one another than they are to the air. The result is a thin film layer with greater cohesiveness on the surface than beneath the surface of the water. This film of strongly attracted molecules produces a surface tension like a stretched rubber membrane that tends to make the surface contract to its smallest possible area. That is why small raindrops are spherical.

An alveolus is so wet that its watery lining would tend to make it collapse if there were not some way to reduce surface tension. That tendency is counteracted by lung surfactant (a surface active agent such as soap), which reduces the surface tension of the water on the inner surface of each alveolus. As a result, the alveolar walls can be thin without collapsing.

A single alveolus looks like a tiny bubble supported by a basement membrane. It is lined with two types of cells: type one—a single layer of squamous cells; and a type two - septal cells (which secrete the detergent-like lung surfactant). They also contain phagocytic alveolar macrophages that adhere to the wall or circulate freely. They ingest and destroy foreign substances that enter the alveoli. The foreign material is moved upward by ciliary action to be expelled by coughing, as it enters the lymphatics to be carried to the lymph nodes at the *hilum* (L. trifle) of the lung.

The walls of the alveoli and the capillary walls that line them are thinner than a sheet of paper. The small size and the large number of capillaries facilitate the gas exchange that takes place here. Capillaries are so small that red blood cells flow through them in single file giving each cell maximum exposure to the walls of the alveoli. There are so many capillaries that at any instant almost a quart of blood is being processed in the lungs.

When the walls harden or are destroyed, it is called emphysema. A major cause of this is inhaling tobacco smoke that is full of poison. The blood carries this poison to the cells instead of oxygen. As the walls of the capillaries and alveoli must be thin for gas exchange, so must we be open to the workings of the Spirit of God. We must always be ready to exchange our will for His will, our ideas for His ideas, our judgment for His. We can do things to deliberately harden our hearts: We can choose to: disobey, watch, read, and hear things originated by the devil.

A group of several alveoli with a common opening into an alveolar duct is called alveolar sac.

Oxygen

The whole purpose of the respiratory system is to exchange oxygen, which is necessary for all body processes. Carbon dioxide, which is a waste product of metabolism, must also be eliminated.

Oxygen and carbon dioxide are gases at ordinary temperatures and pressures. They must be able to dissolve, first in the fluid lining of the alveoli, and then in the water component of the blood. A liquid can usually dissolve more gas at high pressure than it can at lower pressure. That is why a carbonated drink fizzes when it is poured. It is under high pressure in the bottle, but when opened, it is subject only to ordinary atmospheric pressure and the carbon dioxide comes out of solution. Under normal conditions, the lungs and blood are not subjected to great variation in pressure. When they are, it can affect gas exchange. For instance, in high altitudes, the atmospheric pressure is less and there is not enough oxygen to maintain normal respiration. Newcomers to high mountains become sluggish and feel faint, nauseated, or uncomfortable. People who live there compensate for the reduced oxygen pressure by producing more red blood cells. Deep-sea divers may have too much nitrogen dissolved in their blood as a result of the higher atmospheric pressures under water. When they come to the surface too quickly, the nitrogen comes out of solution in the form of gas bubbles. The bubbles may lodge in joints, creating the painful malady known as "the bends" (decompression sickness). If the body is decompressed slowly or breathes a mixture of oxygen and helium, the bends can be avoided.

Gas diffusion is faster when small molecules are crossing a thin membrane than it is when large molecules are crossing a thick membrane. The factors that determine the exchange of oxygen and carbon dioxide across the alveolar-capillary membrane are:

1. The partial pressure on either side of the membrane;
2. The surface area;
3. The thickness of the membrane; and
4. The solubility and size of the molecules. Large molecules diffuse more slowly than small molecules.

Without oxygen the body processes cannot be carried forward. Without God all religion is in vain. *"For through Him we both have access by one spirit unto the Father."*—Ephesians 2:18. *"That the righteousness of the law might be fulfilled in us, who walk not after the flesh, but after the Spirit. For they that are after the flesh do mind the things of the flesh; but they that are after the Spirit the things of the Spirit. For to be carnally minded [is] death; but to be spiritually minded [is] life and peace. Because the carnal mind [is] enmity against God: for it is not subject to the law of God, neither indeed can be. So then they that are in the flesh cannot please God. But ye are not in the flesh, but in the Spirit, if so be that the Spirit of God dwell in you. Now if any man have not the Spirit of Christ, he is none of His."*—Romans 8:4-9.

The oxygen cannot get where it will do any good without the blood. The blood would be useless without the oxygen to carry. The Holy Spirit's job is to testify of Christ. The great sacrifice made by the Heavenly Father and His only Begotten Son is amplified by their Spirits of LOVE working upon our hearts.

When Christ gave us His Holy Spirit of LOVE it was the highest of all gifts that Christ could solicit from His Father for the exaltation of His people. The Holy Spirit of LOVE was to be given as a regenerating agent, and without this, the sacrifice of Christ would have been of no avail. The power of evil had been strengthening for centuries, and the submission of men to this satanic captivity is amazing. Sin could be resisted and overcome only by the mighty agency of the Spirit of LOVE dwelling in the hearts of mankind. It is this that makes effectual what has been wrought out by the world's Redeemer. It is by the Spirit of LOVE that the heart is made pure. Through the Spirit of LOVE the believer becomes a partaker of the divine nature. Christ has given us His Spirit of LOVE as a divine power to overcome all hereditary and cultivated tendencies of evil, and to impress His own character upon His church.

Free Radicals

13 There are molecules in the body that do a great deal of damage. They are called free radicals. They are created by a metabolic process or by an oxygen molecule that does not have enough electrons so it robs one from another cell. This sets up a chain reaction. These processes create excessive heat and cause damage everywhere. *"Beloved, believe not every spirit, but try the spirits whether they are of God: because many false prophets are gone out into the world."*—I John 4:1.

There is a false spirit of selfishness that tries to take the place of the Holy Spirit of LOVE. There are lots of warm feelings created by this false spirit that delude people into thinking that it is the Holy Spirit of LOVE. There is one ingredient missing,. This ingredient is obedience to the Word of God *"Him (Christ) hath God exalted with His right hand [to be] a Prince and a Saviour, for to give repentance to Israel, and forgiveness of sins. And we are His witnesses of these things; and [so is] also the Holy Ghost, whom God hath given to them that obey Him."*—Acts 5:31, 32.

Some will experience false dreams and false visions which contain some truth but lead away from the original faith delivered to the apostles. Every variety of error will be brought out in the mysterious working of Satan. *"Let no man deceive you by any means: for [that day shall not come], except there come a falling away first, and that man of sin be revealed, the son of perdition; Who opposeth and exalteth himself above all that is called God, or that is worshipped; so that he as God sitteth in the temple of God, showing himself that he is God."*—II Thessalonians 2: 3, 4.

This false spirit of selfishness can take on a spirit of an accuser of the brethren. (Revelation 12:10). Satan is the "*accuser of the brethren.*" It is his spirit that inspires men to watch for the errors and defects of the Lord's people and to hold them up to notice while their good deeds are passed by without a mention. Satan is always active when God is at work for the salvation of souls and His servants are rendering Him true homage.

Satan also accuses the sinner. The Holy Spirit of God's LOVE convicts us which includes hope for the forgiveness of sin. The accuser only condemns while trying to remove all hope. *"Even the Spirit of truth; whom the world cannot receive, because it seeth Him not, neither knoweth Him: but ye know Him; for He dwelleth with you and shall be in you."*—John 14:17. When God's Spirit of LOVE dwells within us, He becomes our Comforter. There is comfort and peace in the truth, but no real peace or comfort can be found in falsehood. It is through false theories and traditions that Satan gains his power over the mind. By directing men to false standards, he misshapes the character. Through the Scriptures God's Holy Spirit of LOVE speaks to minds and impresses, truth upon the heart. Thus He exposes error and expels it from the soul.

Respiration

The term respiration has several meanings in physiology. Cellular respiration is the sum of biochemical events by which the chemical energy of foods is released to provide energy for life's processes. In cellular respiration, oxygen is utilized as a final electron acceptor in the mitochondria, the power house of the cell. External respiration, which occurs in the lungs, is the exchange of gases between the blood and the lungs, oxygen from the lungs diffuses into the blood, and carbon dioxide and water vapor diffuse from the blood into the lungs. Internal respiration is the exchange of gases in the body tissues; carbon dioxide from the body cells is exchanged for oxygen from the blood.

14 Because the lungs do not empty completely with each expiration, the air in the lungs contains more carbon dioxide and less oxygen than is in the outside air. Like water flows down hill, gas flows down from where there is a higher concentration to where there is a lower concentration. This is what makes the exchange take place in the alveoli. When the blood comes back from the tissues, it has a lower concentration of oxygen, so the oxygen flows into it. It has a higher concentration of carbon dioxide and the carbon dioxide flows out of the capillaries into the alveoli to be ex-

pelled by the lungs. Blood leaving alveolar capillaries going back to the tissues has lost carbon dioxide and gained oxygen. These processes are taking place simultaneously.

Internal respiration is just the opposite. When the blood containing oxygen gets back to the cell, the cell has become low on oxygen because it has used it for its processes and made carbon dioxide as a waste product. The oxygen enters the cell and the carbon dioxide leaves it. This sounds simple, but in reality is a very complicated chemical process that takes only seconds.

In the spiritual life there is a process that matches this. It is called sanctification and simply means the exchanging of bad for good. The Holy Spirit of LOVE through the Word accomplishes this. As we take in the ideas and standards of God, the worldly ways leave us and are expelled. *"For this is the will of God, even your sanctification."*—I Thessalonians 4:3. *"... because God hath from the beginning chosen you to salvation through sanctification of the Spirit and belief of the truth."* (II Thessalonians 2:13.) If for any length of time this process stops, it is the same as if we stop breathing physically. Our cells begin to starve for oxygen, and the decaying process begins. Sanctification is a lifetime process. One lung in this process is always trying to find out what best pleases God, so that we do not disappoint Him.

Breathing

Breathing is the mechanical process that moves air into and out of the lungs. It includes two phases: inspiration and expiration. Inspiration means breathing in. The spiritual application of this is that it is God's Holy Spirit inspires us. We use the term expiration to mean breathing out. When death occurs, God's Spirit of LOVE (the breath of life) leaves, and we call that expiration.

In a diagram of the lungs notice that the ribs provide a wall around the lungs and heart to protect them. The law of God provides a spiritual wall around us to protect us from the enemy.

There are two sets of muscles in between the ribs: the external intercostals and the internal intercostals. In a picture of the ribs, notice how the direction of the fibers of each are at almost right angles to each other. Here again we have the lesson of two, spiritually representing division and support. Each one helps out the other by doing the job it is assigned. The internal intercostals originate on the upper border of each bony rib (except #1) and insert on the lower border of the rib below. They depress the lower ribs for forced inspiration and contract for expiration. The external intercostals are just the opposite, each originate from the lower border of each bony rib (except #12) and insert in the upper border. They elevate the ribs to increase all diameters of the thorax resulting in inspiration.

The diaphragm does 75% of the work in inspiration. When air pressure inside the lungs becomes less than air pressure outside, air is sucked into the lungs as the dome shaped diaphragm contracts and moves down. At the same time the muscles of the abdominal wall relax. The contraction of the external intercostal muscles raises the ribs which increases the size of the rib cage.

The phrenic nerve that originates in the cervical plexus controls the diaphragm. Remember that the neck is the area of wisdom, If we exercise wisdom, we will be taking in deep inspiration of God's Spirit of LOVE and expelling worldly ideas of selfishness and worldly ambitions.

In expiration, the diaphragm relaxes and is raised to its higher resting position. It arches up into the thorax by the pressure generated in the abdominal cavity by the tense muscles in the abdominal wall. The rib cage becomes smaller when the ribs are lowered, largely by elastic recoil. The rib cage can change its shape and size because of the twisted shape of a typical rib and its cartilaginous connections to the sternum and vertebral column.

The spiritual lessons here are clear: Judgment (muscle) is connected to the ten commandments (ribs), which are connected to the merits of Christ (vertebrae) and to the righteousness of God (the sternum which is raised during inspiration). Cartilage is love that holds the ribs to the sternum. It is all flexible denoting the grace of God in applying to us the protection of the law and the merits of Christ, the inspiration of the Holy Spirit of LOVE, and the expiration of sin and selfishness out of our lives.

Respiratory Center

During normal quiet breathing, the main nerve controls are in the medulla oblongata and pons in the brainstem. According to the information known so far, there are two circuits of neurons. One stimulates inhalation and the other stimulates exhalation. They work alternately. During a period of about two seconds, the firing of the inspiratory neurons causes contraction of the diaphragm and contraction of the external intercostal muscles. When these muscles contract, the rib cage expands and air rushes into the lungs. While the inspiratory neuron circuit is firing, it is also sending impulses to a second circuit consisting of expiratory neurons telling them not to fire. When the inspiratory neurons stop firing, then the expiratory neurons start firing. The diaphragm and external intercostal muscles relax and the rib cage becomes smaller. The increased pressure in the lungs forces air to be expelled. During the three seconds that the expiratory circuit is activated, impulses are sent to the inspiratory neurons to keep them from firing. With these two circuits alternating, rhythmic breathing continues even in periods of unconsciousness.

Two other breath-control areas are in the pons. One is the apneustic area which causes strong inhalations and weak exhalations. The other is the pneumotaxic area that stops breathing actions. The medulla oblongata center can function without the action of these areas in the pons but may be less efficient. The whole system in both areas is called the respiratory center.

What controls this center? Among factors acting upon the control center are chemical components in the blood and sensory receptors in the joints, lung tissues, and muscles. Exercise increases the rate of breathing, because carbon dioxide is being produced more rapidly and the message goes from the blood to the brain to get rid of the carbon dioxide faster. Even a paralyzed person can have their oxygen intake increased if someone else moves their limbs for them. We should not be inactive in our spiritual lives. If we are not going forward, we are sliding backward.

We cannot see the wind, but we know when it is here because we hear it. We can feel it, and we can see the results of its contact with the Earth. We cannot explain it nor predict what it will do. It is the same with the Spirit of LOVE.

The processes of the body all depend upon oxygen. The value of our work is in proportion to the impartation of LOVE into the heart. Christ has promised the gift of His Spirit of LOVE, but like every other promise it is given on conditions. There are many who believe and profess to claim the Lord's promise; they talk about Christ, but receive no benefit. They do not surrender their soul to be guided and controlled by the divine agencies. Through the Spirit of LOVE, God works in His people *"to will and to do of His good pleasure."* (Philippians 2:13). But many will not submit to this. They want to manage themselves. This is why they do not receive the heavenly gifts. Only to those who wait humbly upon God, who watch for His guidance and His grace is the Spirit given. The power of God awaits their demand and reception. Only on these conditions are heavenly gifts bestowed.

The diseases of the respiratory system impair the reception of oxygen, some of these diseases are: asthma, emphysema, hay fever, pneumonia, tuberculosis, lung cancer, pleurisy, etc. These diseases are caused by a variety of factors, most of them by what we take into our bodies voluntarily such as junk food, tobacco, and other toxic drugs. Others are caused by things that are taken in involuntarily such as pollutions and bacteria in the air and chemicals sprayed on the produce and added to the food.

Let us remember all the spiritual lessons of the respiratory system with every breath that we take!

Study Questions

The Skeletal System

1. The skeletal system is the support of the body and spiritually represents _________ and _____________.
2. Each of the seven cervical vertebrae has a pair of openings called ________.
3. T or F. The atlas supports the head and makes it possible for us to say "yes."
4. What ability is given to us by the axis?
5. How many vertebrae are considered to be part of thoracic vertebrae? In your own words give the spiritual connection of each.
6. What are the three parts of the sternum?
7. T or F. There are six lumbar vertebrae.
8. T or F. The sacrum is five bones fused together.
9. T or F. The coccyx serves no purpose and can easily be removed. Support your answer.
10. The pectoral girdle is designed more for ____________ than ____________.
11. T or F. The clavicle acts like a strut to hold the shoulder joint and arm away from the rib cage.
12. The humerus bone spiritually represents the principles and methods of __________________________.
13. There are two bones in the forearm called the ________ and ________.
14. Name the five principles of the everlasting Gospel.
15. Name the 10 Laws of Health.
16. What is the function of the patella?

The Cardiovascular System

1. The cardiovascular system is composed of the ___________ & _________.
2. How many events are recorded in an electrocardiogram?
3. How many times will the heart beat during an average lifetime?
4. Name the three layers of the heart wall.
5. Describe the blood path.
6. What are the two different kinds of arteries?
7. How many gallons of blood are pumped to the hearts own muscle tissue every day?
8. What is an aneurysm?
9. What is arteriosclerosis?
10. Name the two portal systems in the human body and give the reason for each.

The Muscular System

1. What are the three types of muscle tissue? Give an example where each can be found.
2. What are the three basic works of muscle tissue?
3. What is meant by the word "fascia"?
4. What is the difference between tendons and aponeuroses?
5. Give the three different types of joints and how they work.
6. What is tetanization?
7. Why is calcium necessary in the function of muscles?

The Respiratory System

1. The human body can survive without _________ for several weeks and without _________ for several days; but if the _____________ stops for _______ minutes, death is likely.
2. List the seven terms we are to become familiar with in regard to the respiratory system.
3. What are the paranasal sinuses & why are they named the way they are?

4. What are the fauces?
5. Name the three types of tonsils and why are they important?
6. Sound is produced by ____.
7. What are the three veils involved with breathing and swallowing.
8. T or F. Both lungs are approximately the same size and shape.
9. What is the parietal pleura and give its function.
10. Each lung contains ____________alveoli.
11. What is emphysema and how do we prevent it?
12. the factors that determine the exchange of oxygen and carbon dioxide across the alveolar-capillary mem brane are: __________, __________, ___________ and ________________.
13. What are free radicals?
14. T or F. There is more carbon dioxide in the lungs than oxygen. Why?
15. Name the two sets of intercostal muscles and tell what they do.
16. What are the pons? Give their function.

Chapter Thirty

THE DIGESTIVE SYSTEM

Genesis 2:9

"And out of the ground made the LORD God to grow every tree that is pleasant to the sight, and good for food; the tree of life also in the midst of the garden, and the tree of knowledge of good and evil."

In the course work the digestive system is spiritually represented by the tribe of Benjamin. The characteristics that this tribe represents are ones of mental and spiritual nourishment of the body. The raven, a bird of prey, was chosen to represent the tribe of Benjamin. The word raven literally means *"to pluck off, pull in pieces, to supply with food."* This is a good description of the job of the digestive system.

What food is to the body, Christ is to the soul. Food cannot benefit us unless we eat it, unless it becomes a part of our being and Christ can be no value to us if we do not know Him as a personal Saviour. A theoretical knowledge will do us no good. We must feed upon Him and receive Him into our heart. Only then can His life become our life. His love and His grace must be *assimilated.*

The physical digestion and assimilation of food is a mechanical and chemical process that breaks down large food molecules into smaller ones. Unless the small molecules can enter into the cells, they are useless. They must pass through the cells of the intestine into the bloodstream and lymphatic system. This is called *absorption*. There is a saying "you are what you eat," which is not really true. In reality, you are what you assimilate or absorb.

The same is true of spiritual nourishment. It does not do any good to study the Word of God if it does not become part of you. *"Ever learning, and never able to come to the knowledge of the truth."*—II Timothy 3:7.

The perception and appreciation of truth depends less upon the mind than upon the converted heart. Truth must be received into the soul; it claims the homage of the will. If truth could be submitted to the reason alone, pride would be no hindrance in the way of its reception. But it is to be received through the work of grace in the heart and its reception depends upon the renunciation of every sin that the Spirit of God reveals. Man's advantages for obtaining a knowledge of the truth; however great these may be, will prove of no benefit to him unless the heart is open to receive the truth. There must be a conscientious surrender from every habit and practice that is opposed to principles. To those who thus yield themselves to God, having an honest desire to know and to do His will, the truth is revealed as the power of God for their salvation. These enlightened ones will be able to distinguish between truth and error.

The physical body must grow and make repairs to all of its cells. Where does the body get the material it needs for energy and to grow and to make repairs? The food that we eat and the air we breathe supplies the raw material for this energy. It is very important to eat the right food, just like it is important to put the right fuel in the car so that the engine will run correctly. Let's examine two classes of foods that the body assimilates.

Carbohydrates are the easiest compounds for the body to use because the cells have the chemical machinery to break down carbohydrates. God intended that most of our nourishment would be derived from foods that contain high amounts of carbohydrates so He gave the body this ability.

The simplest type of carbohydrate is called a *monosaccharide* (mah-no-SAK-ah-ride). This comes from the Greek word meaning *one sugar*. It has only one link of a chain in it and cannot be broken down into any smaller components. Some monosaccharides are *glucose, fructose* and *ribose*. When there is an *"ose"* on the end of a word, it usually stands for sugar. There are eight different sugars used in the body. The way the atoms are joined together in these sugar molecules give them different names. People in your family have the same kind of atoms but are joined together differently. Some are girls or boys, some are tall and some short, and they have different names. They are different people, but belong to the same family.

When two monosacchrides are joined together, it is called a *disaccharide.* If three are joined together, it is called a *trisaccharide*.

<u>Proteins</u> are the major building blocks in most of the tissues of the body. They are complex compounds and must be broken down into separate *amino acids* to be assimilated. There are twelve amino acids that the body can make, but there are eight more that the body does not make. We must get them from the food we eat. (II Peter 1:5-7 tells us about the eight essential things we need to get from God to build our character.) These twenty different amino acids are linked together by the body to form the necessary material to grow or repair our body cells. They also combat disease and give us energy to do work.

When two amino acids are bound together, they form a protein called a *dipeptide.* Three amino acids bound together are called *tripeptide.* Many bonded together are called *polypeptides. Enzymes* put the chemical chains together.

The work of the digestive system includes five processes:

1. *Ingestion* is the taking in of food into the body by mouth; eating.
2. *Peristalsis* is the muscle contractions that move the food along the digestive tract.
3. *Digestion* is the breaking down of large molecules into small ones.
4. *Absorption* is the passage and uptake of food molecules from the intestines into the blood stream and lymphatic system and on into the cells.
5. *Defecation* is the elimination from the body of the solid wastes from these processes.

<u>The alimentary canal</u> is the digestive tract that is a tube that reaches from the mouth to the anus. It is about 30 feet long in an adult. It consists of eight parts; mouth, pharynx, esophagus, stomach, small intestine, large intestine, rectum and the anal canal. There are eight structures that are associated with the digestive system; lips, teeth, tongue, cheeks, salivary glands, pancreas, liver, and gallbladder. Each of these structures is designed for its specific job.

The walls of the alimentary canal consists of four main layers: (1) The <u>*mucosa*</u> is made up of the *epithelium* (site of interaction between the food and the body, which has large folds), the *supporting layer* of connective tissue which contains lymphatic nodules and many lymphocytes, and the thin *muscular layers* which contain a nerve plexus. The mucosa contains cells that absorb and also secrete mucus, enzymes and various ions. (2) The <u>*submucosa*</u> is connective tissue that contains many blood vessels, nerves; and in certain areas lymphatic nodules. (Submucosal glands are present in the esophagus and duodenum.) (3) The <u>*muscularis externa*</u> is the main muscle layer consisting of an inner circular layer and an outer longitudinal layer of mostly smooth muscle which moves the food by waves of muscular contraction called *peristalsis.* (4) The *serosa* is the outside layer of thin connective tissue.

The mouth is the first part of the digestive tract. This is the area that spiritually represents the Table of Shewbread in the body temple. (Exodus 25:23-30). Notice the table had two crowns. You have two sets of crowns in your mouth: the upper teeth, and the lower teeth. The lower border in the sanctuary was just a hand's breadth wide. Your mandible bode is narrow. The Table of Shewbread was one and one half cubits high. It was the same height as the grate (net) on the Altar of Sacrifice which was in the *midst* of the altar of three cubits (Exodus 27:1-5), which would make it 1½ cubits high. Similarly, the medulla and the mouth are at the same height or level in the head.

The mouth is the place where *physical* nourishment begins. The Table of Shewbread represented *spiritual* nourishment. The bread on the Table of Shewbread was called the bread of *presence* but is more accurately translated bread of *face.* (See Strong's Concordance, number 6440.) *"For God, who commanded the light to shine out of darkness, hath shined in our hearts, to [give] the light of the knowledge of the glory of God in the face of Jesus Christ."*— II Corinthians 4:6. The Word became flesh and dwelt among us, He was the ultimate communication between us and the Heavenly Father.

"I am the bread of life: he that cometh to me shall never hunger; and he that believeth on me shall never thirst." —John 6:35. Read all of John 6, which is the story of earthly bread and heavenly bread. Verse 51 tells us we need to eat of this Bread. Verse 54 calls the flesh of Christ this Bread, and verse 63 explains that the Bread is the word of Christ. The communion service reminds us of this. There are some who believe that the bread literally becomes the body of Christ because they do not understand the true meaning of this scripture. If we do not *live His life,* we have not eaten the Bread.

The life of Christ that gives life to the world is His Word. His Word manifests His power. It was by His Word that Jesus healed disease and cast out demons; by His word He stilled the sea, and raised the dead. The people bore witness that His Word was delivered with power. He spoke the Word of God as He had spoken through all the prophets and teachers of the Old Testament. The whole Bible is manifestation of Christ, and the Saviour desires to fix the faith of His followers on the Word. Without His visible presence, the Word must be our source of power. Like our Master, we are to live *"by every word that proceedeth out of the mouth of God."*—Matthew 4:4. Then we too, will have this power that Christ had.

Leviticus 24:5-9 gives the instruction for the preparation of the communion bread. There were 12 loaves, one loaf representing each of the 12 Tribes of Israel. This object lesson lets us know that the Word of God has nourishment for every member of every tribe.

The ingredients of the bread were the same as the unleavened bread of offerings talked about in Leviticus 2. (Remember that *meat* just means *food* and not flesh of dead animals.) The bread was made of fine flour which can only be made through much grinding and pulverizing. Christ's body was ground into the dirt and pulverized verbally. Another ingredient of the bread was oil, the symbol of God's Holy Spirit. The fat from off an animal was a symbol of sin and men were not to eat the fat. Water represents truth. Review frankincense (His divinity) and salt (His grace) in the Smell study. The bread contained no leaven which was a symbol of sin.

The mouth is also called the *oral cavity.* The opening to the throat, *fauces* (L. throat), contains sensory nerve endings that trigger swallowing. These nerve endings cause gagging if touched by a tongue depressor or other objects.

The lips are the veil that protects the opening of the mouth. The lips contain sensory nerve endings from cranial nerve V (5) that makes it possible for us to identify different textures of food. The lips help place food in the mouth and keep it in a position so it can be chewed. Likewise grace is given to help test that spiritual food we should ingest. *"He that speaketh from himself seeketh his own glory: but he that seeketh the glory of Him that sent him, the same is true, and no unrighteousness is in him."*—John 7:18.

Cheeks help hold the food where the teeth can chew it. The muscles of the cheeks contribute to the chewing process by pushing the food between the teeth. This usually all happens without your even being aware of it unless you bite one of your cheeks. There are four pairs of muscles that produce biting and chewing movements.

The Palate was discussed in the Respiratory Study. The roof of the mouth (palate—PAL-iht) has two sections which are perfectly designed for the functions they perform. (1) The hard palate is formed by a portion of the palatine bones and maxillae. When food is being chewed, the tongue is constantly pushing it against the ridged surface of the hard palate to help crush and soften the food before it is swallowed. (2) The back portion, the soft palate, is elevated during swallowing and the uvula prevents food from entering the nasal pharynx.

The front of the tongue helps move the food around during chewing and the base of the tongue aids in swallowing. There are three pairs of muscles with attachments outside the tongue and three pairs of muscles wholly within the tongue. This makes 12 muscles in all which are innervated by cranial nerve XII. We should have perfect government and judgment over our tongue both in tasting and talking. *"When thou sittest to eat with a ruler, consider diligently what is before thee: And put a knife to thy throat, if thou be a man given to appetite. Be not desirous of his dainties: for they are deceitful meat."*—Proverbs 23:1-3.

"Eat thou not the bread of him that hath an evil eye, neither desire thou his dainty meats: For as he thinketh in his heart, so is he: Eat and drink saith he to thee; but his heart is not with thee. The morsel which thou hast eaten shalt thou vomit up, and lose thy sweet words."—Proverbs 23:6-8. This admonition not only includes rich food that tastes good but makes us unhealthy in the long run. It also represents the modern social gospel that makes us feel good about self but does not cause true spiritual growth.

Salivary glands

There are three pairs of salivary glands which secrete saliva. This saliva contains about 99 percent water and one percent electrolytes and proteins. Saliva lubricates chewed food, moistens the oral walls, buffers chemicals in the mouth with its salts, and contains an enzyme that begins the digestion of carbohydrates. Each of these three different glands secrete saliva that has a slightly different composition.

Salivary secretion is entirely under the control of the autonomic nervous system and it has no hormonal stimulation like the rest of the digestive secretions. If subjected to unpleasant stimuli, such as the smell of rotten food, the mouth will become dry. This can also happen during stressful situations. Salivation slows down during sleep.

Have you ever had a dry mouth or tried to swallow something dry? In the dry desert, God provided water. God constructed our bodies to use water to lubricate to lubricate the digestive tract and to break down the food so that it can be used by the body. When we mix spiritual food with God's thoughts, it will be digestible. *"For My thoughts are not your thoughts, neither are your ways My ways, saith the LORD. For as the heavens are higher than the earth, so are My ways higher than your ways, and My thoughts than your thoughts."*—Isaiah 55:9,10.

Some people have made it a habit to drink fluids with their food. This neutralizes and dilutes saliva and slows down digestion. When we mix worldly ideas with the bread of life, it dilutes our understanding and absorption of the Word. The Jews prided themselves on their knowledge of the Scriptures and had even memorized them. They still did not understand the scriptures. Jesus told them, *"Ye do err, not knowing the scriptures nor the power of God."* —Matthew 22:29.

Peter admonishes that careful understanding of Paul's writings is needed. Peter 3:16 *"As also in all [his] epistles, speaking in them of these things; in which are some things hard to be understood, which they that are unlearned and unstable wrest, as [they do] also the other scriptures, unto their own destruction."*

We must approach Bible study with great humility and much prayer. *"But I keep under my body, and bring it into subjection: lest that by any means, when I have preached to others, I myself should be a castaway."*—I Corinthians 9:27.

Teeth: A tooth has three sections: root, neck, and crown. The crown is the part that shows, the neck is covered by the gums, and the root is embedded in a socket in the bone.

It is interesting that the Hebrew word for tooth comes from a root word that can mean *to teach diligently*. As the teeth grind and crush the food, so we need to rightly divide the word of truth. *"Study to shew thyself approved unto God, a workman that needeth not to be ashamed, rightly dividing the word of truth."*—II Timothy 2:15.

Top teeth are held by the maxilla and bottom teeth by the mandible. Without being rooted by these two elements, the help of the Holy Spirit and discipline, we cannot chew the Word of God. Teeth are held in sockets by connective tissue called periodontal ligaments. Nerve endings in the ligaments monitor the pressures of chewing and relay the information to the brain.

The teeth are composed of four parts: dentine, enamel, cement; and pulp. The surface that comes in contact with the food is enamel which is the hardest substance in the body. We have to be hard with ourselves when we come in contact with the Word of God or else we will turn away from Him when His convictions interferes with our own will and desires.

Acid from decaying food can cause decay in the enamel of teeth. If we hear the Word of God without taking it to heart and acting upon it, it will cause decay in our spiritual enamel, and soon it will cause pain in the inner part.

Dentine is the extremely sensitive yellowish portion that forms the bulk of the tooth. *Odontoblasts* line the innermost surface of the dentine and convey the effects of stimulation to nerve endings in the pulp, which causes the teeth to be sensitive to pain.

Cement is the bonelike covering of the neck and root of the tooth. The *pulp* is the soft core of connective tissue that contains the nerves and the blood vessels of the tooth.

Chewing can be voluntary but most of it is automatic caused by the pressure of food against the teeth, gums, tongue, and hard palate. The pressure causes the jaw muscles to relax and the jaw to drop slightly Then, as opposite muscles contract in an attempt to balance the relaxation, the jaw is pulled up again.

Adults have 32 teeth. There are only 20 baby teeth. Baby teeth are also called "milk" teeth. The Hebrews only partially understood the Levitical priesthood that had been in place since the Exodus from Egypt. They were still using their milk teeth and ingesting milk instead of meat. Paul wanted to talk to the Hebrews about Jesus' priesthood in the order of Melchisedec, but they were still barely able to understand the rudiments or the foundation of spiritual things. The priesthood of Melchisedec deals with victory and perfection. (Read Hebrews 5:1- 6:2.) Paul here likens this to the different nourishment of babies and adults. Babies are not able to chew but must be fed liquids.

According to Hebrews 6:1-2, spiritual milk has seven ingredients, or the basics of salvation which create and nourish a new creature:

1. principles of the doctrine of Christ, how to be saved;
2. repentance from dead works, we cannot save ourselves;
3. faith toward God;
4. baptisms;
5. laying on of hands;
6. resurrection of the dead;
7. eternal judgment.

As it is necessary for a baby to start on milk, so it is necessary for the Christian to start nourishment with these teachings. The first growth must come from the milk. But the plan of God is to progress from milk to solid food.

The whole Book of Hebrews was to try to wean the people from spiritual milk and start them on solid food. The solid food was perfection; obedience from the heart. "*This [is] the covenant that I will make with them after those days, saith the Lord, I will put my laws into their hearts, and in their minds will I write them; And their sins and iniquities will I remember no more.* Hebrews 10: 16, 17. "*For this [is] the covenant that I will make with the house of Israel after those days, saith the Lord; I will put my laws into their mind, and write them in their hearts: and I will be to them a God, and they shall be to me a people: And they shall not teach every man his neighbour, and every man his brother, saying, Know the Lord: for all shall know me, from the least to the greatest.*"—Hebrews 8:10,11 "*Now the God of peace, that brought again from the dead our Lord Jesus, that great shepherd of the sheep, through the blood of the everlasting covenant, Make you perfect in every good work to do his will, working in you that which is wellpleasing in his sight, through Jesus Christ; to whom [be] glory for ever and ever. Amen.*"—Hebrews 13: 20,21. It brings glory to Jesus when we stop sinning and reflect His character.

Sometimes teeth are used for biting people and tongues are used for wounding. *"My soul is among lions; and I lie even among them that are set on fire, even the sons of men, whose teeth are spears and arrows, and their tongue a sharp sword."*—Psalm 57:4 *"But if ye bite and devour one another, take heed that ye be not consumed one of another."* —Galatians 5:15.

Before eruption the teeth are completely surrounded by alveolar bone. Our tools for dividing the word of God should be grounded in principle, not emotion. During eruption, the portion of the alveolar process lying over the crown is absorbed.

Gums

Gums (gingiva; jihn-JYE-vuh) are the firm connective tissues covered with mucus membrane that surrounds the alveolar (hollow) processes of the teeth. The stratified squamous epithelium of the gums is slightly keratinized (tough, from Gr. *horn*) to withstand friction during chewing. The gums are usually attached to the enamel of the tooth somewhere along the crown.

Swallowing spiritually represents believing. Even in the secular world you hear people say, "He swallowed that one, hook, line, and sinker."

After food in the mouth is chewed and moistened by saliva, it is known as a bolus (Gr. *lump).* The first stage of swallowing is voluntary. It is the only voluntary movement until the feces are expelled. So there is a voluntary act at the beginning and at the ending of the digestive process. There is a voluntary act when we decide to accept what we have heard and discard what we cannot use. *"The words of a wise man's mouth are gracious; but the lips of a fool will swallow up himself. The beginning of the words is foolishness; and the end of his talk is mischievous madness."*—Ecclesiastes 10-:12,13. *"Woe unto you, scribes and Pharisees, hypocrites! for ye pay tithe of mint and anise and cumin, and have omitted the weightier [matters] of the law, judgment, mercy, and faith: these ought ye to have done, and not to leave the other undone. [Ye] blind guides, which strain at a gnat, and swallow a camel."*—Matthew 23:23,24.

Pharynx

The pharynx is about 4 or 5 inches long and was studied in the Respiratory System because it is both the air and food passage.

Esophagus

The esophagus is about 10 inches long and consists of three coats of muscle and membrane. Each end of it is closed by a sphincter muscle when the tube is at rest and collapsed. A sphincter (SFINK-tur; Gr. *to bind tight*) is usually in a state of contraction like the tightened drawstrings of a purse. The bottom sphincter opens only long enough to allow food and liquids to pass into the stomach. If there is a digestion problem with it, the acid in the stomach comes into the esophagus causing "heartburn." This can happen when too much food is eaten or wrong combinations of food are ingested.

The cavity below the diaphragm is called the abdominal cavity. It has a double walled sac surrounding it. The two walls (membranes) that make up these sacs are separated by a thin film of serous fluid, a watery fluid, which acts as a lubricant.

Serous membranes, consist of a smooth sheet of simple squamous epithelium and an adhering layer of loose connective tissue containing capillaries, that line the cavity and the viscera contained within the abdominal cavity. The serous membrane of the cavity is the *peritoneum* (per-uh-tuh-NEE-uhm). The *parietal* peritoneum lines the cavity, and the *visceral* peritoneum covers most of the organs in the cavity. Between the two is a space called the *peritoneal cavity.* This cavity usually contains a small amount of serous fluid, which is secreted by the peritoneum and allows for nearly frictionless movement.

When the heart is not pumping right or during grave illness, this cavity fills with fluid. This serous fluid represents the tolerance that we should have in our fellowship with other people. When there is too little fluid, there is condemnation, when there is too much fluid, we are condoning sin. It is a fine line to walk in order to love the sinner and hate the sin, and this can only be done by the Holy Spirit of LOVE dwelling within us.

The *mesenteries* are two fused layers of serous membrane. They suspend some organs from the back abdominal wall. They are also used for the liver, stomach, spleen, most of the small intestine, and the transverse colon of the large intestine. They provide a point of attachment plus convey the major arteries, veins, and nerves of the digestive tract, liver, pancreas, and spleen to and from the body wall.

The peritoneal sac is subdivided into greater and lesser sacs by the stomach and two special mesenteries, the greater and lesser *omenta* (L. fat skin). There are a few cells and some fluid but no organs within these sacs. The greater omentum is folded membrane from the stomach down to the pelvic cavity. It contains large amounts of fat, if too much, a pot belly occurs. It hangs down like a fatty apron to protect and insulate the organs. The lesser omentum extends from the liver to the stomach.

Stomach

Terms: *Fundus* (L. base); *Pylorus* (pie-LOR-us; Gr. gate keeper); *Rugae* (ROO-jee, L. folds); *Gastro* (Gr. *gaster*, belly, which is why the digestive tract is called the gastrointestinal tract); *Follicle* (L. little bag, a small secretory sac or cavity.

There is no organ that changes form or position as much as the stomach, except the uterus during pregnancy. Its normal capacity is four to eight pints, but can be extended to over a gallon. When it is empty, it is all wrinkled and in a different position than when full. It is important to not eat so much that the stomach is overstretched, because then it cannot work efficiently.

As our physical life is sustained by food, so our spiritual life is sustained by the Word of God. And every soul is to receive life from God's Word for himself. As we must eat for ourselves in order to receive nourishment, so we must receive the Word for ourselves. We are not to obtain it merely through the medium of another's mind. We should carefully study the Bible, asking God for the aid of the Holy Spirit of LOVE, that we may understand His Word. We should take one verse, and concentrate the mind on the task of ascertaining the thought which God has put in that verse for us. We should dwell upon the thought until it becomes our own, and we know *"what saith the LORD."* It is better to thoroughly understand one verse than to hurriedly gulp down a whole chapter.

When people say they have a stomach ache, they usually refer to the area around the navel, but actually the stomach is much higher than that. It is located right under the dome of the diaphragm and is protected by the rib cage. It has been divided into four sections.

The stomach wall has four layers. The layer called the *muscularis externa* (outside muscle) contains three layers of muscle running in three different directions. Muscles represent judgment.) They all have a job to perform when the stomach is churning. This represents the judgment that must be exercised when we are dividing the Word of God. It must go in many directions so that we consider the context, the circumstances, and who is making the statement.

For instance, *"Therefore the Jews sought to kill Him, because He not only had broken the Sabbath, but said also that God was His Father, making Himself equal with God."*—John 5:18. If we only look at that verse, we might assume that Christ was a Sabbath breaker, but if we put it together with all the verses that say He kept His Father's commandments and never broke a law of any kind, then we understand that the Jews *thought* that He broke the Sabbath. The Jews did not understand how to keep the Sabbath. Luke, the second chapter, explains the errors of the Pharisees. John 8:46 and 15:10 tell us that no one could accuse Jesus of breaking the law. People still make statements like that today. We must use judgment when we study.

There are four classes of stomach cells. There are six functionally active cell types lining the surface, pits, and glands of the stomach:

1. surface mucous cells,
2. neck mucous cells,
3. parietal cells,
4. zymogenic cells,
5. enteroendocrine cells,
6. enterochromaffin cells.

These all secrete different substances. There are also undifferentiated cells that are able to replace the other glandular cells when they die, making a total of seven cells.

Lining and Glands: The stomach lining is completely renewed every *three* days. It does not get digested along with the food because it is covered with a protective coat of mucus. Mucosa contains dense masses of lymphocytes.

The stomach lining is indented with millions of gastric pits. Extending from each of these pits are 3 to 8 gastric glands. There are 3 types of gastric glands: (1) cardiac glands called this because they are near the opening to the esophagus; the cardiac orifice; (2) pyloric glands in the pyloric canal; and (3) fundic glands; which are the most numerous. Normally over one quart of gastric juice is secreted daily by these cells.

Stomach Action: The stomach has three main functions: (1) secretes acid and enzymes that help digest food and kill bacteria; (2) churns food and breaks it into small particles and mixes them with gastric juices; (3) stores ingested nutrients until released into the small intestine.

In the stomach's smooth muscle cells are pacemaker cells. A few minutes after food enters the stomach, these pacemaker cells generate action in the muscles and slow mixing begins. As actions become stronger, they chop and mix the food until it becomes *chime* (KIME), a soupy liquid mixture.

While the stomach is moving, its glands are secreting hydrochloric acid and enzymes that break down the proteins. This secretion is controlled by three different phases form the head, belly, and intestine.

Head, cephalic phase, is caused by a stimulation of the vagus nerve, (cranial X), which responds to activities such as smelling, tasting, chewing, and swallowing. This causes the cells of the pyloric gland area to secrete *gastrin,* which is carried by the blood back to the cells of the stomach glands and stimulates them to secrete gastric juices. If you are in any way upset by anger, aggression, pain, sadness, depression, etc., this will disrupt the head phase and cause more or less secretions than is needed. If more acid is produced ulcers can occur; if less acid is produced, food cannot be digested. *Be very careful as to what subjects are discussed at the table.

Belly, gastric phase, is caused by fragments of protein and peptide, distension of the stomach, and stimulants such as alcohol and caffeine all of which increase secretion.

Intestinal phase, consists of components that excite and inhibit. These are controlled by conditions in the duodenum.

When the consistency is just right the *gatekeeper*, the pyloric valve, allows some chime to come through into the small intestine. This is a small opening, and the gatekeeper tries to carefully regulate what comes through. For instance, if more than *two teaspoons* of fat is allowed through at once, the small intestine has a hard time trying to take care of it.

The rate of emptying is monitored by the stomach, the duodenum, and the central nervous system which should ensure that the stomach does not become too full and does not empty faster than the small intestine can process the chime. The stomach should empty in 3 to 4 hours after a full meal.

The stomach does not ordinarily absorb. The exception to this is alcohol, which is absorbed by the stomach. Also, recent studies indicate that in some people, milk causes the stomach lining to become porous, allowing some large proteins to enter the blood stream from the stomach. *This is a cause of allergies.*

Small intestine

The small intestine begins at the pyloric opening and ends at the iliocecal valve, where it joins the large intestine. The large and small intestine were named by the size of diameter, not the length. After the chime is passed from the stomach, it enters the small intestine where further contractions continue to mix it with more secretions. Depending upon the content the food should take one to six hours for it to move through this 20 feet of small intestine. This part of the intestine absorbs almost all the digested molecules of food and sends it into the bloodstream and the lymphatics which carry the nutrition on to the cells.

The small intestine is subdivided into three parts:

1. duodenum (doo-oh-DEE-nuhm; Gr. *12 fingers wide*);
2. jejunum (jeh-JOO-nuhm; *fasting intestine*, so named because it was found empty when a corpse was dissected) and
3. ileum (ILL-ee-uhm; L. *groin*).

The wall of the small intestine is composed of the same four layers as the rest of the digestive tract; however, its mucosa has three different features:

1. It has folds that are permanent, unlike the stomach whose folds disappear when it is full;
2. It has millions of long fingerlike protrusions called *villi* (VILL-eye; L. *shaggy hairs*) which increase the absorptive surface;
3. Within the mucosa are simple, tubular glands that secrete several enzymes that aid in digestion.

This entire epithelial surface is replaced about every five days.

The duodenum is only 10 inches long and is divided into four sections. Most of the remaining digestive processes take place here, and most of the absorption occurs here and in the jejunum.

The jejunum is about 8 feet long.

The ileum has regions in the lamina propria where lymphoid tissue is concentrated. Large lymphatic nodules form Peyer's patches, which help destroy microorganisms. (Review the lymphatic study if you do not remember what this is.)

There are many organs that assist in the digestive processes, and we will consider some of those now since they deal directly with the small intestine. When the chime is emptied into the small intestine, juices secreted by the intestinal cells and secretions from the pancreas and liver are added to mix with it.

Liver

The liver is the largest organ in the body, weighing about three to four pounds. It is the most complicated and the most versatile of all organs and without it the body would perish in 24 hours. It performs some 500 functions. The *hepatic cell* is a unique, extraordinary structure that is responsible for nearly all of the vital work of the liver.

The spiritual lesson of the liver is the lesson of faith. *". . . The just shall live by faith."*—Habakkuk 2:4. Of all the organs in the body, probably no other tells of the wondrous grace of God better than the liver. It can regenerate better than any other organ. *"And the apostles said unto the Lord, Increase our faith."*—Luke 17:5. It is enclosed in a network of connective tissue. *"That your faith should not stand in the wisdom of men, but in the power of God."*—I Corinthians 2:5. The Liver is divided into unequal parts. *"For I say, through the grace given unto me, to every man that is among you, not to think of himself more highly than he ought to think; but to think soberly, according as God hath dealt to every man the measure of faith.. For as we have many members in one body, and all members have not the same office."*—Romans 12:3,4

The liver is the main organ of detoxification. *"But the Lord is faithful, who shall stablish you, and keep you from evil."*—I Thessalonians 3:3. The liver sorts out the nutrients, stores them and sends them where they are needed. *"If thou put the brethren in remembrance of these things, thou shalt be a good minister of Jesus Christ, nourished up in the words of faith and of good doctrine, whereunto thou hast attained."*—I Timothy 4:6.

The blood from the intestines passes through the liver before it passes through the heart. *"But without faith it is impossible to please him: for he that cometh to God must believe that His a rewarder of them that diligently seek Him."*—Hebrews 11:6.

If the liver is wounded within its capsule, it can be repaired; but if the protective lining is pierced and the laceration is extensive, death usually occurs from loss of blood. The veins in the liver are contained in rigid canals and are unable to contract, and they have no valves. If our love is damaged and the life of Christ begins to drain out of our spiritual life, faith becomes weak. Then we are not able to digest spiritual food, and death usually occurs.

Faith is the condition upon which God has seen fit to promise pardon to sinners: not that there is any virtue in faith whereby salvation is merited, but because faith can lay hold of the merits of Christ, the remedy provided for sin. Faith can present Christ's perfect obedience instead of the sinner's transgression. When the sinner believes that Christ is his personal Saviour, then according to His unfailing promises, God pardons his sin and justifies him freely. The repentant soul realizes that his justification comes because Christ, as his substitute and surety, has died for him. Christ is his atonement and righteousness.

"Abraham believed God, and it was counted unto him for righteousness. Now to him that worketh is the reward not reckoned of grace, but of debt. But to him that worketh not, but believeth on him that justifieth the ungodly, his faith is counted for righteousness."—Romans 4:3-5.

Righteousness is obedience to the law. The law demands perfect obedience, righteousness, and this the sinner owes to the law; but he is incapable of rendering it. The only way in which he can attain to righteousness is through faith. By faith he can bring to God the merits of Christ, and the Lord places the obedience of His Son to the sinner's account. Christ's righteousness is accepted in place of man's failure, and God receives, pardons, justifies the repentant, believing soul, treats him as though he were righteous, and loves him as He loves His Son. This is how faith is accounted righteousness.

In Gray's Anatomy, the section on the liver states that: There are five surfaces that are considered when examining it. The liver has five ligaments, four that are peritoneal folds plus the round ligament. The liver has five vessels, four blood plus lymphatics. The liver has five fissures on the under and back part which divide it into five lobes, four in the right side plus one in the left (the lobus Spigelii is not mentioned in the later anatomy books so they list only four.) The lobule has five or six sides. It is through these numbers that can see *grace* at work in this organ that represents faith.

Some terms connected with the liver: The name, *liver,* comes from an Anglo-Saxon word meaning *lifer,* because this organ is so vital to life. Anastomoses (Gr. *opening*, and opening where two vessels meet.) Hepatic (Gr. *pertaining to the liver*.) Parenchyma (pahr-EN-kie-mah; Gr. to *pour in beside*.) Polyhedral (Gr. ploys, many + hedra, base, *having many surfaces*.). Sinusoids (L. *sinus,* a hollow + Gr. *eidos,* image, something seen, form), a minute blood vessel in the liver, larger than a capillary.)

Lobules: Each lobule is about the size of a millet seed. In a diagram of the liver notice how the cells are arranged in one-cell-thick plate-like layers that radiate from the central vein to the edge of the lobule. There is a portal area on each corner of the lobule which is composed of branches of the portal vein, hepatic artery, bile duct, and nerve.

Sinusoids

The sinusoids are delicate blood channels between the rows of cells. They transport blood from branches of the portal vein and hepatic artery. 20% of the blood in the liver comes from the hepatic artery and 80% from the hepatic portal vein. All the blood from the gastrointestinal tract comes to the liver via the hepatic portal vein before it goes to the heart. The blood flows from the artery and vein in the portal areas into the *sinusoids* and then to the central vein then drains into the hepatic vein, to the inferior vena cava, and to the heart.

Macrophages

Macrophages are collectively referred to as the reticulo-endothelial system ("network of endothelial cells"). These cells are found in many places, including the interior walls of the sinusoids in the liver. These cells group into large clusters that surround and isolate foreign particles that are too large to phagocytize. They engulf and ingest worn-out red and white blood cells, microorganisms, and other foreign particles that pass through the liver.

Functions: The following functions of the liver are very condensed and each one requires many complicated chemical processes to accomplish. As these functions are considered, keep in mind that all this takes place without conscious effort. The only thing one can do is to feed the liver the right material. *"Faith cometh by hearing and hearing by the word of God."*—Romans 10:17.

The Liver:

1. Removes amino acids from organic compounds.
2. Forms urea from worn-out tissue cells (proteins) and converts excess amino acids into urea to decrease body levels of ammonia.
3. Manufactures most of the plasma proteins, destroys worn-out red blood cells, storing the chemical *hematin* (needed for red blood cells to become mature), manufactures *heparin,* helps synthesize the blood clotting agents from amino acids; all these processes help to keep blood pure.
4. Removes *bilirubin* (a bile pigment produced by the liver) from the blood.
5. Synthesizes certain amino acids.
6. Converts galactose and fructose to glucose.
7. Oxidizes fatty acids.
8. Forms lipoproteins, cholesterol and phospholipids, essential constituents of plasma membranes.
9. Converts carbohydrates and proteins into fat.
10. Detoxifies, by metabolizing drugs, pesticides, herbicides, poisons, etc.
11. Synthesizes vitamin A from carotene.
12. Maintains stable body temperature by raising the temperature of the blood passing through. The liver's many metabolic activities make it the body's major heat producer during rest.
13. Stores glucose in the form of glycogen, and with the help of enzymes, it converts glucose as it is needed by the body. Glucose is the body's main energy source. It also stores the fat-soluble vitamins A, D, E, and K, minerals such as iron, and vitamin B12. The liver can also store fats and amino acids and convert them into usable glucose as required.
14. Secretes over one quart of bile every day. Bile is an alkaline liquid containing water, sodium bicarbonate, bile salts, bile pigments, cholesterol, mucin, lecithin, and bilirubin.

It should be remembered here that the liver performs over 500 other functions too numerous to be listed. Some of these functions even scientists do not know how the liver performs them.

Bile salts are derived from cholesterol. They are part of the bile that is stored in the gallbladder and used in fat digestion and absorption. After they have done their job, they are reabsorbed or taken up into the blood by a special active transport system located in the last part of the ileum. From there they return to the liver, which re-secretes them into the bile.

Bile pigments give bile its color and are derived from the hemoglobin of worn-out red blood cells that have been transported to the liver for excretion. The condition called jaundice, where the skin looks yellowish, is caused by excessive amounts of bilirubin in extracellular fluids.

Pancreas

The pancreas is similar in structure to the salivary glands, but is softer and less complicated. The pancreas is a mixed gland. The weight and mass is 99% exocrine gland which secretes through a duct, and about 1% endocrine gland which secretes directly into the blood stream. The exocrine cells secrete the digestive juices while the endocrine cells secrete hormones. Refer to the Endocrine System for more information on this organ..

The pancreas is five to six inches long and is divided into three parts: head, body, and tail. It tucks its head into a curve in the small intestine and its tail into a curve in the spleen.

The exocrine cells form groups called *acini* (ASS-ih-nye; L. *grapes*), which secrete digestive fluids. These cells are clustered around tiny ducts that pick up the fluid and carry it to the main pancreatic duct which joins the common bile duct. The common duct widens into the *ampulla* (L. *little jar*), which then enters the small intestine. The pancreas produces more digestive fluid than any other digestive organ: over 1½ quarts a day.

Gallbladder

The gallbladder is about four inches long with three layers in its wall. It stores bile that comes from the liver and delivers it when the messenger hormone, *chole-cystokinen*, comes from the small intestine to say that bile is needed in the small intestine. The gallbladder will then contract, squirting bile into the duodenum via the ducts. Bile is rich in cholesterol which may form stones in the gallbladder. The bilary system has three ducts: hepatic, cystic, and common.

Action of the Small Intestine: The main jobs of the small intestine are digestion and absorption which are made possible by muscle movement and chemical action of enzymes. The digestive juices come from different sources. The exocrine glands in the mucosa secrete into the lumen about 1½ quarts of gastric juice containing a dilute salt and mucus. The small intestine synthesizes some digestive enzymes. These enzymes act within the cells or on the borders of the epithelial cells lining the lumen and are not secreted into the lumen.

The pancreas secretes juices containing enzymes, and the liver produces bile that comes via the gallbladder. These substances join and enter the intestine through the common duct and help to change the chemical environment form acidic to alkaline as the digestive process continues. Proteins are broken down into small peptide fragments and some amino acids. Carbohydrates are broken down into disaccharides and monosaccharides. Lipids are completely broken down into their absorbable units (glycerol and free fatty acids). This completes the digestion of the lipids and they are absorbed by the lacteals and carried away in the lymphatics to be emptied into the blood stream.

In the small intestines a brush border of stiff hair like projections is found. The carbohydrates and proteins undergo further digestion within this brush border. There are three categories of enzymes here: (1) *Enterokinase,* which converts the pancreatic enzyme into trypsin, which along with several other enzymes, completes the breakdown of peptides into amino acids. (2) *Disaccharidases,* sucrose, maltase, and lactase, convert the remaining disaccharides into simple monosaccharides completing carbohydrate digestion. (3) *Amiopeptidases* aid in breaking down the rest of the peptides.

Now the finished products go through the absorptive cells and enter the capillaries. All these products, plus most of the ingested electrolytes, vitamins, and water, are absorbed in the small intestine but mostly in the top two parts. The bile salts and vitamin B12 are absorbed mostly in the ileum.

Large Intestine

The large intestine is referred to as the *colon,* and sometimes it is called the *bowel* (L. botulus, *sausage*). After the chime has been in the small intestine for one to six hours, it passes in liquid form through the ileocecal valve. This valve is located between the ileum and cecum. Then the chime enters the large intestine. This ileocecal valve is very important and is often the source of much trouble in the digestive system. It has two segments; the one on the ileum side has the same kind of lining as the ileum; the one on the cecum side has the same kind of lining as the large intestine. It is not uncommon for this valve to be in spasm. If it is in spasm in an open position, it allows a backwash of wastes to return into the small intestine. If in spasm while shut, it does not allow material to enter the colon to be eliminated. Spiritually, this is likened to our determination and willingness to be rid of sin. If our will is not functioning properly, sin will not be eliminated or will be allowed to reenter.

The large intestine is divided anatomically into 8 parts which spiritually shows the superabundant provision God has made to rid the body of waste and sin. The last part of the colon is called the rectum and it is only about 8 inches long.

Besides the difference in the diameter and length, there are three distinct differences in structure between the small and large intestines. In the large intestine there are: (1) Three separate bands of muscle called taeniae coli (TEE-nee-ee-KOHL-eye; L. *ribbons* + Gr. *intestine*). They run longitudinal the full length of the intestine. (2) Because they are not as long as the intestine, they pucker the wall into bulges called *haustra.* (3) It contains fat-filled pouches in the serous layer.

The re-absorption of any remaining water in the chime and some salts takes place in the first part of the colon. The removal of the water along with the action of microorganisms, converts liquid wastes into feces (FEE-seez; L. faex, *dregs*). Bacteria, both living and dead, make up 25 to 50 percent of the dry weight of feces. Intestinal bacteria are harmless as long as they remain where they belong and are eliminated soon. Fat, nitrogen, bile pigments, undigested food, and other waste products from the body also make up feces. The odor is caused by *indole* and *skatole,* byproducts from the decomposition of undigested food residue, unabsorbed amino acids, dead bacteria, and cell debris. The higher the protein intake, the more indole and skatole.

The feces is stored in the colon until ready to be eliminated through the anus. The lamina propria and sub-mucosa contain many lymphoid nodules to help defend against the harmful substances.

The rectum contains strong muscles that form shelves to make it difficult for anything to enter this passage.

Defecation is a voluntary act. When it comes time for elimination, both the internal and external sphincter relax while the peristaltic waves in the colon are stimulated and the rectum and sigmoid colon are contracting. The abdominal muscles and the diaphragm increase abdominal pressure and create a pushing action. So you can see that many structures in the body help us to eliminate waste. Think about the spiritual significance of this. What a wonderful body God created for us!

The average length of time for food to be in the digestive tract is called the *transit* time and should not be over 36 hours from the time the food eaten until it is eliminated. But the transit time for people on a wrong diet and who do not exercise or drink enough water can be up to 90 hours or more. If the waste is not eliminated out of the colon rapidly, it can cause poison in the whole body and irritation that leads to cancer.

The Bible has a great deal to say about the wastes that need to be eliminated from the Body Temple. The longer the wastes are there, the more poisoning occurs.

"Mortify therefore your members which are upon the earth; fornication, uncleanness, inordinate affection, evil concupiscence, and covetousness, which is idolatry: For which things' sake the wrath of God cometh on the children of disobedience: In the which ye also walked some time, when ye lived in them. But now ye also put off all these; anger, wrath, malice, blasphemy, filthy communication out of your mouth. Lie not one to another, seeing that ye have put off the old man with his deeds; And have put on the new [man], which is renewed in knowledge after the image of him that created him:"—Colossians 3: 5-10. *"But fornication, and all uncleanness, or covetousness, let it not be once named among you, as becometh saints; Neither filthiness, nor foolish talking, nor jesting, which are not convenient: but rather giving of thanks. For this ye know, that no whoremonger, nor unclean person, nor covetous man, who is an idolater, hath any inheritance in the kingdom of Christ and of God."* Ephesians 5: 3-5.

"Being filled with all unrighteousness, fornication, wickedness, covetousness, maliciousness; full of envy, murder, debate, deceit, malignity; whisperers, Backbiters, haters of God, despiteful, proud, boasters, inventors of evil things, disobedient to parents, Without understanding, covenantbreakers, without natural affection, implacable, unmerciful: Who knowing the judgment of God, that they which commit such things are worthy of death, not only do the same, but have pleasure in them that do them."—Romans 1:29-32.

By looking constantly to Jesus with the eye of faith, we shall be strengthened. God will make the most precious revelations to His hungering, thirsting people. They will find that Christ is a personal Saviour. As they feed upon His word, they find that it is spirit and life. The word destroys the natural, earthly nature, and imparts a new life in Christ Jesus. The Holy Spirit of LOVE comes to the soul as a *Comforter*. By the transforming agency of His grace, the image of God is reproduced in the disciple; he then becomes a new creature. LOVE takes the place of hatred, and the heart receives the divine similitude. This is what it means to live *"by every word that proceedeth out of the mouth of God."* This is eating the Bread that comes down from Heaven. All who receive Him will partake of His nature and be conformed to His character. This involves the relinquishment of our cherished ambitions and requires the complete surrender of ourselves to Jesus. We are called to become self-sacrificing, meek, and lowly in heart. We must walk in the narrow path traveled by the Man of Calvary if we would share in the gift of life and the glory of Heaven.

Chapter Thirty-One

THE URINARY SYSTEM

Psalms 119:9

"BETH. Wherewithal shall a young man cleanse his way? by taking heed [thereto] according to thy word."

The urinary system is spiritually represented by the tribe of Manasseh. The urinary system, also known as the *renal system* (L. *renes*, kidneys), removes waste products (cleanses) through the production of urine and regulates the water content of the body. It consists of six structures: 2 kidneys, 2 ureters, the urinary bladder, and the urethra. In a diagram of this system, notice how the kidneys are located on each side of the great blood vessels. When given their inheritance in Canaan, the tribe of Manasseh was divided into two sections each located on opposite sides of the Jordan River (Numbers 34:14,15 and Joshua 12:7).

Homeostasis (ho-mee-oh-STAY-siss; Gr. *homois,* same + *stasis,* standing still) is a state of inner balance and stability in the body. This is a process that entails thousands of chemical reactions going on at the same time, usually on a sub-conscious level. We are not aware of all this activity unless something goes wrong.

The elimination of the waste products of metabolism and the removal of surplus substances from body tissues is called *excretion* (L. *excernere,* to sift out). Excretion contributes greatly to maintaining homeostasis. Excretion is performed in several ways by several organs. The lungs excrete carbon dioxide and water. During heavy sweating, the skin releases some salts, urea, and ammonia. The large intestine excretes the solid wastes. But the prime system for regulating balance between water and other substances is the urinary system.

Remember that water represents truth. The other substances that are in the plasma of the body represent other principles that make the truth more useful to us, such as love, joy, peace, patience, gentleness, goodness, faith, meekness, temperance, justice, and mercy. All these are needed in the right amounts to keep the truth in balance and nourish the body.

The body must maintain a balance between acids and bases (alkalines). The *higher* the concentration of *positively charged* hydrogen ions, the more *acidic* a solution is. The numerical scale that measures acidity and alkalinity is called the pH scale. It runs from 0 to 14, with 7 being neutral, which indicates the concentration of free hydrogen ions in water. Each whole number on the pH scale represents a tenfold change in acidity. For instance, a cup of black coffee at pH 5 has ten times the acidity as peas, at pH 6. If the concentration of the $H+$ ions is high, the concentrations of negatively charged hydroxyl ions ($OH-$), is low. When the number of $H+$ ions equals the number of $OH-$ ions the solution is considered neutral.

Food and drink have either an acid or alkaline pH. The body is designed to operate most efficiently on mildly acidic substances such as fruits and vegetables.

Metabolism of ingested substances leaves byproducts to be disposed of like the burning of fuel in a fireplace or stove. The smoke that is produced dissipates into the air, but then the ashes must be discarded. The smoke would represent the alkaline waste and the ashes the acid waste. It takes more work for the body to get rid of the acid wastes.

Too much of a strong acid or a strong alkaline (base) can destroy the stability of cells and too sudden a change in the pH balance can be destructive. Therefore, the body has *buffer systems* to help the blood and other body fluids resist changes in pH when even small amounts of strong acids or bases are added. When a cup of acid coffee is ingested, immediately the body has to go to work to neutralize it; all this is going on without conscious thought of the individual. This is true of too much acid foods, drugs, alcohol, etc. It is a strain on the body to stabilize the pH when extreme substances are ingested. This buffer system is one of the body's major ways of maintaining pH homeostasis. The kidneys maintain acid-base or pH balance. They also release some hormones and alter others.

Our blood is the river of life, as we learned in the Body Fluids study. It needs to be cleansed and balanced in order to keep us alive. The kidneys are the means of blood purification. They spiritually represent works. They have the help of the liver which represents faith (see the Digestive System study). Did you ever wonder why the kidneys were offered to God in the sacrifices that Israel preformed in the Sanctuary service? Obviously, it was because the kidneys are used by the body to purify the blood. The word for *offering* (Strong's 7133), means "something brought near the altar, a *sacrificial present:* "And thou shalt take all the fat that covereth the inwards, and the caul [that is] above the liver, and the two kidneys, and the fat that [is] upon them, and burn [them] upon the altar."—Exodus 29:13.

There were five occasions upon which the kidneys were to be offered and burned on the altar of sacrifice: Two were during a consecration service for the priests, Leviticus 8: verses 14-16, *a sin offering*, and verses 22-25, *a ram of consecration*. The other three occasions are found in Leviticus 9: verses 8-11, *sin offerings for the priest*, verse 15, *sin offerings for the people*, and verses 18-21, *peace offerings for the people.*

During the burnt offering, the whole animal was burnt on the altar. Our works are to be brought as an offering to God. *"Commit thy works unto the LORD, and thy thoughts shall be established."*—Proverbs 16:3.

The word kidneys and reins are both translated from the same word *kilyah* (Strong's 3629). In the modern translations of the Bible, instead of this word being translated as *reins* as in the King James Version, it is translated as *heart* in some places and *mind* in others; but the literal translation is *kidney.* It might be assumed that there was a mistake somewhere, but even in the Greek, the word *nephros* from which we get our word *nephritis* meaning "a kidney infection" is translated reins. *"And I will kill her children with death; and all the churches shall know that I am He which searcheth the reins and hearts: and I will give unto every one of you according to your works."*—Revelation 2:23. Here the LORD Himself tells us that we will be judged by our works and motives by using the words *reins* and *hearts.* We will give a few more examples of this important principle.

"Oh let the wickedness of the wicked come to an end; but establish the just: for the righteous God trieth the hearts and reins."—Psalm 7:9.

"Examine me, O LORD, and prove me; try my reins and my heart." Psalm 26:2.

"But, O LORD of hosts, that judgest righteously, that triest the reins and the heart, let me see thy vengeance on them: for unto thee have I revealed my cause."—Jeremiah 11:20.

"I the LORD search the heart, I try the reins, even to give every man according to his ways, and according to the fruit of his doings."—Jeremiah 17:10.

"But, O LORD of hosts, that triest the righteous, and seest the reins and the heart, let me see thy vengeance on them: for unto Thee have I opened my cause."—Jeremiah 20:12.

"Thus my heart was grieved, and I was pricked in my reins."—Psalm 73:21.

"Righteous art thou, O LORD, when I plead with Thee: yet let me talk with Thee of Thy judgments: Wherefore doth the way of the wicked prosper? Wherefore are all they happy that deal very treacherously? Thou hast planted them, yea, they have taken root: they grow, yea, they bring forth fruit: Thou art near in their mouth, and far from their reins."—Jeremiah 11:20.

Satan has always sought to shut out from men a knowledge of God, to turn their attention from the temple of God, and to establish his own kingdom. Through heathenism, Satan has for ages turned men away from God. By contemplating and worshipping their own conceptions, humans lose a knowledge of God and become more and more corrupt. The principle that man can save himself by his own works lies at the foundation of every heathen religion. Satan implanted this principle and wherever it is held, men have no barrier against sin. In ourselves we are incapable of doing any good thing; but that which we cannot do will be wrought by the power of God in every submissive and believing soul. It is through *faith* that spiritual life is begotten, and we are enabled to do the works of righteousness. It is interesting to note that, if it were not for the help of the *liver,* the kidneys could not do their job.

Kidneys

The kidneys are about the size of a large bar of soap, are bean-shaped, and are reddish-brown in color. They are located in the back wall of the abdominal region, one on each side of the vertebral column from Thorax 12 to Lumbar 3. They are partially protected by the last pair of ribs.

Each kidney has three layers of tissue that cover and support them. The *renal capsule* is the innermost layer of a tough, fibrous material. It is covered by the *adipose capsule* (composed of fat), and it gives the kidney a protective cushion. The outer layer is called the *renal fascia,* composed of connective tissue and surrounded by another layer of fat. The renal fascia attaches the kidney firmly to the posterior abdominal wall.

There are three distinct regions of the kidney:

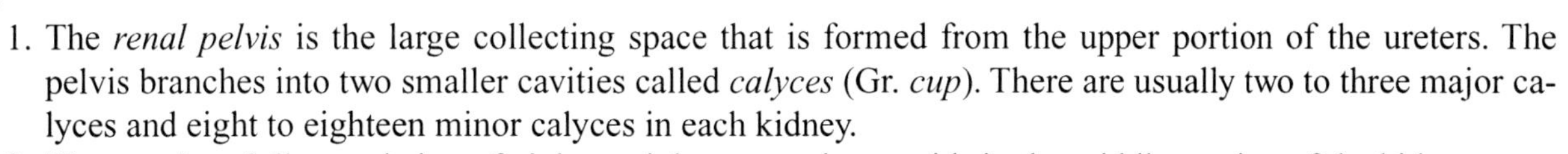

1. The *renal pelvis* is the large collecting space that is formed from the upper portion of the ureters. The pelvis branches into two smaller cavities called *calyces* (Gr. *cup*). There are usually two to three major calyces and eight to eighteen minor calyces in each kidney.
2. The *renal medulla* consisting of eight to eighteen renal pyramids in the middle portion of the kidney.
3. The *renal cortex* has a granular, textured appearance caused by the spherical bundles of capillaries and the associated structures of the nephrons. Cortical tissue forms *renal columns* which are composed mainly of collecting tubules and are located between the renal pyramids. The pyramids and the cortical tissue around them are called lobes. Each minor calyx receives urine from a lobe of the kidney.

Spiritual Application

There are three distinct features of works.

(1) They must be ordained by God to be acceptable. *"For we are His workmanship, created in Christ Jesus unto good works, which God hath before ordained that we should walk in them."*—Ephesians 2:10. *"And Moses said, Hereby ye shall know that the LORD hath sent me to do all these works: for I have not done them of mine own mind."*—Numbers 16:28.

(2) They must be accomplished by God and man working *together. "LORD, thou wilt ordain peace for us: for Thou also has wrought all our works in us."*—Isaiah 26:16. *". . . work out your own salvation with fear and trembling. For it is God which worketh in you both to will and to do of His good pleasure."*—Philippians 2:12,13.

(3) They must glorify God. *"Let your light so shine before men, that they may see your good works, and glorify your Father which is in Heaven."*—Matthew 5:16. *"Herein is my Father glorified, that ye bear much fruit; so shall ye be my disciples."*—John 15:8.

The kidneys receive more blood in proportion to their weight than any other organ in the body. Approximately one quart of blood passes through the kidney every minute, and there are 1440 minutes in a day. That is a lot of filtering! *"And let us note weary in well doing: for in due season we shall reap, if we faint not."*—Galatians 6:9.

Blood comes to each kidney directly from the abdominal aorta through the *renal artery.* The renal artery branches into the *interlobular arteries* inside the kidneys. The interlobular arteries pass through the renal columns. At the bases of the pyramids these arteries turn and run parallel to the base and are called *arcuate arteries* (L. *arcuare,* to *bend like a bow).* The interlobular arteries branch off of the arcuate arteries, and then into small afferent arterioles which carry the blood to the glomeruli.

The efferent arteriole branches, which form a second capillary bed made up of *peritubular* (around the tubules) capillaries. These empty into the interlobular veins, on to the arcuate veins, on to the interlobular veins, and finally to the renal vein that carries blood from each kidney to the inferior vena cava.

In a diagram of the kidney notice the vasa recta, which is the name for the straight arteries that are extensions of the efferent arterioles. They surround the loop portion, and only about one to two percent of the total renal blood flows through these vessels. Even so, they play an important role in the concentration of urine.

The smooth muscle in the walls of the arterioles permits these vasa rectas to constrict or dilate in response to nervous or hormonal stimulation. The filtration rate is regulated by the blood pressure in the glomerulus which is controlled by this constriction or dilation of the arterioles. Muscle represents judgment.

The capillaries of the glomerulus eventually empty into a single efferent arteriole which carries blood away from the glomerulus. Because the efferent arteriole is narrower than the afferent arteriole, resistance increases blood pressure in the glomerulus, which forces the filtrate into the capsular space and produces efficient filtration.

Lymphatics

The lymphatic vessels are not shown in many pictures of the liver, but they accompany the larger renal blood vessel and are more prominent around the arteries than the veins in the kidney. These lymphatics all meet in the renal sinus region and form several large vessels that leave the kidney at the renal hilus.

Nerves

The nerves for the smooth muscles in the afferent and efferent arterioles come from the renal plexus of the autonomic nervous system. These nerves communicate the messages that control the constriction and dilation of the arterioles which changes the pressure in the glomerulus. This affects the blood distribution throughout the kidney and is affected by changes in physical activity, stress, and changes in posture and gravity. During a kidney transplant these nerves are severed permanently.

Nephrons

Each kidney contains one million nephrons. This is the structural and functional unit of the kidney. These nephrons are where the urine is formed. If all the tubes in all the nephrons in both kidneys were stretched out, they would have a combined length of 50 miles. A nephron contains the *renal corpuscle,* the *proximal tubule,* the *loop of henle,* the *distal tubule,* the *collecting tubule,* plus the *collecting duct,* six structures in all that do the actual work of the kidneys.

The renal corpuscle consist of the *glomerulus* (glow-Mare-yoo-luhss; L. ball), which is a tightly coiled ball of capillaries. The capsule called *Bowman's capsule* (named after the man who discovered it) surrounds the renal corpuscle. This is where the blood is filtered.

There are two walls of the Bowman's capsule, and in between the two layers is a cavity called the *capsular space.* The outer parietal layer is composed of simple squamous epithelial cells with basement membranes. The second layer is the inner visceral layer, which is composed of specialized epithelial cells called *podocytes* (POH-doh-sites: *foot like cells).*

Blood enters the renal capsule through the capillaries. Filtration of the blood takes place across three layers. First is the endothelium of the capillary which contains tiny pores called *fenestration* (windows).

Next this filtrate passes through the middle layer of basement membrane of the glomerulus.

Third is the visceral layer of the glomerular capsule and the podocytes. The podocytes are relatively large cells that have several processes which branch to form many smaller fingerlike processes called foot processes or *pedicels* (PEHD-ih-sehlz; L. *little feet).* The pedicels are in contact with the glomerular capillaries. The filtrate passes into the capsular space.

The cellular components of the blood, such as white and red blood cells and the large proteins, do not normally pass through this membrane. Glomerular filtrate is composed of water and dissolved substances such as electrolytes, sugars, urea, amino acids, and polypeptides that have passed from the blood into the capsular space.

The epithelial cells with their surfaces are different in cach of the sections of the tubes in the nephron. Their structure changes as their function changes. This is what allows water to move in and out freely in some places and not in others.

Proximal tubule

From the renal capsule the filtrate drains into the proximal tubule. It is called proximal because it is nearest to the glomerular capsule. It is here that many substances are reabsorbed into the blood. These substances are water, electrolytes, glucose, some amino acids and polypeptides. Because most reabsorption takes place here, this part of the tube has many microvilli which are fingerlike projections that increase the surface of the lining for absorption.

Loop of Henle

This loop has a thin descending limb and a thick ascending limb. It is designed this way to generate a difference in the concentration of a substance on either inside or outside of the tube causes that substance to move in or out to equalize the pressure—called a *"concentration gradient."* Moving "down" the concentration gradient in the kidneys is a way of moving a molecule from an area of high concentration to one of lower concentration. Water cannot go through the walls of the ascending limb which means that if sodium and chloride ions move out, water cannot move in. This is what happens in the other portion of the tubule. This is why 99% of the water is able to return to the body. The reason for this is that the epithelial cells are different in this portion. Without getting into all the chemistry happening here, suffice it to say that if the tubules were not designed in this particular way, it would not be possible to keep the blood balanced.

Distal tubule

This tubule is similar in design to the proximal tubule but has very few microvilli, because not much absorption takes place here. The cells in this tubule have many mitochondria near their basal surfaces.

Collecting duct

The collecting duct is lined by two types of epithelial cells, the light cell, containing cilia, and the dark cell which has flaps. The distal tubules of many nephrons empty into a single collecting duct which travels through the medulla. The collecting ducts join to form larger and larger tubes reaching to the minor calyx. By now the filtrate is called urine and drains into the renal pelvis down the ureters and into the bladder.

To say it simply, the kidneys produce and modify the glomerular filtrate that finally is excreted as urine. This involves three processes:

1. Glomerular filtration: The kidneys filter blood in the glomerulus into filtrate. The rate of glomerular filtration depends on:

    a. Effective filtration pressure,
    b. Stress,
    c. Capillary permeability,
    d. Release of renin,
    e. Total surface area available for filtration, and
    f. Auto regulation by conditions in the kidney itself.

2. Tubular reabsorption: As the filtrate passes through the tubules, useful substances are put back into the capillaries surrounding the tubules.
3. Tubular secretion: Unwanted and excess substances may pass from the blood in the peritubular capillaries into the filtrate in the tubules.

Spiritual Application

How can works keep our life purified? We know that we are not saved by our works, but we are judged by them. Does this sound like a contradiction? When we understand the relationship between the *means* of salvation and the *results* of salvation, this becomes clear.

"Ye shall know them by their fruits. Do men gather grapes of thorns, or figs of thistles? Even so every good tree bringeth forth good fruit; but a corrupt tree bringeth forth evil fruit. A good tree cannot bring forth evil fruit, neither [can] a corrupt tree bring forth good fruit. Every tree that bringeth not forth good fruit is hewn down, and cast into the fire. Wherefore by their fruits ye shall know them. Not every one that saith unto me, Lord, Lord, shall enter into the kingdom of heaven; but he that doeth the will of my Father which is in heaven. Many will say to me in that day, Lord, Lord, have we not prophesied in thy name? and in thy name have cast out devils? and in thy name done many wonderful works? And then will I profess unto them, I never knew you: depart from me, ye that work iniquity. Therefore whosoever heareth these sayings of mine, and doeth them, I will liken him unto a wise man, which built his house upon a rock: And the rain descended, and the floods came, and the winds blew, and beat upon that house; and it fell not: for it was founded upon a rock. And every one that heareth these sayings of mine, and doeth them not, shall be likened unto a foolish man, which built his house upon the sand: And the rain descended, and the floods came, and the winds blew, and beat upon that house; and it fell: and great was the fall of it."—Matthew 7:16-27

When the nations are gathered before God, there will be but two classes, and their eternal destiny will be determined by what they have done or have neglected to do for Him for the poor and suffering. In that day Christ does not present before men the great work He has done for them in giving His life for their redemption. He presents the faithful work His children have done for the Father. Even among the heathen are those who have cherished the spirit of kindness and who worship God ignorantly. Though ignorant of the written law of the Father, they have heard His voice speaking to them in nature, and have done things that the law required. Their works are evidence that His Spirit of LOVE has touched their hearts, and they are recognized as the children of God. Every deed of kindness done to uplift a fallen soul, every act of mercy, is accepted as done to Him.

Those on the left hand of Christ were unconscious of their guilt. Satan had blinded them; they had not perceived what they owed to their brethren. They had been self absorbed, and cared not for others' needs. It is because this work is neglected that so many young disciples never advance beyond the mere alphabet of Christian experience. The light which was glowing in their own hearts when Jesus spoke to them, *"Thy sins be forgiven thee,"* they might have kept alive by helping those in need. Those who minister to others will be ministered unto by the angels and by Christ Himself. They themselves will drink of the living water, and will be satisfied. They will not be longing for exciting amusements, or for some change in their lives. The great topic of interest will be how to save the souls that are ready to perish. Love to *man* is the outward manifestation of love to *God.* The King of glory became one with us to implant this love. When we love the world as He has loved it, then for us His mission is accomplished. We are fitted for heaven; for we have heaven in our hearts.

The people that hear the words *"depart from me, ye that work iniquity"* thought they were doing God's will. What did they lack? There are those today who would be surprised if Jesus said that He did not know them. The people here described apparently are weaving Jesus into all their doings but are at war with the sayings and laws of God. Christ calls them workers of iniquity, because they are deceivers, having on the garments of righteousness but disobeying. Satan in these last days will work with all deceivableness of unrighteousness by working miracles that seem to help and do good. He will claim to be Christ Himself. He gives power to those who are aiding him in his deceptions. The Lord tells us that if possible he will deceive the very elect. How will we tell the difference? The sheep's clothing seems so real, so genuine, that the wolf cannot be discerned except by one thing. If he is at war with the commandments of

God, we know that he has no light. When these people are shown the word of God they have one excuse or another for not obeying it as it is written.

Juxtaglomerular apparatus

Juxta means near in Latin. The juxtaglomerular apparatus is not shown in any picture of the kidneys, but we will mention it here because it manufactures a hormone named rennin that helps regulate blood pressure. Located in the space between the afferent and efferent arteriole and the distal tubule are some smooth muscle cells called juxtagomerular cells. These cells are in contact with the macula densa. The cells of the macula densa are longer and narrower than the epithelial cells of the rest of this tubule; also their nuclei are closer together. Together the juxtaglomerular cells, macula densa, and the afferent and efferent arterioles make up this glomerular apparatus.

Only one percent of the water in the filtrate is excreted in the urine. The other 99% is returned to the blood.

Over two quarts of water is lost each day, half of this being excreted in urine and the rest by lungs and skin. An adult should take in at least two quarts of water besides the water ingested in food. All food contains varying amounts of water; the highest percentage of water in any food is watermelon at 97% water. The lowest is baked sunflower seeds at 5% water.

Urine passes from the colleting ducts in the pyramid to the minor calyces, major calyces, and renal pelvis.

Ureters

Connected to the renal pelvis in each kidney is a tube called a *ureter,* which carries urine to the bladder. There are three layers of tissue in the walls of the ureters. The outermost layer is connective tissue that holds the tube in place. When these tubes reach the bladder, they function as a check valve so that urine cannot return to the kidney to cause infection.

Urinary Bladder

This is a hollow, muscular organ that is the size of a walnut when it is empty but can expand to the size of a fist when it is full. It has three main layers. The lining is transitional epithelium which allows the bladder to stretch and contract. When it is empty it has folds called rugae (ROO-gee; L. folds). The middle layer consists of meshed smooth muscle. The outer layer is derived from the peritoneum.

The openings of the ureters and urethra into the bladder outline a triangular area, the known as the *trigone.* The smooth muscle in the bladder wall forms a bundle at the place where the urethra leaves the bladder. The muscles are in a spiral, longitudinal, and circular pattern which contract to prevent the bladder from emptying before it should.

Urethra

This is the final tube of the urinary tract. It is smooth muscle lined with mucosa. It has a voluntary skeletal muscle that forms the external urethral sphincter that holds back the urine until urination. In the female, this tube is only about 1½ inches long. In the male it is about 8 inches long and passes through three different regions: the prostatic portion, the membranous portion, and the spongy portion.

Kidney Problems

Acute kidney failure is the total or nearly total stoppage of kidney functions. Substances that are normally eliminated from the body are retained, which begins to poison all the cells. Causes for this condition vary.

Some of them are:

A diminished blood supply to the kidneys caused by a heart attack, an injury, or a blood clot. Think of the spiritual lesson in this. If we do not have a sufficient amount of knowledge of the truth, Jesus' sacrifice, and the work of the Holy Spirit of LOVE, our good works die. This can be caused by a problem in the church, hurt feelings, or a blood clot (jealousy) that impedes our intake of the blood of Christ.

A high level of toxic materials builds up in kidneys faster than they can be eliminated. Albeit sometimes the motives for our works become impure. Instead of doing work for the love of God and man, we do it for *"brownie points"* or *show.*

Obstructions like kidney stones cause kidney problems. Spiritually, wrong attitudes, stress, and malice toward other people can cause kidney problems.

Chronic kidney failure develops slowly and progresses over periods of years and sometimes the kidneys are damaged beyond repair before the person is aware of it. Likewise, spiritual "burnout" sometimes sneaks up on a person. It usually happens when people become so involved in working for the church that they do not have time for personal bible study and prayer. In other words we are working so hard for heaven's sake that we are no earthly good. Let us keep our spiritual kidneys functioning properly. Works without faith are useless, and faith without works is dead.

Chapter Thirty-Two

THE REPRODUCTIVE SYSTEM

Genesis 1:26-28

"And God said, Let us make man in our image, after our likeness: and let them have dominion over the fish of the sea, and over the fowl of the air, and over the cattle, and over all the earth, and over every creeping thing that creepeth upon the earth. So God created man in his [own] image, in the image of God created he him; male and female created he them. And God blessed them, and God said unto them, Be fruitful, and multiply, and replenish the earth, and subdue it: and have dominion over the fish of the sea, and over the fowl of the air, and over every living thing that moveth upon the earth."

God took Adam's rib and created the woman named Eve. This would indicate that the woman was part of the original being (man) that God created. This is why there is no record of God forming her (the woman) from the dust and breathing into her the breath of life. They walked and talked with God and received from God all the directions and tasks that were to be performed in the garden. They were in perfect agreement at all times. They were perfectly equal with different roles to perform. This state remained until sin entered and then a change was necessary. Because they changed leaders (from God to Satan) sin would cause disagreement. Can two walk together unless they agree? Obviously not! Havoc, disorder, and chaos are the result. Genesis 3:16 tells us that the woman was put under the authority of the man. This is not because God is a male chauvinist, but because it was the woman that was beguiled. Man was to be the High Priest of the home and love his wife as his own flesh.

"And Adam said, This [is] now bone of my bones, and flesh of my flesh: she shall be called Woman, because she was taken out of Man."—Genesis 2:23. Bone here means substance or selfsame; woman was his second self showing the close union and the affectionate attachment that should exist in this relationship.

"Wives, submit yourselves unto your own husbands, as unto the Lord. For the husband is the head of the wife, even as Christ is the head of the church: and he is the savior of the body. Therefore as the church is subject unto Christ, so [let] the wives [be] to their own husbands in every thing. Husbands, love your wives, even as Christ also loved the church, and gave himself for it; That he might sanctify and cleanse it with the washing of water by the word, That he might present it to himself a glorious church, not having spot, or wrinkle, or any such thing; but that it should be holy and without blemish. So ought men to love their wives as their own bodies. He that loveth his wife loveth himself. For no man ever yet hated his own flesh; but nourisheth and cherisheth it, even as the Lord the church: For we are members of his body, of his flesh, and of his bones. For this cause shall a man leave his father and mother, and shall be joined unto his wife, and they two shall be one flesh. This is a great mystery: but I speak concerning Christ and the church. Nevertheless let every one of you in particular so love his wife even as himself; and the wife [see] that she reverence [her] husband."—Ephesians 5:22-33.

The act of two becoming one is related here as a symbol of the spiritual togetherness that Christ wishes with His people. The devil (Satan) determined that the lesson would be lost and so he set out to pervert marital relationships. Instead of sexual intercourse being a holy symbol, Satan wanted it to be a symbol of self gratification. It became perverted and misused. This perversion became the central issue of many heathen religions. Sexual body parts came to be worshipped and idolized. This was the origin of many symbols that are still used today: the Egyptian ankh, the obelisk, the church steeple, the neck tie, the wedding ring, are all examples of sexual symbols. An interesting project might be to research the origin of these things.

The confusion that exists on the subject of male and female existed even in the days of Christ. The religious teachers of His day did not understand what the scriptures had to say about it. *"The same day came to him the Sadducees, which say that there is no resurrection, and asked him, Saying, Master, Moses said, If a man die, having no children, his brother shall marry his wife, and raise up seed unto his brother. Now there were with us seven brethren: and the first, when he had married a wife, deceased, and, having no issue, left his wife unto his brother: Likewise the second also, and the third, unto the seventh. And last of all the woman died also. Therefore in the resurrection whose wife shall she be of the seven? for they all had her. Jesus answered and said unto them, Ye do err, not knowing the scriptures, nor the power of God. For in the resurrection they neither marry, nor are given in marriage, but are as the angels of God in heaven."*—Matthew 22:23-30. *"And Jesus answering said unto them, The children of this world marry, and are given in marriage: But they which shall be accounted worthy to obtain that world, and the resurrection from the dead, neither marry, nor are given in marriage: Neither can they die any more: for they are equal unto the angels; and are the children of God, being the children of the resurrection."*—Luke 20:34-36. Apparently in heaven the division of male and female will not exist. We will each be recreated to reflect the image of God. The Bible does not elaborate on the time when we will inherit a new perfect world except that in referring to the *new Earth,* the Bible says, *"The wolf also shall dwell with the lamb, and the leopard shall lie down with the kid; and the calf and the young lion and the fatling together; and a little child shall lead them. And the cow and the bear shall feed; their young ones shall lie down together: and the lion shall eat straw like the ox. And the sucking child shall play on the hole of the asp, and the weaned child shall put his hand on the cockatrice' den. They shall not hurt nor destroy in all my holy mountain: for the earth shall be full of the knowledge of the LORD, as the waters cover the sea."*—Isaiah 11:6-9. This would seem to indicate that there will be "young" animals and "little" children. How this will occur we do not know at this time.

The story of Hosea is of his relationship with his wife, who was a prostitute and God likened it to His relationship with His people. It breaks His heart when we commit adultery with Satan and love the world more than we love our Savior. This is why the LOUD CRY must be sounded and swelled. The church must come out from this adulterous love affair relationship with the world and "be separate." *"Know ye not that your bodies are the members of Christ? shall I then take the members of Christ, and make [them] the members of an harlot? God forbid. What? know ye not that he which is joined to an harlot is one body? for two, saith he, shall be one flesh. But he that is joined unto the Lord is one spirit. Flee fornication. Every sin that a man doeth is without the body; but he that committeth fornication sinneth against his own body. What? Know ye not that your body is the temple of the Holy Ghost [which is] in you, which ye have of God, and ye are not your own? For ye are bought with a price: therefore glorify God in your body, and in your spirit, which are God's."*—I Corinthians 6:15-20.

This union between two people is so important that we are admonished to run away from the misuse of it because it is a sin against our own body. When we become joined to someone physically, we become one with that person. If sexual relations are enjoined with many different people, we can see ramifications of broken relationships. Satan refers to these adulterous relationships as *love* when in actuality these are the result of *lust* or *selfishness.* Fornication is the acting out of sexual relations outside of marriage. In the Bible when a man "lay" with a woman they became *one flesh,* and she became his wife. Any subsequent relationship was termed *adultery.* This is the reason why God says that the marriage relationship is a life-long union and can never be broken. *"And said, for this cause shall a man leave father and mother, and shall cleave to his wife: and they twain shall be one flesh? Wherefore they are no more twain, but one flesh. What therefore God hath joined together, let not man put asunder."*—Matthew 19:5,6. When we are baptized and declare ourselves to be Christians, we are married to Christ. Any departure from the Christian lifestyle from that time on is an act of adultery with the world and a violation of the third and seventh commandments.

As we study the body parts for this system, think about the spiritual significance and how the joining of the male and female can create a new creature. Compare this to the new creature that God creates when we turn our lives over to His Holy Spirit of LOVE.

Terms:

Gonads: (Gr. *gonos,* offspring) sex organs, the testes in males, the ovaries in females.

FSH - follicle-stimulating hormone and LH—luteinizing hormone: hormones that come from the anterior pituitary gland and regulate gonad activity.

Gene, genetic, geno: when you see any of these words you know it has something to do with producing (Gr. gennan, *to produce*). For instance, *spermatogenesis* means to produce sperm.

Peristalsis (peri, around + stalsis contraction): the progressive wavelike movement that occurs in the hollow tubes of the body caused by involuntary muscular contractions.

Male Reproductive System

The reproductive role of the male is to produce sperm and deliver them to the vagina of the female. There are four different structures that make this possible:

1. The testes, which produce sperm and testosterone;
2. Accessory glands that furnish the fluid for transport;
3. Accessory ducts which store and carry secretions; and
4. The penis, which deposits the semen.

This is a complicated system and each structure will be discussed separately.

The Scrotum

The scrotum (SCROH-tuhm; L. *scrautum,* a leather pouch for arrows) is an external sac of skin that holds the testes and part of the spermatic cords. It has two layers. The first is very thin skin of a brownish color and has folds. The second is a thin layer of loose reddish tissue which is able to contract. This layer divides the scrotum into two compartments, one for each testis. The division line is know as the median septum and can be seen on the outside as a ridge of skin.

The Testes

The testes (TESS-teez, plural; testis, singular. L. witness) are the organs which produce sperm. They are also known as testicles or gonads (GO-nads; Gr. *gone,* seed). In the fetus they are formed in the abdominopelvic cavity, and by the third month they have descended to the inguinal canal. During the seventh month they pass through this canal that leads to the scrotum. Each testis measures about 1.75 inches long and one inch wide and has three layers:

1. The tough tunica albuginea (al-byoo-JIHN-ee-uh; L. albus, white) which is lined by
2. the tunica vasculosa which contains a network of blood vessels, and
3. the tunica vaginalis which is a continuation of the membrane that lines the abdominopelvic cavity.

Each testes contains over 800 seminiferous tubules which are tightly coiled in compartments called lobules. If the tubules in both testes were stretched out they would reach about 750 feet. These tubules produce thousands of sperm each second. The walls are lined with germinal tissue which contains two types of cells:

1. Spermatogenic cells consist of three classes:

   a) spermatogenia,
   b) spermatocytes, and
   c) spermatids.

2. Sustentacular cells provide nourishment for the sperm as they mature and secrete a fluid for the sperm to swim in. They also secrete a protein that bids to the male hormones and carries them into the fluid where they are available to the maturing sperm. They also secrete a hormone, *inhibin*, which inhibits the secretion of another male hormone, FSH.

Between the tubules are clusters of endocrine cells. *Leydig* cells, which secrete three male sex hormones called androgens: *testosterone, dihydrotestosterone,* and *androstenediaone.* The production and secretion of testosterone are controlled by FSH (follicle stimulating hormone) and LH (luteinizing hormone). They are part of a complicated system that bring on puberty, change of voice, growth of hair on the face and under arms and in the pubic area.

Removal of the testes is called *castration,* which causes a diminishing of the male characteristics. In the Bible, men who were designated as *Eunuchs* were subjected to this operation.

Sperm

The sperm (Gr. *sperma,* seed) is one of the smallest cells in the body, only 0.05mm from one end to the other. Each sperm requires more than two months for its complete development. Sperm have three parts: head, body, and tail. The *acrosome* at the tip of the head contains enzymes that help the sperm penetrate the egg. All of the genetic material (chromosomes) of the donating male are contained in the center of the head in the nucleus. Coiled mitochondria, which supply ATP to provide the energy for movement, are located in the middle piece or body. The beating movement of the tail drives the swimming sperm.

Epidiymis

The sperm travels to the epididymis (ep-ih-DIHD-ih-mus; Gr. *upon the twin*) for its final maturation. This tightly coiled tube would be about twenty feet long if straightened out, but it is bunched up to about 1.5 inches long. It has three parts: a head, a body, and a tail. It has three functions: (1) It stores sperm, (2) it is a duct system for the sperm to travel in, and (3) it contains circular smooth muscle to help propel the mature sperm.

The sperm mature as they travel along the epididymis; this journey may take from ten days to five weeks. They are nourished by the lining of the epididymis. After they are mature, they will remain fertile for about a month, and if not ejaculated by that time, degenerate and are reabsorbed by the body. They wait either in the epididymis or the *ductus deferens.*

Ductus deferens

The ductus deferens (L. *deferre,* to carry away) is a dilated extension of the epididymis, and is the main carrier of the sperm. It contains three thick layers of smooth muscle, which, when stimulated by the sympathetic nerves, produce peristaltic contractions that move the sperm along. The ductus deferens is covered by the spermatic cord as it leaves the tail of the epididymis. When the ductus deferens enters the inguinal canal, it becomes free of this cord and passes behind the urinary bladder. Here it travels alongside the seminal vesicle, but just before reaching this point it widens into an enlarged portion called the *ampulla.* This is where sperm are stored. The ampulla is joined by the duct of the *seminal vesicle* at the ejaculatory duct which is about one inch long. These ducts receive secretions from the seminal vesicles, and more secretions from the prostate as they pass through there, and finally join the urethra.

Seminal vesicles

The seminal vesicles provide the bulk of the seminal fluid, which is secreted by the mucous membrane lining. The seminal fluid provides an energy source for the sperm and helps neutralize the natural acidity of the vagina. It consists mostly of water, fructose, prostaglandins and vitamin C. Stimulation during sexual excitement causes the fluid to be emptied into the ejaculatory ducts by contractions of the smooth muscle layers.

Prostate gland

The prostate gland surrounds the first portion of the urethra and is bout the size of a chestnut. The smooth muscles of the prostate contract like a sponge and squeeze the prostatic secretions through tiny openings into the urethra. Most of these secretions join the seminal fluid to help neutralize the vaginal acidity, but some are also released with urine. This fluid comes from three types of glands: (1) The inner mucosal glands secrete mucus and are the glands that become inflamed in prostatitis, (2) the middle submucosal glands, and (3) the main prostatic glands which are the ones involved in cancer of the prostate. These supply most of the secretions which contain mainly water, acid phosphatase, cholesterol, buffering salts, and phospholipids.

Bulbourethral glands

Bulbourethral *(bulb root + urethra)* glands are bout the size of a pea. They secrete clear alkaline fluids into the urethra to neutralize the acidity of any remaining urine and act as a lubricant. These glands are stimulated by sexual excitement.

Semen

Semen is made up of all the above secretions together with the sperm, which only makes up about 1% of the semen. The odor of semen is caused by amines which are produced in the testes. The average ejaculation contains 300 to 500 million sperm.

Penis

Besides the urethra, the penis (L. tail) consists of three layers of erectile tissue; the *corpus spongiosus* (*corpus,* body + *sponge*), which contains the *urethra*, and two *corpus cavernosa.* They are surrounded by a dense, connective tissue called the tunica albuginea. The corpus cavernosa contains many vascular cavities full of veins.

The corpus spongiosus contains the urethra and extends beyond the corpus cavernosa. The part expanded into the tip of the penis is called the *glans penis.* This is a sensitive area containing many nerve endings which are most concentrated in the corona. The *prepuce* (foreskin) is the loosely fitting skin that is folded forward over the glans. This is the skin removed in circumcision.

Spiritual Application

There are six structures involved in the male reproductive system:

1. The Testes,
2. The epididymis and its extension, the Ductus deferens tube,
3. the seminal vesicles,
4. the prostate,
5. the bulbourethral gland, and
6. the penis.

There are six structures in the female reproductive system:

1. The ovaries,
2. the uterine tubes,

3. the uterus,
4. the vagina,
5. the vulva, and
6. the mammary glands.

When joined together they become 12, to represent perfect government. When they join as God intended, they are head to head—mentally joined, heart to heart—emotionally joined—and creative parts to creative parts—physically joined. Anything less than this will leave a marriage unfulfilled.

God's definition of love is totally different than the cheap sentimentalism (lust) that is being put forward today as love. Instead of marriage and commitment, now the big word is relationship. This word is supposed to cover up fornication and adultery and try to make it acceptable. Great hurt to the spirit and the soul comes from these "relationships" as people pass from one to another never finding a true belonging. This is promoted in all the media that assaults the senses, and unless the Word of God is the foundation for decision making, soon it seems acceptable to do what the world is doing. This was the great problem with the Israelites, and they finally ended up at the foot of Baal having sex orgies with the heathen. (Read Numbers 25)

Female Reproductive System

Females produce egg cells, and after fertilization, they nourish, carry and protect the developing embryo. They are able to nurse the infant after it is born. As we study the body parts for this system, think about the spiritual significance and how the joining of the male and female can create a new creature. Compare this to the new creature that God creates when we turn our life over to His Holy Spirit of LOVE.

Ovaries

Ovaries (L. *ovum,* egg) are about the size and shape of an unshelled almond. They produce ova and female hormones. They are covered by a layer of specialized epithelial cells called the germinal layer. Beneath it is the *stroma* (STROE-mah; Gr. mattress) that actually contains ova in many different stages of development.

Vesicle means a little bladder and follicle means a little bag. The follicles in the cortex of the ovary are the actual centers of egg production or oogenesis (oh-oh-GEN-eh-sis). There are two classes of follicles: (1) The primordial follicles which are not yet growing. Each follicle contains an immature ovum called a *primary oocyte* (OH-oh-sit). Baby girls are born with about 400,000 primordial follicles in each ovary (total 800,000) which is reduced to about 200,000 at the time of puberty. (2) Vesicular ovarian follicles are those which are almost ready to release an ovum, which is called a *secondary oocyte.* When this ovum is released, the process is called *ovulation.*

A high concentration of FSH and LH comes from the pituitary and causes the mature follicle to rupture. After puberty, about 20 follicles mature each month, but usually only one actually ruptures to release the mature ovum. (Fraternal twins are the result of two separate oocytes being released and fertilized.) The site of the rupture heals leaving scar tissue on the surface of the ovary.

Hormones regulate the activities of the ovaries. Follicles secrete estrogen. After the follicle has released the egg, it becomes the *corpus luteum* (yellow body), which generates *estrogen* and *progesterone.* These hormones, in the right concentration, stop additional ovulation and cause the uterine wall and mammary glands to become ready for pregnancy. The high concentration of progesterone stops the uterus from contracting. Within 14 days, if the body has not become pregnant, the corpus luteum degenerates into the *corpus albicans,* a form of scar tissue. If pregnancy occurs, the corpus luteum works for about two to three months until the placenta takes over. Besides the other hormones, they both produce relaxin, a hormone which promotes the softening of the cervix and ligaments of the symphysis pubis and the relaxation of the birth canal.

Uterine Tubes

The uterine tubes are not directly connected to the ovaries. The ovum is gathered up by the feathery fimbrae (FHM-bree-ee; L. threads). Each uterine tube is about four inches long. There are three distinct portions of the tubes:

1. the *infundibulum* (L. funnel) that is near the ovary and contains the fimbria,
2. The thin walled *ampulla,* which is where the ovum is usually fertilized, and
3. the *isthmus,* which opens into the uterus.

The wall of the uterine tube is made up of three layers:

1. The outer *serous membrane,*
2. the middle *muscular layer*, which is made up of two layers and
3. the inner *mucous membrane,* which is lined with two alternating epithelial cells; the *secretory* to nourish the oocyte and the *ciliated* for propelling it.

The oocyte is unable to move on its own, so it is carried along the tube toward the uterus by the peristaltic contractions of the tube and the waving movements of the cilia. If the ovum is not fertilized, it will degenerate in the uterine tube. The fertilized ovum is called a *zygote* (Gr. *zygotos,* yoked), and it continues on to the uterus. The journey from the ovary to the uterus takes from four to seven days. (If for some reason, the zygote plants itself in the tube it is called a tubal or *ectopic* pregnancy and will cause serious problems.)

Uterus

The uterus is about the size and shape of a pear. It can increase to six times its size. It is divided into three sections: The fundus, the body and the cervix (L. neck). It is made up of three layers of tissue:

1. The outer *serosal layer,*
2. The middle muscular layer, or *myometrium* (Gr. *myo,* muscle + *metra,* womb) which has three layers of smooth muscle fiber and
3. The *endometrium,* the lining which is deep and velvety in texture and contains many blood vessels. The endometrium is divided into two layers: The *stratum functionalis* (functional) and the *stratum basalis* (foundation). Every month the endometrium is built up to prepare for the possible implantation of a zygote, and if this does not occur, the stratum functionalis is shed along with blood and glandular secretions. This passes through the cervical canal and vagina and makes up the menstrual flow. The stratum basalis layer is permanent and forms a new stratum functionalis for the next monthly cycle.

Vagina

The *vagina* (L. sheath) is a muscle-lined tube about three to four inches long. It performs three functions:

1. It is the site where semen from the penis is deposited during sexual intercourse,
2. It is the birth canal for the baby, and
3. It is the channel for removal of menstrual flow.

The wall is composed mostly of smooth muscle and fibroelastic connective tissue which is lined with mucous membrane containing many folds. The mucus that lubricates the vagina comes from glands in the cervix of the uterus. The vagina has an acidic environment which helps prevent infection. This acidity would kill sperm if it were not for the alkaline fluid in the semen, and because at the time of ovulation, the cervix secretes a more alkaline mucus.

External genital organs

The external genital organs consist of five main parts:

1. The *mons pubis* (L. mountain + pubic) is a mound of fatty tissue that covers the *symphysis pubis,* and at puberty this area becomes covered with hair;
2. The *labia majora* (major lips) form the outer borders of the vulva. They contain fat, smooth muscle, areolar tissue, oil glands, and sensory receptors;
3. The *labia minora* (minor lips), which are two smaller folds of skin that lie next to the labia majora. Along with the majora, the minus surround the vaginal and urethral openings. They contain only oil glands and blood vessels, and no fat or hair;
4. The *vestibule* (entryway) is the space between the labia minora (pl. minus). Its floor contains the openings for the urethra and vagina and two glands that secrete an alkaline mucus during sexual excitement, and
5. The *clitoris* (KILT-ur-rihss; Gr. small hill) is a small erectile organ at the upper end of the vulva. It contains many nerve endings and has two corpora cavernosa that can fill with blood during sexual stimulation, causing it to become enlarged. It is capped by a sensitive glans.

The *perineum* (per-uh-NEE-uhm) is the name given to the region of the vulva and anal organs. The upper portion is called the *urogenital triangle,* and the bottom portion, the *anal triangle.*

The reproduction system in the female is controlled by eight hormones. The hormonal pattern in the monthly cycle has eight steps. The uterus is held by eight ligaments. (Review the number eight in the number study).

Forming a new creature

Each cell contains 46 chromosomes. Cell division is called *mitosis* (Gr. *mitos,* thread) and involves four stages. When a sperm and egg come together as a zygote, the zygote can only have 46 chromosomes, so it is necessary that the egg only have 23 and the sperm 23 chromosomes. How does this occur? The process is called *meiosis* (Gr. *meioun,* to diminish), and has six phases. Both the egg and the sperm must go through the process of being divided into only a portion of their original cell. The ovum does not get rid of its 23 until the sperm penetrates it. (Keep this in mind as we discuss the spiritual aspect of the new creature.) When sperm and the ovum combine they have the total chromosome count again of 46.

In the male and female sets of chromosomes, 22 always match in size and shape and determine the same traits. These 22 matching pairs are called *autosomes.* The other pair are sex chromosomes and they determine the sex of the new individual. In the female, all sex chromosomes look alike, so they are designated XX. The sex chromosomes do not all look alike in males, and are designated XY. Half of the sperm contains an X chromosome and half contain a Y chromosome. If an ovum is fertilized by an X chromosome bearing sperm, the baby will be a girl; if the ovum is fertilized by a Y chromosome bearing sperm, the baby will be a boy. The sex of the child is determined by the father at the moment of conception, just as the Holy Spirit of God plays the same part in our spiritual development. *"But all these worketh that one and the selfsame Spirit, dividing to every man severally as He will."*—I Corinthians 12:11.

In order to produce a new physical life it takes both *male* and *female.* In order to produce a *spiritual* life, it takes two beings, *God* and *human.* When we try to do this on our own, it is called self-righteousness. This is symbolized by masturbation, which creates a totally different electrical charge than intercourse, and it is harmful to the body. When two males or two females physically join or come together, it produces nothing but *trouble.* God created Adam and Eve, not Adam and Steve. (See Leviticus 18:22 and Romans 1:26, 27.)

Spiritual Application

The Father in heaven knows all about us before we are formed. He has a plan for our lives that will make us happy and satisfied. *"For thou hast possessed my reins: thou hast covered me in my mother's womb. I will praise thee; for I am fearfully [and] wonderfully made: marvellous [are] thy works; and [that] my soul knoweth right well. My substance was not hid from thee, when I was made in secret, [and] curiously wrought in the lowest parts of the earth. Thine eyes did see my substance, yet being unperfect; and in thy book all [my members] were written, [which] in continuance were fashioned, when [as yet there was] none of them. How precious also are thy thoughts unto me, O God! how great is the sum of them!"*—Psalm 139:13-17. The secret parts of the earth mentioned here are the *creative* parts in Hebrew. The word for *"curiously wrought"* is embroidered, which is what the DNA and RNA look like in the fertilized egg. Our heavenly Father has loving and caring thoughts for us long before we know anything about Him.

"Before I formed thee in the belly I knew thee; and before thou camest forth out of the womb I sanctified thee, [and] I ordained thee a prophet unto the nations."—Jeremiah 1:5. *"Listen, O isles, unto me; and hearken, ye people, from far; The LORD hath called me from the womb; from the bowels of my mother hath he made mention of my name."*—Isaiah 49:1.

God knows, sadly enough, whether we are going to hearken to His plans or not. *"Yea, thou heardest not; yea, thou knewest not; yea, from that time [that] thine ear was not opened: for I knew that thou wouldest deal very treacherously, and wast called a transgressor from the womb."*—Isaiah 48:8. Not because it was ordained, but because God can see into the future. Hear the sadness in our Creator's voice in John 3:3-8: *"Jesus answered and said unto him, Verily, verily, I say unto thee, Except a man be born again, he cannot see the kingdom of God. Nicodemus saith unto him, How can a man be born when he is old? can he enter the second time into his mother's womb, and be born? Jesus answered, Verily, verily, I say unto thee, Except a man be born of water and [of] the Spirit, he cannot enter into the kingdom of God. That which is born of the flesh is flesh; and that which is born of the Spirit is spirit. Marvel not that I said unto thee, Ye must be born again. The wind bloweth where it listeth, and thou hearest the sound thereof, but canst not tell whence it cometh, and whither it goeth: so is every one that is born of the Spirit."*

Jesus' birth was the example of the spiritual rebirth that must occur in order for us to join the family of God; so let us examine that process. *"Now the birth of Jesus Christ was on this wise: When as his mother Mary was espoused to Joseph, before they came together, she was found with child of the Holy Ghost."*—Matthew 1:18. Mary was impregnated by God's Holy Spirit of LOVE—God gave us the power to become the sons and daughters of God: *"But as many as received Him, to them gave He power to become the sons of God, even to them that believe on His name: Which were born, not of blood, nor of the will of the flesh, nor of the will of man, but of God."*—John 1:12-13. *"Of His own will begat He us with the word of truth, that we should be a kind of firstfruits of His creation."*—James 1:18. When we truly become converted, we will put off the old creature of *selfishness* and put on HIS character of *LOVE. "Therefore if any man be in Christ, he is a new creature: old things are passed away; behold, all things are become new."*—II Corinthians 5:17.

Just as the moment that the sperm meets the egg is not something the conscious mind is aware of, we are sometimes not consciously aware of God implanting the spirit of LOVE in our lives. The zygote takes nine months to become a full term baby, so like the person grows in the knowledge of God's *LOVE*-ing character. *"To whom God would make known what is the riches of the glory of this mystery among the Gentiles; which is Christ in you, the hope of glory."*—Colossians 1:27. *"Being born again, not of corruptible seed, but of incorruptible, by the word of God, which liveth and abideth for ever."*—I Peter 1:23. Selfishness will die—God's LOVE never dies. *"Whosoever is born of God doth not commit sin; for his seed (love) remaineth in him: and he cannot sin, because he is born of God."*—John 3:9. The *seed* is LOVE. The One who *implants* it is God.

The seed can fall on four types of ground. The sperm enters four different situations. By the *wayside,* there is no egg to fertilize or it is never put inside the body. On *stony ground,* there is a miscarriage in the early stages. *Among thorns,* there is a miscarriage in the late stages. On *fertile ground*—a baby is born. *"Thy people shall be willing in the day of thy power, in the beauties of holiness from the womb of the morning: thou hast the dew of thy youth."*—Psalm 110:3. There are four ways to be spiritually born: full term, premature, miscarriage, stillborn.

Mammary glands

The *mammary glands* (L. *mammae,* breasts) are present in male and female, but are never developed in the male (unless peculiar circumstances occur). The actual size of the breast, which depends on how much adipose tissue is present, does not affect how much mammary tissue is present. Each breast is composed of 15 to 20 lobes of compound areolar glands that look like bunches of grapes. These clusters have lactiferous (milk carrying) ducts that carry milk from the glands to the lactiferous sinus where the milk is stored. An extensive drainage system, which is made up of many lymph vessels, is present in the breasts. *"And in the midst of the seven candlesticks one like unto the Son of man, clothed with a garment down to the foot, and girt about the paps with a golden girdle."*—Revelation 1:13. Paps means breasts. Here we have Jesus represented as having breasts which were covered by a golden bra. What does this mean? One of the names of God is the Almighty, *EL SHADDAI.* It comes from the root word *shad,* which means a woman's breasts. A nursing mother is the symbol of nourishment, love, self-sacrifice, protection, giving, comfort, and all the other terms that describe that relationship to the child. This is what is implied in the title, ALMIGHTY. God uses this title when He comes to speak to Abram in Genesis 17. God gives Himself to Abram, and then Abram gives himself to God. Abraham then becomes fruitful. God adds something to Abram's name, the letter—He, the chief letter of His own name, which can only be uttered by an out breathing. This symbolizes God giving Abraham something of His own nature of LOVE. God promises Abraham that he will be a blessing to all nations, that out of a dead womb will the promised son come. Then Abraham yields himself to God and the ALMIGHTY seals it with the act of circumcision which signifies removal of selfishness and instilling LOVE into our hearts. God invented sexual intercourse, and it was a giving act rather than a selfish act. Today it is often times an act of LUST.

"Even by the God of thy father, who shall help thee; and by the ALMIGHTY, who shall bless thee with blessings of heaven above, blessings of the deep that lieth under, blessings of the breasts, and of the womb."—Genesis 49:25.

When Balaam had the vision and poured forth the promises of blessings on Israel, it was the "vision of the ALMIGHTY." (Numbers 24:4.) *"He that dwelleth in the secret place of the most High shall abide under the shadow of the ALMIGHTY."*—Psalm 91:1. EL SHADDAI will protect us with more fury than a mother protects her baby. *"Howl ye; for the day of the LORD is at hand; it shall come as the destruction from the ALMIGHTY."*—Isaiah 13:6.

There is another side to the nourishment of God. If a baby is not ready to receive the mothers milk and it is flowing too fast, he may choke on it. The same sun can shine on clay and wax; it hardens the clay and melts the wax. It is so with the blessings of God. When the SON shines on the *selfish heart,* it becomes combative or like a stone. When the SON shines on the *loving heart,* it becomes selfless.

The details of this study could go on and on as there is a spiritual "parable" to every part of the growth of the baby, in the order of development, the length of time, the different sections of original tissue. If we covered all these lessons, the length of this book would be prohibitive. Perhaps someone would like to write a book or prepare a dissertation on these subjects.

Chapter Thirty-Three

THE INTEGUMENTARY SYSTEM

Exodus 26:14

"And thou shalt make a covering for the tent [of] rams' skins dyed red, and a covering above [of] badgers' skins."

God instructed Moses to cover the sanctuary in the wilderness with two coverings; one of rams' skins dyed red and the other of badgers skins which were 'hairy'. God has covered His sanctuary (our bodies) with a covering of two layers the dermis (which when exposed is red) and the epidermis (which in most cases is covered with hair). This covering is called *integumentary* (in-teg-yoo-MEN-tah-ree, L. covering). That is exactly what this system does; it covers everything on the outside or the body. Naphtali represents this system. (Review the lesson on Touch.) This system consists of the skin, hair, fingernails, toenails, and several types of glands. It covers and protects, helps regulate body temperature, eliminates some waste material, helps make vitamin D from the sun's rays, and perceives stimuli such as pain, pleasure, and temperature.

Skin

The skin is the covering of love that holds in the fluids, flesh, and bone of the body. When people look at us, they should not see us but should only see the reflection of God's character of LOVE. This was what made up the "covering" that Adam and Eve were clothed in before sin. It was only when they sinned and began thinking of self in place of God's LOVE that this covering was removed and they felt naked. Keep this in mind as you study the different layers and functions of skin. This coincides with the different layers and functions of love. The tribe of Naphtali was lacking in common sense and practicality. Sometimes it looks as though love does not have much common sense, especially as we see Christ on the cross dying for a world that rejected Him. That fact makes no sense at all to the worldly minded person.

The skin is the largest system in the body. It occupies about 21 square feet of area, of course depending upon your size. It can vary in thickness from less than 0.5mm on the eyelids to more than 5mm in the middle of the upper back. It is multi-layered like love which has many facets. There is love that protects, love that provides, love that comforts, love that disciplines, and love that never seeks her own. We should read I Corinthians 13 every day until the meaning of love is clear and becomes a part of us.

The skin has three main parts: the epidermis, dermis and hypodermis (subcutaneous).

Epidermis

The outer layer of the skin is called the epidermis (Gr. *epi,* over + *derma,* skin) and is stratified (layered) with squamous epithelium. It contains three layers in most parts of the body. In the thick epidermis on the palms of the hands and the soles of the feet there are five layers. Remember that there are also five bones there, of grace in your works and your ways.

Starting with the outermost layer the layers are:

1. Stratum corneum (L. horny layer). This flat is a thick layer of dead cells that contains keratin (Gr. *keras,* horn), a fibrous protein produced by epidermal cells called *keratinocytes.* The keratin in the skin is soft compared to the keratin in the fingernails and toenails. The epidermis cells are arranged in parallel rows and protect the cells underneath from being exposed to the air and drying out. Everything in the body that is not bathed in fluid is dead. If we are not immersed in the truth of God, we eventually dry up spiritually and die. These cells are constantly being shed and replaced with new cells. The cells on the soles of the feet are completely replaced every month or so. Albeit truth is progressive. *"But the path of the just is as the shining light, that shineth more and more unto the perfect day."*—Proverbs 4:18. The next two layers appear only in the palms of the hands and the soles of the feet.
2. Stratum lucidum (L. *lucidus,* bright, clear) This consists of flat, translucent layers of dead cells.
3. *Stratum granulosum* (L. *granum,* grain) is next and is usually 2 to 4 cells thick. The cells contain granules of *keratohyaline.* The process of *keratinization,* which is associated with the dying of cells, begins in this layer. The next two layers combined are known as the *stratum germinativum* (jer-mih-nuh-TEE-vuhm; L. *geminare,* to sprout) because they generate new cells.
4. Stratum spinosum (L. *spinousus,* spiny) produces some new cells, which are pushed to the surface to replace the cells that have shed off. This layer is composed of several layers of polyhedral (many sided) cells that have delicate spines protruding that interlock with their neighboring cells to help support this binding layer. *"I drew them with cords of a man, with bands of love . . ."* (Hosea 11:4.) Love is binding and draws us to Him and to each other.
5. Stratum basale (buh-SAY-lee; L. *basis,* base) consists of a single layer of columnar or cuboidal cells. It also produces new cells to replace the old ones that are shed off. This represents the basis of our Christian experience, love to God in obedience and love to man in service.

Dermis

The dermis (Gr. *derma,* skin) is composed of a strong, flexible connective tissue containing collagenous, reticular and elastic fibers. The toughness of the skin is because the collagenous fibers are very thick. The reticular fibers provide a supporting network and the elastic fibers give the skin elasticity. Thank God for the lesson of love in the supporting skin. Let us group together the blessed assurances of His love that we may look upon them continually:

1. The Son of God leaving His Father's throne, clothing His divinity with humanity, that He might rescue man from the power of Satan.
2. His triumph in our behalf, thereby opening heaven to men.
3. Revealing to human vision the presence chamber where the Deity unveils His glory.
4. Rescuing us from the pit of ruin into which sin had plunged us.
5. Bringing mankind again into connection with the infinite God, and having endured the divine test through faith in our Redeemer.
6. Then clothing us in the righteousness of Christ and exalting us to His throne.

These are the pictures given on which God would have us contemplate.

The dermis is composed of two layers, the thin papillary layer and the reticular layer. The *papillary layer* (L. *papula,* pimple) is named because of the tiny projections that join it to the ridges of the epidermis. Most of the papillae contain loops of capillaries and some have special nerve endings called *Meissner's corpuscles,* which are corpuscles of touch. Fingerprint patterns are formed by these papillae as the epidermis follows these corrugated contours. The palms of the hands and the souls of the feet have a double row of papillae that produce ridges that help keep the skin from tearing. These ridges improve the gripping surfaces and produce distinctive patterns in the lines there which are different in every person. No two people, not even identical twins, have the same patterns.

Every individual has a life distinct from all others, and an experience differing essentially from all others. The relations between God and each soul are as distinct and full as though there were not another soul upon the earth to share His watchcare; not another soul for whom He gave His beloved son. When we seem to doubt God's love and distrust His promises, we dishonor Him and cause Him grief. How would a mother feel if her children were constantly complaining about her, as though she did not mean them well when her whole life's effort had been to forward their interests and to give them comfort? Suppose they should doubt her love; it would break her heart. And how can our heavenly Father regard us when we distrust His love which has led Him to give his only-begotten son that we may have life?

The reticular layer (netlike) is the part of the animal's skin that is processed to make leather, so you can see how tough it is. It is made up of dense connective tissue with coarse collagenous fibers and fiber bundles that crisscross to form a strong elastic network. If you have ever used a chamois to dry your car, think of how you can stretch it. This represents how enduring love is.

There is a dominant pattern to these layers and different directional patterns are found in each area of the body. These lines of tension are known as cleavage lines. If a cut or incision is made parallel with the directional lines, it heals faster because it only slightly disrupts the collagenous fibers. If cut crosswise, it is harder to heal. If we accept our trials as learning experiences, they are much easier to cope with. If we murmur and complain, it takes us longer to learn the lesson intended.

Embedded in the reticular layer are many blood and lymphatic vessels, nerves, free nerve endings, fat cells, gland, and hair roots.

Glands

There are two different classes of glands in the skin, *sudoriferous* (L. *sudor,* sweat) glands and *sebaceous* (L. *sebum,* tallow) glands. There are two different types of sudoriferous glands: *eccrine* and *apocrine.*

Eccrine Glands

Eccrine (EK-rihn; Gr. to exude, secrete) glands are small sweat glands that are distributed over nearly the entire body surface, and work to keep the body temperature regulated. Most of these glands secrete sweat through perspiration when the external temperature of the body rises. The word perspiration means "to breathe through" because it was believed that the skin breathes. The constant combination of perspiration and evaporation keeps the body from overheating. During one hour of vigorous physical activity, over a quart of sweat is produced.

Sweat is a colorless fluid that holds neutral fats, albumin and urea, lactic acid, sodium and potassium chloride, and traces of sugar and ascorbic acid. Sweat produced by inactive persons is lost by simple diffusion. This occurs waking or sleeping and is called insensible perspiration. Perspiring can also be caused by psychological stress and occurs mostly on the palms, fingers, and soles. Jesus was under such stress in the garden that He actually sweat blood through the pores of His face. The face represents spirit and character. As you look at the picture of the skin, see what a strain takes place to put blood in the sweat pores. *"And being in an agony He prayed more earnestly: and His sweat was as it were great drops of blood falling down to the ground."*—Luke 33:44. The humanity of the Son of God trembled in that trying hour. He prayed not then for His disciples, that their faith might not fail, but for His own tempted agonized soul. The awful moment had come —that moment which was to decide the destiny of the world. The fate of humanity trembled in the balance. Christ could have even then refused to drink the cup apportioned to guilty man. He might have wiped the bloody sweat from His brow, and left man to perish in his iniquity. He could have said, "Let the transgressor receive the penalty of his sin, and I will go back to my Father. I will not drink the bitter cup of humiliation and agony and be separated from my Father in consequence of the curse of sin." But Jesus saw the helplessness of man and the power of sin. The woes of a doomed world rose before Him and He chose to save man at any cost to

Himself. *"Ye have not resisted unto blood, striving against sin?"*—Hebrews 12:4. How much strain have we ever put on ourselves to keep from hurting Him?

Apocrine glands

Apocrine glands are larger and more deeply situated than eccrine glands. They are found in the armpits, the *areolae* (dark regions around the nipples), and the outer areas of the genital and anal regions. These glands become active at puberty and respond to stress by secreting sweat with a characteristic odor. The smell that we call "body odor" is produced when bacteria on the skin decompose the secretions as they feed on them. The odor becomes stronger if toxins are present in the body. Love tries to get rid of sin and keep the body pure. If our works are not pure, with a motive of love, we stink and the world notices, which makes Christianity unattractive.

Ceruminous glands are also apocrine glands and are located in the outer ear canal. They secrete ear wax which helps trap foreign substances before they can enter the deeper portions of the ear. *Mammary* glands that produce milk are modified apocrine glands.

Sebaceous Glands

In the body the oil glands are connected to hair follicles. These glands are alveolar (hollow) and their main functions are lubrication and protection. The part that secretes is made up of a cluster of cells and the oily secretion called *sebum* is produced by the breaking down of interior cells. Sebum is composed almost entirely of lipids. This helps to keep our skin soft, acts as a permeability barrier, and protects against bacteria and fungi. When these oil glands become inflamed and plugged, a blackhead results. A pimple or even a *sebaceous cyst* can develop if they are not unplugged. The black color is not caused by dirt, but by the sebum being exposed to air.

Openings from these glands are found all over the surface of the skin except the palms and soles. When you take a bath, the underside of your toes and fingers absorb a great deal of water because this area is not coated with sebum, so water soaks through. The epidermis expands, but not the dermis, so the epidermis becomes wrinkled. After you get out of the bath, this water evaporates quickly.

Hypodermis

The *hypodermis* (Gr. *hypo,* under + *derma,* skin), the layer under the dermis, is *subcutaneous* (L. *sub,* under + *cutis,* skin); it is composed of loose, fibrous connective tissue, is supplied with lymphatic and blood vessels and nerves, and, where extra padding is needed, thick sheets of fat cells. The distribution of fat in this layer creates the characteristic body curves of the female.

Skin Color

The color of the skin is caused by three factors: the color of the blood reflected through the dermis, the accumulation of the yellow pigment *carotene,* and the presence of *melanin.*

Melanin (Gr. *melas,* black) is a dark pigment produced by specialized cells called *melanocytes,* which are usually located in the deepest part of the stratum basale. There is hardly any melanin in the palms and soles. There is more in the areas where skin is darker: the external genitals, the nipples and areolae, the armpits, and the anal region. Melanin is also present in hair and in the iris and retina of the eyes.

Melanin screens out excessive ultraviolet rays. Suntan is caused by a darkening of the melanin and it moves to the outer skin layers. Freckles are caused by spots of melanin. Darker races have slightly more melanocytes than light skinned people and produce more melanin. There is also a wider distribution of melanin into the higher levels of the epidermis. If there is no melanin produced, the person is known as an albino (L. *albus,* white).

Cancers of the skin are caused, in part, by the wrong kind of fats and an overabundance of fat in the diet. Add to that exposure to UV rays of sunlight and there is bound to be a problem. *Malignant melanoma* is cancer of the pigment producing melanocytes. It usually starts as small, dark growths resembling moles that gradually become larger, change color, become ulcerated, and bleed easily.

The sun helps the skin to manufacture vitamin D, but on the other hand, it can cause a terrible burned condition of the skin. The sun can be a blessing or a curse.

When Israel strayed away from God, it was always to worship the sun god in one form or another. Satan set up sun worship as his form of religion to steal the worship of the Son to which only the Creator is entitled. The people of God tried to mix the two worships together and brought the symbols of sun worship right into the Temple of God. The early Christians were careful not to use pagan symbols. Some of these pagan symbols did not come into the church until around 300 AD. Some of these symbols are: The fish, the cross, the steeple, Sunday keeping, and the sun disk (halo) around people's heads seen in many religious paintings. We should take seriously the second commandment (Exodus 20:4-6). The fourth plague of Revelation 16 comes because man insisted upon worshipping the sun god.

Just as our outward actions indicate how pure our love is, the skin shows the condition of the body. Skin provides many clues about the general health of the body. The skin color can be an indication of abnormal conditions such as yellow-orange skin indicates jaundice or liver impairment. This is caused by too much bile pigment in the blood. The blood becomes impure, the complexion sallow, and eruptions frequently appear. That which colors the skin and makes it dingy also clouds the spirits and destroys cheerfulness and peace of mind. Every habit that injures the health reacts upon the mind. Those who obey the laws of health will give time and thought to the needs of the body and to the laws of digestion. The results of this obedience will be a clearness of mind and skin.

Cyanide poisoning turns the skin blue, indicating a lack of oxygen in the blood. When the cortex of the adrenal gland is underactive, the skin appears a bronze tone. A greenish tinge sometimes indicates cancer.

Rashes on the skin indicates problems in the liver or an overworked lymph system. Dry skin and loss of skin color may come from a protein deficiency. A lack of essential fatty acids can cause patchy baldness and eczema. Dry, thickened skin can be a problem with the thyroid gland or lack of vitamin A.

Functions of the Skin

Protection: The skin acts like a shield to prevent harmful microorganisms from entering the body. The surface of the skin is acidic which restricts the growth of microorganisms. These little creatures are removed before they can enter the body as the outer layer of skin is constantly shed and replaced. There are several harmful substances that can be absorbed by the skin and enter the bloodstream. They are: Nickel, mercury, many pesticides and herbicides, and chemicals in the poison plants such as poison ivy. *"He shall cover thee with His feathers, and under His wings shalt thou trust: His truth shall be thy shield and buckler."*—Psalm 91:4

Prevent loss of body fluid: The skin is designed with its many layers to perfectly impede the entrance and loss of fluid. When all the layers are burned, fluids begin to leak out. If this occurs over a large area, all systems of the body are affected. The lymph and blood circulation is hampered and severe loss of body water, plasma, and plasma proteins produce shock. An infection at a burn site is a major concern. The neutrophils are the skin's specialized infection fighting cells. When circulation is hampered, the neutrophils are immobilized and instead of rushing to the infection site and releasing their disease fighting chemicals, they release them before they get there. Then the chemicals interfere with the chemical signal from the infection site, and they do not know where to go. *"If a man love Me, he will keep My words: and My Father will love him, and We will come unto him, and make Our abode with him. He that loveth Me not keepeth not My sayings."*—John 14: 23,24.

Temperature regulation: A cooling effect occurs when sweat is excreted through the pores and then evaporates from the surface of the skin. On cold days your skin acts as a sheet of insulation that helps retain body heat. There is a dense bed of blood vessels in the dermis contract to hold heat from the blood. Also when you are frightened, these blood vessels in the skin contract, causing the skin to become cool. That is where the expression "getting cold feet" come from. This refers to the fear of a situation. When it is warm the opposite reaction takes place. The vessels dilate and heat radiation from the blood is increased then the body heat is lost.

Excretion: The skin has many sweat pores. Small amounts of waste materials such as urea are excreted through these pores. This happens especially when we do not have regular bowel movements and toxins build up in the body. Excretion of these toxins is what causes body odor.

Syntheses: The skin lets in some ultra violet rays, which convert a chemical in the skin into vitamin D. This vitamin is vital to the normal growth and maintenance of bones and teeth. If it is lacking, it affects the absorption of calcium from the intestine into the bloodstream. If children are deprived of sunshine, they can develop rickets, which is a disease that may deform bones permanently. *"For this is the love of God, that we keep His commandments: and His commandments are not grievous."*—I John 5:3.

Sensory reception: The sensory receptors in the skin respond to heat, cold, touch, pressure, and pain. The many nerve endings keep us aware of changes in the environment. In this way the body can respond appropriately, such as not touching a hot object, sensing a sharp blade, feeling a heavy weight, a tight binding, etc. An itch or a tickle is a low-level pain response. Pain and pleasure travel the same identical pathways to the brain. *"Whosoever believeth that Jesus is the Christ is born of God: and every one that loveth Him that begat loveth Him also that is begotten of him. By this we know that we love the children of God, when we love God, and keep His commandments."*—I John 5:1,2.

Spiritual Application

The Bible gives a prophecy concerning the skin (love), the flesh or muscles (judgment), and the bones (principles and laws), about what was going to be happening in the churches in the last days: *"And many false prophets shall rise, and shall deceive many. And because iniquity shall abound the love of many shall wax cold."*—Matthew 24:11,12. *"And I said, Hear, I pray you, O heads of Jacob, and ye princes of the house of Israel;* (leaders in the church) *[Is it] not for you to know judgment? Who hate the good, and love the evil; who pluck off their skin from off them, and their flesh from off their bones; Who also eat the flesh of my people, and flay their skin from off them; and they break their bones, and chop them in pieces, as for the pot, and as flesh within the caldron."*—Micah 3:1-3.

The caldron in which they are placed is the ecumenical movement where love, judgment, and law is interpreted to suit man and not according to the Word of God. This is more tragic than the story in I Kings 4:38-41. One went out into the field (the world) and gathered wild gourds (worldly philosophy), and shredded them into the great pot (mixed error with the Word of God that was going to be fed to the sons of the prophets). It caused death in the pot, but it was cured. In the prophecy in Micah, the people of God get cooked. Albeit cooked food is basically dead food.

How a wound heals: The healing of a wound follows the same steps as spiritual healing. Blood vessels are severed by the wound. (David's connection with God was severed or he would not have added sin to sin.) Platelets and a blood clotting protein called *fibrinogen* help start a blood clot. A network of fibers containing trapped cells forms, and the edges of the wound begin to join together. Tissue forming cells called *fibroblasts* begin to approach the wound. The spiritual application is: The prophet comes with the law to point out David's sin. If God's people would recognize His dealings with them and accept His teachings, they would find a straight path for their feet and a light to guide them through darkness and discouragement. David learned wisdom from God's dealings with him and bowed in humility beneath the chastisement of the Most High. The faithful portrayal of his true state by the prophet Nathan made David acquainted with his own sins and aided him to put them away. He accepted counsel meekly and humiliated himself before God. *"The law of the Lord,"* he exclaims, *"is perfect, converting the soul."*

A scab forms to cover up the wound so that no more bacteria can enter. David was repentant and began to fast and pray. Repentant sinners have no cause to despair, because they are reminded of their transgressions and warned of their danger. These very efforts on their behalf show how much God loves them and desires to save them. They have only to follow His counsel and do His will to inherit eternal life. God sets the sins of His erring people before them that they may behold them in all their enormity under the light of Divine Truth. It is then their duty to renounce them forever.

The injured cells die in the immediate area and release *lysosomes,* which cause more damage. Sin never just affects us, but all around us suffer. Uriah, a faithful servant and a mighty soldier for David, died because of David's sin. All wrong done to others reaches back from the injured one to God.

Phagocytes from the blood come into the wound area and ingest microorganisms, cellular debris and other foreign material. Epidermal cells at the edge of the wound begin to divide and start to build a bridge across the wound. The phagocytes die as a result of having to clean up the wound; this creates pus. The spiritual application is that Jesus' death pays for the sin, but the results of the sin go on even after they are forgiven.

When a totally new epidermal surface has been formed, the protective scab is sloughed off. When David knew he could not change the results of the sin, he washed and ate and got on with his life.

Fibroblasts build scar tissue with collagen. Sin leaves scars, as the results of a departure from God and can go on for years after. *"Now therefore the sward shall never depart from thine house."*—II Samuel 12:10. David paid fourfold and lost four of his sons, and one of them, Absalom, slept openly with David's wives, as God predicted in II Samuel 12:11.

David was of a contrite spirit, as can be seen in the psalms he wrote. God called him a man after His own heart because he was humble and sorry for his sin, but that did not change the results of a wounded life. We need always to watch and pray lest by a moment of carelessness we might yield to temptation and cause a terrible wound to God, to those around us, and to ourselves.

Scrupulous cleanliness is essential to both physical and mental health. Impurities are constantly thrown off from the body through the skin. Its millions of pores are quickly clogged unless kept clean by frequent bathing, and the impurities that should pass off through the skin become an additional burden to the other eliminating organs. A cool or tepid bath every day fortifies against cold because it improves the circulation, the blood is brought to the surface, and a more easy and regular blood flow is obtained. The mind and body are invigorated. The muscles become more flexible, the intellect is made brighter, and the nerves are soothed. Bathing helps the bowels, the stomach, and the liver, giving health and energy to each, and it promotes digestion. Spiritual bathing in the water of life, the Truth of God, and brings health to our spiritual life.

It is important also that the clothing be kept clean. The garments worn absorb the waste matter that passes off through the pores. If they are not frequently changed and washed, the impurities will be reabsorbed. This is true also of our spiritual covering. *"That He might sanctify and cleanse it with the washing of water by the word. That He might present it to Himself a glorious church, not having spot, or wrinkle, or any such thing; but that it should be holy and blameless."*—Ephesians 5:27.

Man, as created in the image of God, had a covering of light (Psalm 104:2). A beautiful soft light, generated by God's Spirit of Love, enshrouded the holy pair. This robe of light was a symbol of their spiritual garments of heavenly innocence. Had they remained true to God, it would ever have continued to cover them. But when sin and selfishness entered, they severed their connections with God and the light that had encircled them departed. Naked and ashamed, they tried to supply the place of the heavenly garments by sewing together fig leaves for a covering. Genesis 3:7.

This is what the transgressors of God's law have done ever since the day of Adam and Eve's disobedience. They have sewed together fig leaves of selfishness to cover the nakedness caused by transgression. They have worn the garments and head coverings of their own devising, and by works of their own, they have tried to cover their sins and make themselves acceptable to God. But this they can never do! Nothing can man devise to supply the place of His robe of perfect LOVE.

Only the covering, which Christ Himself has provided, can make us fit to appear in God's presence. This robe was woven in the loom of heaven from God's love and has in it not one thread of selfishness. Christ in His humanity wrought out a perfect character He offers to impart it to us. *"And we are all as an unclean thing, and all our righteousness are as filthy rags; and we all do fade as a leaf; and our iniquities, like the wind, have taken us away."* —Isaiah 64:6. Everything that we of ourselves can do is defiled by sin. *"Whosoever committeth sin transgresseth also the law: for sin is the transgression of the law. And ye know that he was manifested to take away our sins; and in him is no sin"*—I John 3:4,5. By the Son of God's perfect obedience, He has made it possible for every human being to obey God's commandments. When we submit ourselves to Christ, the heart is united to His heart, the will is merged in His will, the mind becomes one with His mind, the thoughts are brought into captivity to Him, and we live His life. This is what it means to be clothed with the garment of His righteousness. Then as the Lord looks upon us, He sees, not the nakedness and deformity of sin, but His own robe of righteousness which is perfect obedience to God's law of Love.

"I will greatly rejoice in the LORD, my soul shall be joyful in my God; for He hath clothed me with the garments of salvation, He hath covered me with the robe of righteousness, as a bridegroom decketh himself with ornaments, and as a bride adorneth herself with her jewels."—Isaiah 61:10. In Revelation 3:18, Christ counsels us to be sure we put on this robe so that the shame of our nakedness might not appear. Clothes represent character. Christ was stripped naked of His character and took ours and nailed it to the cross that we might have His character and be presentable to God. *"For He hath made Him to be sin for us, who knew no sin; that we might be made the righteousness of God in HIM."*—II Corinthians 5:21.

In pictures of Christ on the cross, He is always portrayed as having on a covering. But the soldiers had divided His underwear into four parts and cast lots for His seamless robe. (John 19:23,24). He hung naked, in open shame, for you and me. The spotless pure King of the Universe died naked on the cross, the symbol of pagan sexual rites, so that I might be able to wear His spotless character. Wonder of Heavens! Marvel O Earth, what manner of LOVE!

Besides our spiritual clothing, God is concerned about the clothing we wear on our bodies. The garments of the priests were specified in every detail and were called holy garments (Exodus 28). We are a royal priesthood (I Peter 2:9) and our garments should reflect that. Not that we dress in the priests' robes, but that we reflect our standing as a child of God, by wearing clean, modest apparel. Read I Timothy 2:9; I Peter 3:3; Deuteronomy 22:5. If we do not want to be peculiar people, and insist on following the world's standards of dress, God will count us as belonging to the world and so will the world. This is the reason that the LOUD CRY must be given to God's people everywhere to come out from the world and be separate.

Hair

Hair represents experience. In the following passages, you can see how hair was an outward sign of the experience occurring.

The first mention of human hair was in connection with the test for leprosy; Leviticus 13:1-3 depicts cleansing of the leper in Leviticus 14. Part of the ceremony was to shave all the hair from the body denoting a cleansing of past experience and a new start. Sores were a symbol of sin (Isaiah 1:6). Jesus did the impossible and healed the lepers just like He cleanses us from sin.

The second occasion for hair to be a sign was Numbers 6, in the case of a Nazarite vow. The person was to let his hair grow all the days of the vow and at the end, make a sacrifice, shave his hair and put it in the fire of sacrifice showing that all of our experience in life should be an offering unto God.

The third occasion was Judges 13–16, the story of Samson, who was to be a Nazarite from his birth for all his life. The sign of this experience was his hair, and when he did not treasure the experience of being especially consecrated to God, he betrayed his calling by disclosing the secret of his strength. There was no strength that came from his hair, but it was the sign of experience of being especially called of God for a certain job. When he lost his connection with God, he lost his strength.

Another interesting story is about Absalom and his hair. Read II Samuel 13-18. He had so much hair that he cut it once a year and weighed it. It weighed about five pounds. When you read the story of Absalom, you will see that his experience was in the ways of the world. The ambitious person sometimes sacrifices principles to get ahead, but God measures a man's worth by his obedience to Him.

Ezra became so upset with what the people had done that he pulled his hair out, which we still say today to denote frustration. (Ezra 9:3.)

In Daniel 3:27, not one of the hairs of Shadrach, Meshach, and Abednego was even singed, showing that the king's punishment could not change their experience with the Lord.

The sign that Jerusalem had disobeyed the Lord was to *"Cut off thine hair, O Jerusalem, and cast it away, and take up a lamentation on high places; for the LORD hath rejected and forsaken the generation of His wrath."* —Jeremiah 7:29.

God promises that baldness will replace well set hair on those people who are caught up in the experience of self love and display. (Isaiah 3:24.)

Face represents spirit or character. Jesus had the hair on His face pulled out by people who were objecting to His experience of a holy life. (Isaiah 50:6.)

Jesus tells us that we cannot make one hair of our head black or white, showing that we cannot change our experience without the help of God. (Read Matthew 5:36 and Jeremiah 13:23.) Of course, today we have hair dye and makeup that can fool people. We cannot fool God with a pretended experience (Timothy 2:9 and I Peter 3:3.)

Our experience will be weighed and divided and rewarded one way or the other. Read Ezekiel 5.

In a picture of the hair shaft, it shows three layers of epithelial cells: The medulla, the cortex, and the cuticle, which are all composed of soft keratin. A strand of hair is stronger than a strand of copper the same size.

The central core of the hair is the medulla. The medulla contains loosely arranged cells separated by air or liquid in its extracellular spaces.

The next layer is the cortex which is the thickest layer and consists of several layers of cells. Pigment in the cortex gives hair its color. When the medulla becomes completely filled with air and hair pigment fades from the cortex, the hair loses its color and becomes white. This can also be accomplished by bleaching the hair with hydrogen peroxide.

The cuticle is the outer layer and is made up of thin squamous cells that overlap. This cuticle can be softened or even dissolved by chlorine in pool water.

Hair has two sections: The shaft is the part that can be seen, and the root is the portion embedded in the skin. The lower portion of the root is called the bulb, which is composed of a matrix of epithelial cells. The entire bulb is enclosed within a little bag called a follicle. It has three sheaths. The hair projects at an angle which makes the hair cover the scalp more efficiently. The shape of the follicle opening on the skin determines whether the hair is straight (round opening) or curly (spiral opening).

Each hair has an *arrector pili* muscle, which contracts and forces oil from a gland. It also pulls the follicle and its hair to and erect position, elevates the skin above, and produces a goose bump. This muscle is what makes the hair stand up when you are frightened.

We shed hair when its growth is complete. The root of the hair becomes hard and detaches from the matrix. The root bulb begins to move along the follicle until it stops near the level of the oil gland, the papilla atrophies, and the outer root sheath collapses. Then there is a resting period for the new matrix to develop and a new hair starts to grow up the follicle.

No area of the body sheds all its old hair at the same time because each hair has its own life cycle. Hair grows faster at night and in warm weather. Hair growth varies with age, the fastest growth rate being in women between 16 and 24. Coarse black hair grows faster than does fine blond hair. Each person has their own experience.

Eyebrow and eyelash hairs last about 10 weeks and scalp hairs about 3 to 5 years. Some parts of our experience last longer than others. If it was not for the growth and rest cycle, scalp hair could grow about 24 feet in a lifetime, but ordinarily does not reach a length more than 3_ feet before it dies. Hair does not grow after death, but as the skin shrinks, the hair appears longer. There is no experience after death until Jesus returns to resurrect us. *"Whatsoever thy hand findeth to do, do it with thy might; for there is no work, nor device, nor knowledge, nor wisdom, in the grave, whether thou goest."*—Ecclesiastics 9:10.

Men's hair is coarser and more obvious, so it appears that they have more body hair. But both men and women have about the same number of hair follicles, around 21 million. I Corinthians 11 indicates that men and women should have a different kind of hair arrangement symbolizing that their experience in life is different. Today there is a big movement to obliterate that difference instead of being content with the gender we are and complementing each other's attributes.

The average scalp contains about 125,000 hairs, but only God knows how many each person has. *"But the very hairs of your head are all numbered."*—Matthew 10:30. Keep your wants, your joys, your sorrows, your cares, and your fears before God. You cannot burden Him; you cannot weary Him. He who numbers the hairs of your head is not indifferent to the wants of His children. His heart of love is touched by our sorrows, and even by our utterance of them. Take to Him everything that perplexes the mind. Nothing is too great for Him to bear, for He holds up worlds, He rules over all the affairs of the universe. Nothing that in any way concerns our peace is too small for Him to notice. There is no chapter in our experience that is too dark for Him to read. There is no perplexity too difficult for Him to unravel. No calamity can befall the least of His children, no anxiety harass the soul, no joy cheer, no sincere prayer escape the lips of which our heavenly Father is unobservant or in which He takes no immediate interest.

Not until a person reaches puberty does he get hair under the arms or in the pubic area. Not until a Christian is mature does he have real experience in his works or his production for God. In Ezekiel 16 is the story of God's love relationship with His people. When they became mature and their hair grew (verse 7), they offered themselves to other gods. Just about the time we have matured enough to be of some use to God, we often decide to spend our time and talents on worldly things.

Nails

Nails are made of hard keratin. This part that shows is the body of the nail and consists of dead cells. The nails appear pink because the nail is translucent, and the red color of the blood in the tissues underneath shows through. The nail rests on an epithelial layer of skin called the nail bed. The root is hidden under the skin folds of the nail groove. The thicker layer of skin beneath the nail root is the matrix where new cells are made for nail growth and repair. If a nail is injured severely enough to cause it to fall off, a new one will grow if the matrix is not damaged.

The nails protect our fingers and toes, they and are useful to scratch with and they allow us to pick up small objects and do delicate work.

Nails spiritually represent experience. They are harder than hair and tell of a harder experience. When a woman was taken as a captive from the enemies of Israel, she was to shave her head and pare her nails (Deuteronomy 21:10-12). This was a symbol that her whole experience of life was changed.

When Nebuchadnezzar forgot that God was in charge of his kingdom and that it was God that had given him the all that he had. Nebuchadnezzar had to go out and eat grass with the cattle. He went crazy, and his hair and nails testified of that experience. Read Daniel 4. It is well for us to remember what Jesus told us: *"For without Me ye can do nothing."*—John 15:5.

As we comb our hair, trim our nails, bathe our skin, and put on our clothing, let us keep in mind the wonderful lessons of this system.

"Peace I leave with you, my peace I give unto you: not as the world giveth, give I unto you. Let not your heart be troubled, neither let it be afraid."—John 14:27

Study Questions

The Digestive System

1. What food is to the ___________ Christ is to the _____________.
2. Physical digestion and assimilation of food is a _________________ and _______________ process that __________________________________.
3. T or F. The perception and appreciation of truth depends less upon the mind than upon the converted heart.
4. God intended that most of our nourishment would be derived from foods that contain high amounts of ________________.
5. Research question: Name the eight essential carbohydrate sugars that our bodies must have to facilitate intercellular communication.
6. What is a "dipeptide"?
7. Name the five processes of our digestive system.
8. Where is the alimentary canal located and give its function.
9. What does the palate have to do with digestion?
10. The saliva contains _________% water and ___% _________________.
11. Why is it dangerous to drink with our meals?
12. What is a bolus?
13. The esophagus is about ________ long and consists of ________ coats of muscle and membrane.
14. What are the mesenteries and what do they do?
15. Name the four classes of stomach cells.
16. What is "chime." Write a short paragraph in your own words how it is derived and describing its function.
17. The stomach is divided into three parts: Name them and tell their function.
18. T or F. The liver is the largest organ inside the body and is the most complicated and versatile of all body organs. Explain why.
19. Explain how the Law of God and Grace are compatible.
20. Give ten functions of the liver.
21. What does the gallbladder do?
22. What does the pancreas do?
23. T or F. Defecation is an involuntary act. Explain your answer. How many times a day should it take place?

The Urinary System

1. The Urinary System consists of what six structures?
2. What is "homeostasis"?
3. The body must maintain a balance between ________ and ________. Why?
4. Where are the kidneys located? Name the three distinct regions of the kidneys.
5. T or F. The kidneys receive more blood in proportion to their weight than any other organ of the body. Why is this?
6. What is a "nephrons"? What do they do?
7. Glomerular filtrate is what?
8. Where is the "Loop of Henle" found? What is its function?
9. When the nations are gathered before God, there will be but two classes, and their eternal destiny will be determined by what?
10. Give the function of the "juxtaglomerular apparatus."
11. What are some of the causes for kidney problems.

The Reproductive System

1. Men ought to love their wives as ___________________.
2. Why is the Biblical story of Hosea important?
3. What is the reproductive role of the male?
4. T or F. The sperm mature as they travel along the epididymis; this journey may take 2 to 5 days.
5. What is the function of the Prostrate gland?
6. What is the reproductive role of the female?
7. Research question: Is it wise for some women to take estrogen? Why?
8. T or F. The uterine tubes are not directly connected to the ovaries.
9. The wall of the uterine tube is made up of what three layers?
10. Name the three functions of the Vagina.
11. Cell division is called ______________. Elaborate.
12. Each female breast is composed of ____ to ____ lobes of compound areolar glands.

The Integumentary System

1. The skin is the _____________ organ in the body and occupies about _______ sq. ft of area.
2. Describe the outer layer of skin and how it functions.
3. What three layers of connective tissue compose the dermis? What other two layers are in the dermis?
4. What are the sudoriferous glands? Give some reasons for these important glands.
5. What are sebaceous glands? Give some reasons for these important glands.
6. Why is there different colors of skin?
7. What are cancers of the skin caused by?
8. What are the various functions of the skin?
9. Describe how a wound heals in your own words.
10. Why is it important to keep our skin clean?
11. Name the three layers of epithelial cells in our hair follicles
12. T or F. Hair grows faster at night and in warm weather.
13. What are the nails used for, and tell how to protect them from disease.

The Story Never Ends

Forever and Ever and Beyond!

Dear Reader, you have come to the conclusion, even the last chapter of this book.

It is not important, or our purpose, that you should remember or memorize all the medical terms and complicated names found within the pages of this book. Indeed, we the authors do not remember all of these medical names nor do we understand fully how our bodies function. Only the Creator can completely understand the created. The most important concepts to remember are these: #1. God, through His Holy Spirit, dwells within our "Body Temple", and #2. He is desperate to save us; He has pulled out all stops to help us understand how much He loves us. He even wants us to come home and live forever with Him in the beautiful home He has provided in His Holy City, the New Jerusalem. God is leading His jewels through the destructive mine fields laid out by the enemy to destroy this homecoming. After man's sinful fall from the state of perfection, and until this present time, God has been leading His dear children along through famine, fire, and flood to full restoration. This is the very reason why we have studied how God fashioned the human body and how He is the source of all energy that keeps life activated.

We have studied how the Skeletal System represents God's laws to provide support for the government of God and for the support of the body.

We can see how the Immune and Lymphatic Systems represent God's forgiveness by working together to expunge disease and sin from the body and restore complete victory through restoration of health and life.

The Muscular System represents judgment by contracting a muscle and then relaxing an opposing muscle, thus representing the phases of judgment and mercy.

The Circulatory or Cardiovascular System in our study represents God's righteousness. This main transportation system carries life giving blood and even God's righteousness to all parts of the body.

The Respiratory System demonstrates the role of God's Holy Spirit in the same way that oxygen actuates in the lungs and, indeed, the whole body. Without oxygen the body cannot live or function.

The Nervous System teaches us about the activities that are going on in Heaven for our soon arrival at the celestial homecoming. God is sending assistance through the brain and nervous system of the body to every expectant one that looks to Him for eternal life.

The Digestive System is a lesson linking us to mental and spiritual nourishment. It reminds us what food is to the body, Christ is to the soul. We must feed upon Him and assimilate Him into the "Body Temple."

In the Urinary and Endocrine Systems, God has placed the power to eliminate toxins and wastes that have accumulated there. These systems represent, in our study, God's Holy Spirit of Love. The Spirit of Love permeates every system, organ, and atom of the body.

The Reproductive System is a most awesome system of the human body. This system gives us insight in the story that never ends; the story of forever and ever and beyond. God gave mankind the power of procreation. Within the act of marriage, God caused the man and woman to become one flesh. This one flesh concept represents the spiritual oneness that Christ will have with His bride.

Finally we view the Integumentary System. Nestled within the glue of the skin is the covering of love that holds together the fluids, flesh, and bones of the body. All of the body's systems, organs, cells, and molecules are held together by a protein molecule called ***laminin***. Laminin derives its name from the word to laminate. Webster's definition of laminate is "to overlay or construct with a thin plate of coat of laminae." To be sure then, laminin molecules overlay and literally hold the body together. They are the glue or biological life; they are the protein molecules found in the outer sheet or skin of the body. These glyco-proteins are numbered in the trillions and are an integral part of the scaffolding in almost every human tissue. From the smallest molecule to the largest organ, the skin of the human body, God has declared His saving message of grace, salvation, and redemption. For you see, the laminin molecule is formed in the shape of the cross! **Please see the diagram*. What an AWSOME GOD we serve! We are literally bound together in Christ!

The Apostle Paul declared in Holy Writ centuries ago that *"all things have been created through Christ, and that He is before all things and in Christ all things hold together."* {Colossions1: 16,17}

It is written- *"Store up these words of mine in your heart and in your soul. Bind them together as a token on your hands and let them be as a forehead band between your feet." "Teach them to your children when sitting at home and when walking on the road, when you lie down and when you get up."* (Deuteronomy 11:18,19)

We are writing this conclusion chapter of this book in the sleepy pre-renewing month of November. We are living in the winter of our years seeing more of our times on earth as *past* than we have *before* us. It is a great reflective time as we see more and more every day how our Heavenly Father superimposes His Regal Presence on everything we are and do. He continually implores us to choose life and not death. When we choose to live eternally we become grafted into His royal family where we will live forever with Him and all our loved ones who made that same choice. When we decide this—AT THAT VERY MOMENT, we are now part of that love story that never ends and lives forever and ever and beyond!

MEDUNA HEALTH INTERNATIONAL

ME = MESSIAH DU= DUANE NA= NANCY

If after reading this book on "The Body Temple" you would like to schedule a book signing or health seminar in your area you may contact the Doctors McEndree by writing to them at 7745 Riata Place, Zephyrhills, FL 33541.

Some of the POWER POINT SEMINAR PRESENTATIONS are listed below.

IN THE BEGINNING A study of Genesis 1 as it relates to God's Ten Ten Laws of Health. God created mankind to live forever. There was no need for sickness or disease in His created beings. We still have the ability to live in optimum health through the guidelines set forth in God's Word. This power point presentation will inspire all of us to walk a little closer with Our Father.

America in Crises – *"State of Health in America"*
Up to the minute health headlines across America – Statistics concerning health care from JAMA (Journal American Medical Association) What is happening among 21st Century Christians *What we can do to help prevent sickness and disease. God's answer to man's problems.*

The Answer for Cancer – *"God's Plan for Renegade Cells"*
Cancer Statistics – Early Warning Signs – Screening Tests – Leading Carcinogens – Power of whole plant foods – Natural remedies and how to use them to combat cancer. *How our diet and lifestyle can help prevent this ravishing disease. Steps we can take to rid ourselves of the problem.*

Heart Health – *"For the love of Mike!"*
Hypertension - Progression of Coronary Heart Disease - Atherosclerosis – The need for good cholesterol and the perils of to much cholesterol build up – Strokes *How to adjust our lifestyle and diet to prevent heart problems.*

Cholelithiasis

Cholecystitis

Gall Bladder Cancer

Gallstone Pancreatitis

Jan 1:29 Dm 1

CPSIA information can be obtained at www.ICGtesting.com
Printed in the USA
LVOW031425140212

268668LV00002B/1/P